MW01014614

Integrated Petroleum Reservoir Management

▼ ▼ ▼

A Team Approach

Integrated Petroleum Reservoir Management

▼ ▼ ▼

A Team Approach

Abdus Satter, Ph.D.
Research Consultant
Texaco E&P Technology Department
Houston, Texas

Ganesh C. Thakur, Ph.D.
Manager—Reservoir Simulation Division
Chevron Petroleum Technology Company
La Habra, California

PennWell Books
PENNWELL PUBLISHING COMPANY
TULSA, OKLAHOMA

Disclaimer

This text contains statements, descriptions, procedures, and other information that have been prepared as general information. There are many variable conditions in reservoir management and related situations, and the authors have no knowledge or control of their interpretation. The information in this text is believed to accurately represent actual situations and conditions, and the user should determine to what extent it is practicable and advisable to follow them. The text is intended to supplement and not to replace the user's judgment, and the authors and the publishers assume no liability for injury, loss, or damage of any kind resulting directly or indirectly from the use of the information contained in this book and other related materials.

PennWell Publishing Company
1421 South Sheridan/P.O. Box 1260
Tulsa, Oklahoma 74101

Library of Congress Cataloging-in-Publication Data

Satter, Abdus.
Integrated petroleum reservoir management: a team approach/
Abdus Satter, Ganesh C. Thakur.
p.cm.
Includes bibliographical references.
ISBN 0-87814-408-0
1. Oil fields—Management. 2. Petroleum industry and trade—
Production control. I. Thakur, Ganesh C. II. Title
TN870.S283 1994
622'.3382'068—dc20 93-44927
CIP

Printed in the United States of America

4 5 6 7 8 05 04 03 02 01

Contents

Foreword

Although elements of petroleum reservoir management have been practiced almost since reservoirs were first recognized, the concept of an integrated approach took form only within the past one or two decades. Many individual articles and at least one manual on the subject have been published in the open literature, and it is probable that proprietary presentations of reservoir management concepts are to be found within the internal libraries of some oil- and gas-producing companies. In the present volume, the authors present the first treatment of the subject to be published in book form.

The roots of petroleum reservoir management are to be found in reservoir engineering, taken in its braodest sense as the technology that deals with the movement of fluids into, out of, and through the geological formations of the earth by means of wells and well systems. However, reservoir management and reservoir engineering are not identical. The latter is one of the elements that enters into achieving the former. Reservoir management infers the existence of goals toward which the reservoir technology is directed. At the heart of petroleum reservoir management is the idea that goals and the technological implementations to acheive those goals will be specific to each individual reservoir. This point is emphasized throughout the book by means of examples and case histories.

The book brings out the importance of the reservoir model and discusses its uses, not only as a tool for integrating the total database that is available on a reservoir, but also for predicting the consequences of alternate future constraints and/or implementation procedures that might be invoked. In short, the individual reservoir has its formalized expression in the model that is constructed to simulate it.

The book also emphasizes that petroleum reservoir management depends upon teamwork and continuous interactions among team players whose expertise may include specializations such as geology, geophysics, drilling, logging, well behavior, recovery mechanisms, subsurface fluid behavior, production operations, facilities engineering, economics, decision strategy, environmental issues and other applicable subjects. The manner in which the team players are brought together, the formulation of a plan, its periodic revision, and its evaluation are presented as parts of the reservoir management process to ensure effective use of the technological spectrum that is available to the team.

Integrated Petroleum Reservoir Management is a book that will serve persons who come to petroleum reservoir management from different backgrounds. Its aim is to present the integrated picture. It provides an overview of a very complex subject, discussing the overall framework of concepts without dwelling in depth on subject matter details that lie within the provinces of individual experts who make up a reservoir management team. It is a timely addition to the literature of petroleum technology.

John C. Calhoun, Jr.
Distinguished Professor Emeritus of Petroleum Engineering, Texas A&M University

Acknowledgments

The authors wish to acknowledge the support and permission of Texaco and Chevron to prepare this book. We also would like to thank our many coworkers and students from whom we have learned about many aspects of reservoir management. Their contributions to this book are gratefully acknowledged.

Our special thanks go to Kathy Gough, Donna Dismuke, and Rose Mary Burditt for their patience and hard work in producing this book.

Lastly, we owe our wives, Betty Satter and Pushpa Thakur, and our families sincere gratitude for their patience, understanding, and encouragement during the long period of planning and writing this book.

Abdus Satter *Ganesh C. Thakur*

CHAPTER 1

▼ ▼ ▼

Introduction

Integrated petroleum reservoir management has received significant attention in recent years. Various panel, forum, seminar, and technical sessions have provided the framework for information sharing and the exchange of ideas concerning many practical aspects of integrated, sound reservoir management.[1–7] The need to enhance recovery from the vast amount of remaining oil and gas-in-place in the United States and elsewhere, plus the global competition, requires better reservoir management practices.

Historically, some form of reservoir management has been practiced when a major expenditure is planned, such as a new field development or waterflood installation. The reservoir management studies in these instances were not integrated (i.e., different disciplines did their own work separately). During the past 20 years, however, a greater emphasis has been placed on synergism between engineering and geosciences. However, despite the emphasis, progress on integration has been slow.

SOUND RESERVOIR MANAGEMENT

A reservoir's life begins with exploration that leads to discovery, which is followed by delineation of the reservoir, development of the field, production by primary, secondary, and tertiary means, and finally to abandonment (see Figure 1–1).[8] Integrated, sound reservoir management is the key to a successful operation throughout a reservoir's life.

A vast amount of hydrocarbon remains unrecovered in the United States and elsewhere in the world. The good news is that many leading-edge technological advances have now been made in geophysics, geology, petrophysics, production, and reservoir engineering. Mainframe super computers, more powerful personal computers, and workstations are providing ever increasing computing power. Integrated life cycle database-management

FIGURE 1–1. Reservoir Life Process *(Copyright © 1992, SPE, from paper 22350[8])*

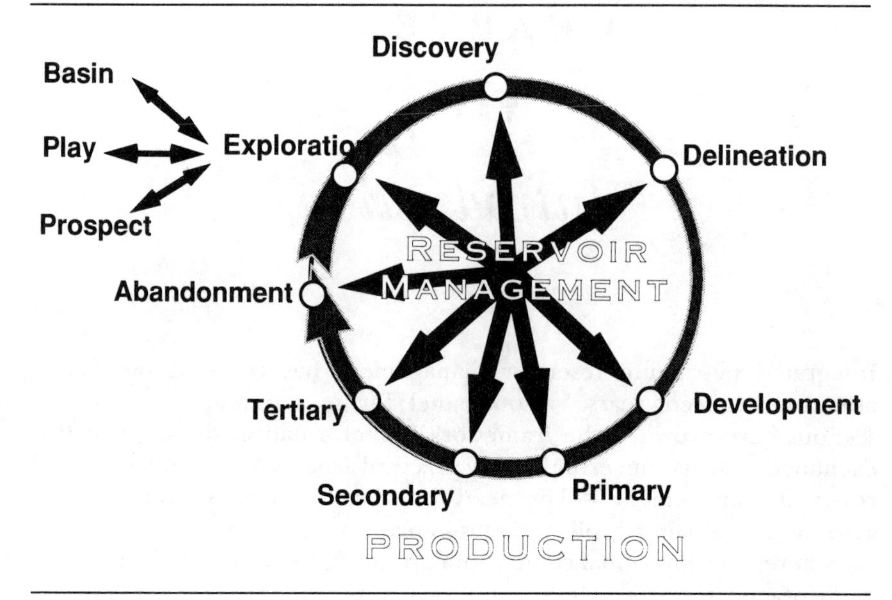

systems are being developed. The technological advances in computer tools and software will provide opportunities for enhancing economic recovery of hydrocarbons.

U.S. reserves have been declining (see Figure 1–2).[9] Even a small percentage increase in recovery efficiency due to sound reservoir management would amount to significant additional reserves (see Figure 1–3). These incentives and challenges should provide the motivation to practice better reservoir management.

SCOPE AND OBJECTIVE

This book is written for practicing engineers, geologists, geophysicists, field operation staffs, managers, government officials, and others involved with reservoirs. College students in petroleum engineering, geoscience, economics, and management can also benefit from this book. We have placed into one book pertinent reservoir management literature along with our own years of practical experience in reservoir studies, operation, and management. The book's focus has been placed on reservoir management as a whole entity by integrating the technologies

FIGURE 1–2. **Why Need Sound Reservoir Management?** *(Copyright © 1990, SPE[9])*

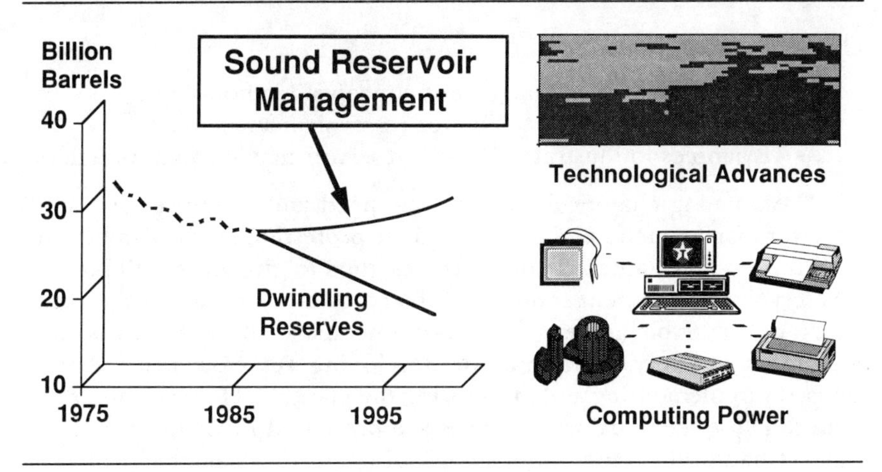

FIGURE 1–3. **Why Need Sound Reservoir Management?** *(Copyright © 1992, SPE, from paper 22350[8])*

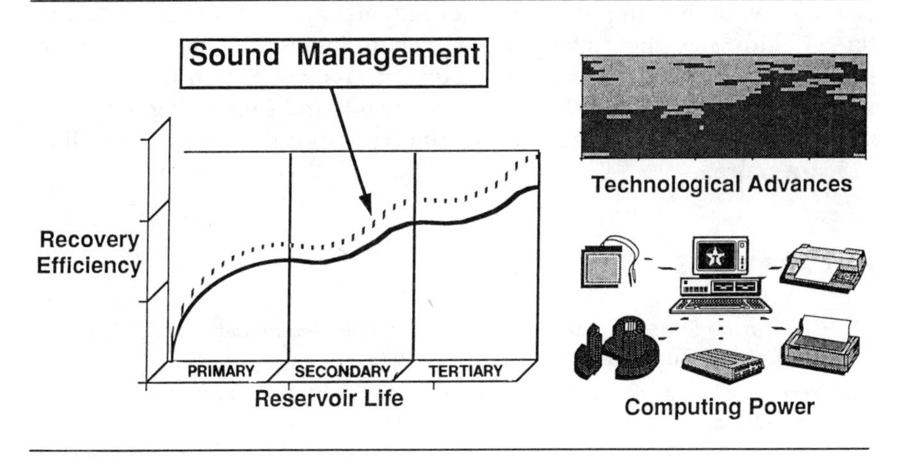

and activities of the many disciplines involved. The objective is to provide a better understanding of the practical approach to asset management using multidisciplinary, integrated teams. This will promote better resource management practices by enhancing hydrocarbon recovery and maximizing profitability.

ORGANIZATION

This book presents:

- Sound reservoir management concepts/methodology.
- Technology needed for better reservoir management.
- Examples to illustrate effective reservoir management practices.

The opening chapter focuses on the need and incentive for sound reservoir management, and it identifies the professionals who can benefit from this book. Chapters 2 and 3 are designed to give an overall picture of reservoir management concepts and methodology. Data management, integrated reservoir model, production rate and recovery forecasts, and economics are very important for developing reservoir management plans, for implementing and monitoring the plans, and for evaluating the results. These technical perspectives are presented in Chapters 4, 5, 6, and 7. Chapter 8 presents an overview of water, thermal, chemical, and miscible flood processes. Chapter 9 presents case histories of selected reservoir management projects with overall analyses including key achievements, inadequacies, and future operating plans. Chapter 10 presents an integrated reservoir management plan for a newly discovered field and an example of an ongoing primary-to-tertiary management program. Chapter 11 addresses the state-of-the-art technologies, the importance of integrated reservoir management, current challenges and areas of improvement, outlook, and the next step. Details and supporting materials are presented in the Appendices for the benefit of those who would like to learn more.

REFERENCES

1. SPE Forum Series V: *Advances in Reservoir Management and Field Applications,* Mt. Crested Butte, CO, August 13–18, 1989.
2. Reservoir Management Panel Discussion, SPE 65th Ann. Tech. Conf. & Exb., New Orleans, LA, September 23–26, 1990.
3. Reservoir Management Practices Seminar, SPE Gulf Coast Section, Houston, TX, April 26, 1991.
4. SPE Forum Series III: *Application of Reservoir Characterization to Numerical Modeling and Reservoir Management,* Mt. Crested Butte, CO, July 28–Aug. 2, 1991.
5. Reservoir Management Panel Discussion, SPE 66th Ann. Tech. Conf. & Exb., Dallas, TX, October 6–9, 1991.

6. Reservoir Management Sessions, Int'l. Mtg. on Pet. Engr., Beijing, China, March 24–27, 1992.
7. Reservoir Management Practices Seminar, SPE Gulf Coast Section, Houston, TX, May 29, 1992.
8. Satter, A., J. E. Varnon and M. T Hoang. "Reservoir Management: Technical Perspective." SPE Paper 22350, SPE Int'l. Mtg. on Pet. Engr., Beijing, China, March 24–27, 1992.
9. Satter, A. "Reservoir Management Training: An Integrated Approach." SPE 65th Ann. Tech. Conf. & Exb., New Orleans, LA, September 23–26, 1990.

CHAPTER 2

▼ ▼ ▼

Reservoir Management Concepts

This chapter presents a historical review of reservoir management practices and discusses technological advances made and computer tools developed in recent years to facilitate better reservoir management. It also provides a reservoir management definition, discusses synergy and teamwork, examines the integration of geoscience and engineering, and analyzes the timing for reservoir management.

DEFINITION OF RESERVOIR MANAGEMENT

There are many reservoir engineers, geologists, and geophysicists who realize that the maximum coordination of their disciplines is essential to the future success of the petroleum industry. With this in mind, they follow the principles of reservoir management for maximizing economic recovery of oil and gas.

One of the objectives of this section is to define *reservoir management.* The *Webster Dictionary* defines management as the "judicious use of means to accomplish an end." Thus, the management of reservoirs can be interpreted as the judicious use of various means available to a businessman in order to maximize his benefits (profits) from a reservoir.[1]

Reservoir management has been defined by a number of other authors.[2–5] Basically, sound reservoir management practice relies on the utilization of available resources (i.e., human, technological and financial) to maximize profits/profitability index from a reservoir by optimizing recovery while minimizing capital investments and operating expenses (see Figure 2–1). Reservoir management involves making certain choices. Either let it happen, or make it happen.[2] We can leave it to chance to generate some profit from a reservoir operation without ongoing deliberate planning, or we can enhance recovery and maximize profit from the same reservoir through sound management practice.

FIGURE 2–1. What Is Reservoir Management? *(Copyright © 1992, SPE, from paper 22350[2])*

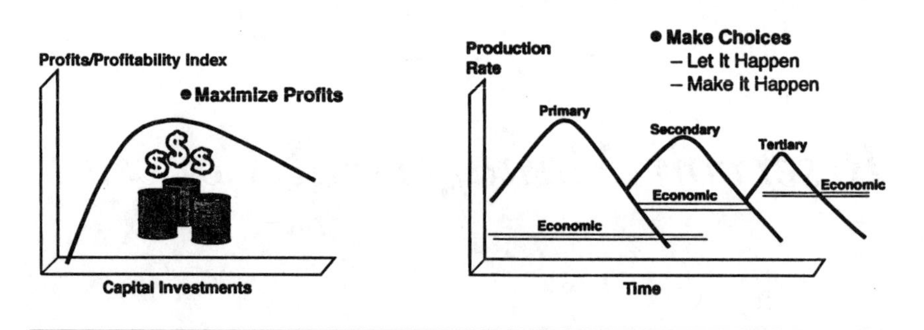

HISTORY OF RESERVOIR MANAGEMENT

Most people considered reservoir management synonymous with reservoir engineering. As recently as the early 1970s, reservoir engineering was considered the most important technical item in the management of reservoirs. However, after understanding the value of geology, synergism between geology and reservoir engineering became very popular and proved to be quite beneficial.

Reservoir management has advanced through various stages in the past 30 years. The techniques are better, the background knowledge of reservoir conditions has improved, and the automation using mainframe computers and personal computers has helped data processing and management. The developmental stages of reservoir management could be described as the following:

Stage 1—Before 1970, reservoir engineering was considered the most important technical item in the management of reservoirs. In 1962, Wyllie emphasized two key items: (1) clear thinking utilizing fundamental reservoir mechanics concepts and (2) automation using basic computers.[6] In 1965, Essley described "reservoir engineering" and concluded that in spite of the technical advancement of reservoir engineering, vital engineering considerations are often neglected or ignored.[7]

Stage 2—This covers the time period of the 1970s and 1980s. Craig et al. (1977) and Harris and Hewitt (1977) explained the value of synergism between engineering and geology. Craig emphasized the value of detailed reservoir description, utilizing geological, geophysical, and reservoir simulation concepts.[8] He challenged explorationists, with the knowledge of geophysical tools, to provide a more accurate reservoir description to be used in engineering calculations. Harris and Hewitt presented a geological perspective of the synergism in reservoir management.[9] They explained the reservoir heterogeneity due to complex variations of reservoir continuity, thickness patterns, and pore-space properties (e.g., porosity, permeability, and capillary pressure).

FUNDAMENTALS OF RESERVOIR MANAGEMENT

Although the synergism provided by the interaction between geology and reservoir engineering has been quite successful, reservoir management has generally been unsuccessful in recognizing the value of other disciplines (e.g., geophysics, production operations, drilling, and different engineering functions).

The prime objective of reservoir management is the economic optimization of oil and gas recovery, which can be obtained by the following steps:

- Identify and define all individual reservoirs in a particular field and their physical properties.
- Deduce past and predict future reservoir performance.
- Minimize drilling of unnecessary wells.
- Define and modify (if necessary) wellbore and surface systems.
- Initiate operating controls at the proper time.
- Consider all pertinent economic and legal factors.

Thus, the basic purpose of reservoir management is to control operations to obtain the maximum possible economic recovery from a reservoir based on facts, information, and knowledge.

In 1963, Calhoun described the engineering system of concern to the petroleum engineer as being composed of three principal subsystems:

- Creation and operation of wells.
- Surface processing of the fluids.
- Fluids and their behavior within the reservoir.[10]

The first two subsystems depend on the third because the type of fluids (i.e., oil, gas, and water) and their behavior in the reservoir will dictate how many wells to drill and where, and how they should be produced and processed to maximize profits.

Since the goal is to maximize profits, neglecting or de-emphasizing any of the previous items could jeopardize our objective. For example, we could do well in studying the fluids and their interaction with rock (i.e., reservoir engineering), but if the proper well and/or surface system design is not considered, then recovery of oil and/or gas will not be optimized. Most people can cite examples of mistakes made in our business where we thoroughly studied various aspects of the reservoir and made decisions that resulted in too many wells being drilled, improper application of well completion technology, inadequate surface facilities available for future expansions, and so forth.

The suggested reservoir management approach emphasizes interaction between various functions and their interaction with management, economics, proration, and legal groups. The reservoir management model that involves interdisciplinary functions has provided useful results for many projects.

The following question-and-answer section provides reservoir management philosophies:

1. When should reservoir management start?

The ideal time to start managing a reservoir is at its discovery. However, it is never too early to start this program because early initiation of a coordinated reservoir-management program not only provides a better monitoring and evaluation tool, but also costs less in the long run. For example, a few early drill stem tests (DST) could help decide if and where to set pipe. Sometimes these data can also provide the same type of information normally available by complex and expensive cased hole, multiple zone testing. An extra log or an additional hour's time on a DST may provide better information than could be obtained from more expensive core analysis.[7] Sometimes it is possible to do some early tests that can indicate the size of a reservoir. If it is of limited size, drilling of unnecessary wells can be prevented.

We can draw an analogy between reservoir and health management.[10] It is not sufficient for the reservoir management team to determine the state of a reservoir's health and then attempt to improve it. To be most effective, the team must maintain the reservoir's and its sister subsystems' health from the start.

Most often reservoir management is not started early enough, and the reservoir, wells, and surface systems are ignored for a long time. Many times we consider reservoir management at the time of a tertiary recovery operation. However, it is critical and a prerequisite for an economically successful tertiary recovery operation to have a good reservoir management program already in place.

In the Permian Basin, carbon dioxide (CO_2) flooding is receiving more and more attention. An efficient reservoir management program for CO_2 flooding (with a $2 per barrel injectant cost) is even more critical compared to waterflooding (with a 5–10¢ per barrel cost for water). Thus, it is very important that all injected CO_2 be properly utilized in displacing oil to the production wells.

2. What, how, and when to collect data?

To answer this question, we must follow an integrated approach of data collection involving all functions from the beginning. Before collecting any data, we should ask the following questions:

- Are the data necessary, and what are we going to do with these data? What decisions will be made based on the results of the data collection?
- What are the benefits of these data, and how do we devise a plan to obtain the necessary data at the minimum cost?

The reservoir management team must prepare a coordinated reservoir evaluation program to show the need for the data requirement, along with their costs and benefits. Amyx et al. provides a detailed review of data evaluation for reservoir engineering calculations.[11]

It must be emphasized that early definition and evaluation of the reservoir system is a prerequisite to good reservoir management.[12,13] The team members must convince the management to obtain necessary data to evaluate the reservoir system. In addition, the team should participate in making operating decisions.

3. What kinds of questions should be asked if we want to ensure the right answer in the process of reservoir management?

Some example questions follow:

- What does the answer mean?
- Does the answer fit all the facts; why or why not?
- Are there other possible interpretations of the data?
- Were the assumptions reasonable?
- Are the data reliable?
- Are additional data necessary?
- Has there been an adequate geological study?
- Has the reservoir been adequately defined?

The modern reservoir management process involves goal setting, planning, implementing, monitoring, evaluating, and revising plans.[2] Setting a reservoir management strategy requires knowledge of the reservoir, availability of technology, and knowledge of the business, political, and environmental climate. Formulating a comprehensive management plan involves depletion and development strategies, data acquisition and analyses, geological and numerical model studies, production and reserves forecasts, facilities requirements, economic optimization, and management approval. Implementing the plan requires management support, field personnel commitment, and multidisciplinary, integrated teamwork. Success of the project depends upon careful monitoring/surveillance and thorough, ongoing evaluation of its performance. If the actual behavior of the project does not agree with the expected performance, the original plan needs to be revised, and the cycle (i.e., implementing, monitoring, and evaluating) reactivated.

SYNERGY AND TEAM

Successful reservoir management requires synergy and team efforts. It is recognized more and more that reservoir management is not synonymous with reservoir engineering and/or reservoir geology. Success requires multidisciplinary, integrated team efforts. The players are everybody who has anything to do with the reservoir (see Figure 2–2).[2] The team members must work together to ensure development and execution of the management plan. By crossing the traditional boundaries and integrating their functions, corporate resources are better utilized to achieve the common goal.

All development and operating decisions should be made by the reservoir management team, which recognizes the dependence of the entire system upon the nature and behavior of the reservoir. It is not necessary that all decisions be made by a reservoir engineer; in fact, a team member who considers the entire system, rather than just the reservoir aspect, will be a more effective decision maker. It will help tremendously if the person has a background knowledge of reservoir engineering, geology, production and drilling engineering, well completion and performance, and surface facilities. Not many people in an organization have knowledge in all areas. However, many persons develop an intuition for the entire system and know when to ask for technical advice regarding various elements of the system.

The team effort in reservoir management cannot be overemphasized. It is even more necessary now than it has ever been because the current trend of the oil industry is not one of expansion. Most companies are carrying on their production activities with a staff much smaller than had existed just five years ago.

Also, with the advent of technology and the complex nature of different subsystems, it is difficult for anyone to become an expert in all areas. Therefore, it is obvious that the reduction of talent and the increasingly complex technologies must be offset by an increase in quality, productivity, and emphasis on the team effort.

A team approach to reservoir management can be enhanced by the following:

- Facilitate communication among various engineering disciplines, geology, and operations staff by: (a) meeting periodically, (b)

FIGURE 2–2. Reservoir Management Team *(Copyright © 1992, SPE, from paper 22350[2])*

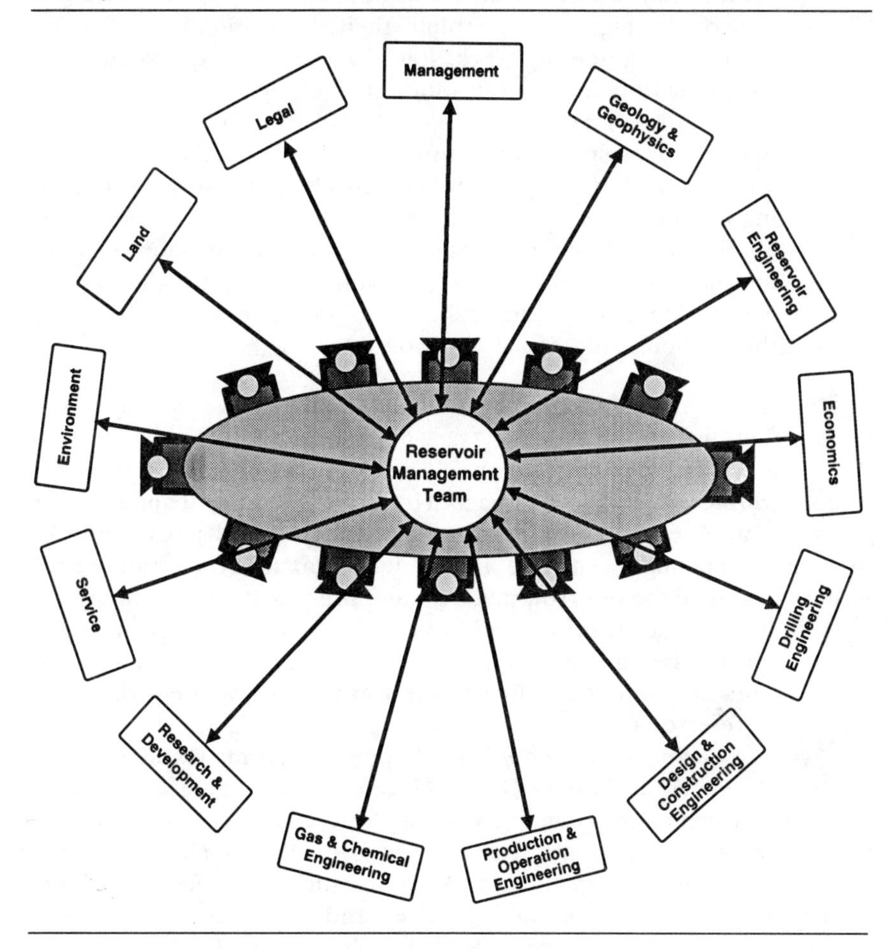

interdisciplinary cooperation in teaching each other's functional objectives, and (c) building trust and mutual respect. Also, each member of the team should learn to be a good teacher.

- To some degree, the engineer must develop the geologist's knowledge of rock characteristics and depositional environment, and a geologist must cultivate knowledge in well completion and other engineering tasks, as they relate to the project at hand.
- Each member should subordinate their ambitions and egos to the goals of the reservoir management team.
- Each team member must maintain a high level of technical competence.

- The team members must work as a well-coordinated "basketball team" rather than a "relay team." Reservoir engineers should not wait on geologists to complete their work and then start the reservoir engineering work. Rather, a constant interaction between the functional groups should take place. For example, it is better to know early if the isopach and cumulative oil/gas production maps do not agree rather than finalize all isopach maps and then find out that cumulative production maps are indicating another interpretation of the reservoir. Using an integrated approach to reservoir management along with the latest technological advances will allow companies to extract the utmost economic recovery during the life of an oil field. It can prolong the economic life of the reservoir.[1]

In summary, the synergism of the team approach can yield a "whole greater than the sum of its parts."

Today, it is becoming common for large reservoir studies to be integrated through a team approach. However, creating a team does not guarantee an integration that leads to success. Team skills, team authority, team compatibility with the line management structure, and overall understanding of the reservoir management process by all team members are essential for the success of the project. Also, most reservoir management teams are being assembled only at key investment times. Missing today are ongoing multidisciplinary reservoir management efforts for all significant reservoirs.[2]

Synergy is not a new concept. Halbouty, chairman and CEO of Michael T. Halbouty Energy Co. in Houston, a long-time advocate of synergy and team approach, recognized this concept as basic to future petroleum reserves and production.[14] According to Sneider, "synergy means that geologists, geophysicists, petroleum engineers and others work together on a project more effectively and efficiently as a team than working as a group of individuals."[15] Talash advocated that teamwork between reservoir and production/operation engineers is essential to waterflood project management.[16] The team approach to reservoir management is essential and involves the interaction between management, engineering, geoscience, research, and service functions.[1] We emphasize again that the team members should work as a well-coordinated "basketball team" rather than a "relay team."

Some independents, such as the Apache Intl. Corp., have used an innovative team approach to increase production on properties acquired from majors. A full-time interdisciplinary team is assigned to one mature acquisition after another to analyze and identify ways to extract additional production. "This keeps the team hungry," Apache CEO William Johnson said recently. "We just feed them one field after another. They always welcome a new challenge."[17]

Limited staff and other resources have caused most independents to take a multidisciplinary approach since the early 1970s. Halbouty said recently, "In our shop, we exclusively use this constructive approach, and it has proved highly valuable in every way. . . From the discovery and early appraisal phases through planning, development, and reservoir management phases, geologists, geophysicists, and engineers work closely together."[17]

Major producers also have used the integrated approach for years. As one early example, Amoco Intl. Oil Co. used a multidisciplinary approach on the East Unit of the North Sea Leman field from the time the field came on stream in 1968. The field contained more than 10 Tcf [280×10^6 m^3] of gas—then the world's largest producing offshore gas field.[8]

Working in a complex fault system, the company's reservoir engineers worked closely with geologists to "produce an accurate *a priori* reservoir description." The team tested the description against field performance in a 2D fine-grid, single-phase model and refined it with measured pressures from the first six years of production. The team gained valuable insight into fault configurations and the relationships of gas in place, permeability, and reserves.

Geologists reviewed the locations of faults and reservoir boundaries on the historical map. The resulting model successfully predicted pressure for an additional two years. The proven accuracy of the model led to confident planning of future platform and compression requirements, providing more than three years' lead time to install equipment.

In the more recent case of the Amoco Production Co. Teak field offshore Trinidad, the key to reversing production declines was a comprehensive data base created at the beginning of field reevaluation.[18] Amoco assigned a reservoir engineer, geologist, and geophysicist to work closely with operations staff in 1988 to improve the economics of the field.

Since 1972, Teak has produced 226 million bbl (36×10^6 m^3] of oil. Production peaked in 1975 at 58,000 B/D [9220 m^3/d], but it dropped 50% by 1988. The team compiled data from the field's 16-year history and updated it throughout and beyond the study. Amoco used the data to produce detailed studies beginning with the shallow producing sands downward. The team developed conformable, detailed sand structure maps.

Along with structure maps, the team used deviated-well cross-sections, true-vertical-depth correlation panels, fault-plane maps, and dipmeter interpretations. The integrated reservoir modeling results and production data helped the team boost Teak production 49,000 B/D [7790 m^3/d] in 1990, the highest level in five years.

Traditionally, oil and gas companies tend to manage reservoirs like relay races. Geologists hand the field to drilling engineers, who hand it to production engineers, who hand it to reservoir engineers. Unfortunately, employees at any stage could drop the "baton."

For example, sometimes pressing lease deadlines can cause incomplete data capture, conversion, and integration. As a result, engineers and geoscientists who do not communicate with each other may study only a fraction of the available data. The quality of the reservoir model can suffer, which adversely affects drilling decisions and production plans throughout the life of the reservoir.

The continental United States can be cited as examples of problems with conventional "relay-team" reservoir management.[17] Major producers are abandoning what they consider "mature" fields in the lower 48 states. First Boston Corp. reported recently that the five largest U.S. oil companies earned an average return of only 4% on investments in the continental United States over the last five years. As a result, $14 to $18 billion in oilfield assets are for sale. Many are "mature" reservoirs.

Better tools and data, along with the advent of new technology, such as workstations and integrated software packages, can minimize remaining barriers between reservoir management disciplines. August 1991 market research from Landmark/Concurrent Solutions Inc. of Houston found that geologists and engineers spend up to one-half of their time sifting through mountains of information and converting data formats, thus wasting hundreds of man-hours every year.[17]

Integrated data storage and retrieval systems that use workstations and interactive technologies provide a solution. Engineers and geoscientists can enter archival or current information compiled in their areas into these shared systems. Information available for these data bases includes everything from speculative data and production histories to recently shot seismic and old maps and well logs. Access to these data helps the team build and refine a subsurface model from the latest available data and interpretations.

Input from petrophysicists and reservoir engineers can help the team check and validate seismic and geologic interpretations. Team members can correct contradictions as they arise, preventing costly errors later in the field's life. They also can view information in multiple formats at any time, tailoring the data to specific project needs. Adding other technologies, such as 3D-seismic techniques, can help teams find and exploit new reserves in existing fields. Halbouty has advocated these approaches for many years as keys to locating the "subtle traps" left behind in earlier exploration efforts.

Walter Ritchie of Geophysical Service U.S.A. of Dallas mentioned that multidisciplinary groups from companies, including Hunt Oil Co., Occidental E&P Co., and Chevron U.S.A. Inc., have used workstations with 3D-seismic data to model reservoirs in the varied conditions of Peru,

Alaska, the Netherlands, and China.[17] He singled out the East Painter reservoir in the Rocky Mountain overthrust as a particularly successful application of 3D-seismic processing. Companies drilled 13 wells on the structure after survey completion. All were successful wells, with an average cost of $4 to $5 million each. The East Painter survey cost $1.6 million.

"Multidisciplinary teams using the latest technology from the first discovery of a reservoir provide the best possible description of a reservoir's potential recovery," stated Houston consultant Joseph G. Richardson. "Continued efforts to describe the reservoir with 3D-seismic data and to enhance its performance are most valuable. For example, improved seismic data can aid where surveillance of production operations in mature projects has revealed a lack of continuity between wells. As reservoirs are developed and projects mature, continuing efforts by multidisciplinary teams are necessary to provide the best, most practical techniques of accurate field descriptions to achieve optimal production."[17]

So far we have discussed synergy and team approach; however, the organization and management of the reservoir management team are also very critical. As previously discussed, sound reservoir management requires a multidisciplinary team effort. The important question is who should set goals and make reservoir management decisions? Should the production manager make the decisions or should they be made by the team? (See Figures 2–3, 2–4a and 2–4b for various types of team organization.) Figure 2–3 shows team members working under functional heads and a production manager, where the functional heads provide functional guidance and perform evaluations, and the production manager provides project direction and focuses on business needs. Figures 2–4a

FIGURE 2–3.

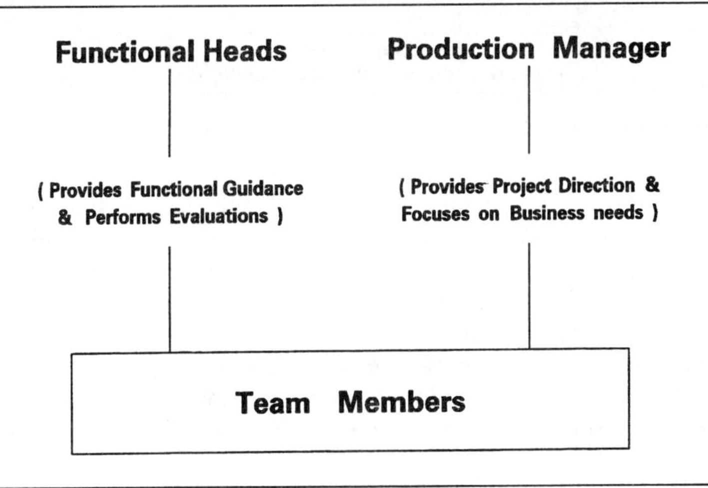

FIGURE 2–4a. Old System-Conventional Organization

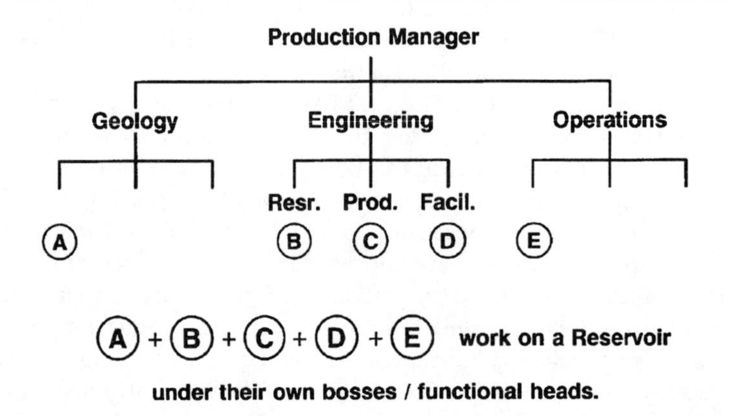

FIGURE 2–4b. New System-Multidisciplinary Team

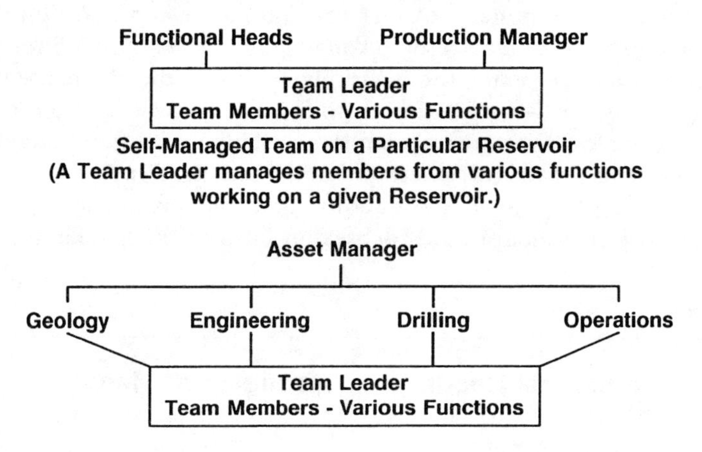

and 2–4b contrast the old and the new system. In the old (conventional) system, various members of the team (i.e., geologist, reservoir/production/facilities engineers, operations staff, and others) work on a reservoir under their own bosses/functional heads; whereas, in the new/multidisciplinary team approach, the team members from various functions work on a given reservoir under a team leader and sometimes operate as a self-managed team. The team leader generally provides day-to-day guidance, and occasional functional guidance is provided by the so-called function-

al "gurus." The team (or its members) do not administratively report to functional "gurus," rather administrative and project guidance are provided by the production manager or the asset manager. The asset management concept emphasizes focusing on a field as an asset, and all team members have a primary objective of maximizing the short- and long-term profitability of the asset. The team members concentrate on their duties more like generalists than functional specialists.

In order to make effective decisions, the production manager has to recognize the dependence of the entire system upon the nature and behavior of the reservoir. However, it is rare to have someone with expertise in all required areas. In our experience, we have observed that reservoir-management plan development and implementation are most effective if the team members work together and are involved in decision making.

Organization and management of the reservoir management team requires special attention. Formation of the team, selection of team members, appropriate motivational tools, and composition of the team (as the needs of the reservoir change) should be carefully considered. Other aspects such as team leadership, establishment of team goals and objectives, and performance appraisals of the team members are some matters that play a key role in effective reservoir management.

Once a team is formed and begins to function, another item of significance is how to sustain team effort. It is easy to get excited when the teams are set up, at times of major expenditures, development effort, 3-D seismic program, and so forth; however, to get the ongoing attention by a multidisciplinary team for all major reservoirs requires great commitment by the operating company.

One model of the team approach follows:

- Functional management nominates team members to work on a project team with specific tasks in mind.
- The team reports to the production manager for this project. Also, the team selects a team leader, whose responsibility is to coordinate all activities and keep the production manager informed.
- The team members consist of representatives from geology and geophysics, various engineering functions, field operations, drilling, finance, and so forth.
- Team members prepare a reservoir management plan and define their goals and objectives by involving all functional groups. The plan is then presented to the production manager; and after receiving the manager's feedback, appropriate changes are made. Next the plan is published and all members follow the plan.

- The team members' performance evaluation is conducted by their functional heads with input from the team leader and the production manager. The performance appraisal, in addition to various dimensions of performance, includes teamwork as a job requirement.
- Teams are rewarded recognition/cash awards upon timely and effective completion of their tasks. These awards provide an extra motivation for team members to do well.
- As the project goals change (e.g., from primary development to secondary process), the team composition changes to include members with the required expertise. Also, this provides an opportunity to change/rotate team members with time.
- Approvals for project AFE's (Appropriation For Expenditures) are initiated by the team members; however, the engineering/operations supervisor and/or production manager have the final approval authority.
- Sometimes conflicting priorities for the team members develop because they essentially have two bosses (i.e., their functional heads and the team leader). These conflicts are generally resolved by constant communication among the team leader, functional heads, and the production manager.

INTEGRATION OF GEOSCIENCE & ENGINEERING

Halbouty stated in 1977: "It is the duty and responsibility of industry managers to encourage full coordination of geologists, geophysicists, and petroleum engineers to advance petroleum exploration, development, and production."[14] Despite the emphasis, progress on integration has been slow.

Sessions and Lehman presented the concept of increased interaction between geologists and reservoir engineers through multifunctional teams and cross-training between the disciplines.[19] They stated that production geology and reservoir engineering within the conventional organization function separately, and very seldom does a production geologist get in-depth experience in reservoir engineering and vice versa. They advocated cross-exposure and cross-training between disciplines. Integrated reservoir management training for geoscientists and engineers offered by many major oil and gas companies is designed to address these needs.[20]

Sessions and Lehman presented Exxon's three case histories where the geology-reservoir engineering relationship was promoted through both a team approach and an individual approach. The results of the three cases (project-based approach, team-based approach, and multi-skilled individual approach) were very positive.

Synergy and team concepts are the essential elements for integration of geoscience and engineering.[2] It involves people, technology, tools, and data (see Figure 2–5). Success for integration depends on:

- Overall understanding of the reservoir management process, technology, and tools through integrated training and integrated job assignments.
- Openness, flexibility, communication, and coordination.
- Working as a team.
- Persistence.

Reservoir engineers and geologists are beginning to benefit from seismic and cross-hole seismology data. Also, it is essential that geological and engineering ideas and reasoning be incorporated into all seismic results if the full economic value of the seismic data is to be realized.

Perfectly conscientious and capable seismologists may overlook a possible extension in a proven area because of their unfamiliarity with the detailed geology and engineering data obtained through development.

FIGURE 2–5. Reservoir Management *(Copyright © 1992, SPE, from paper 22350[2])*

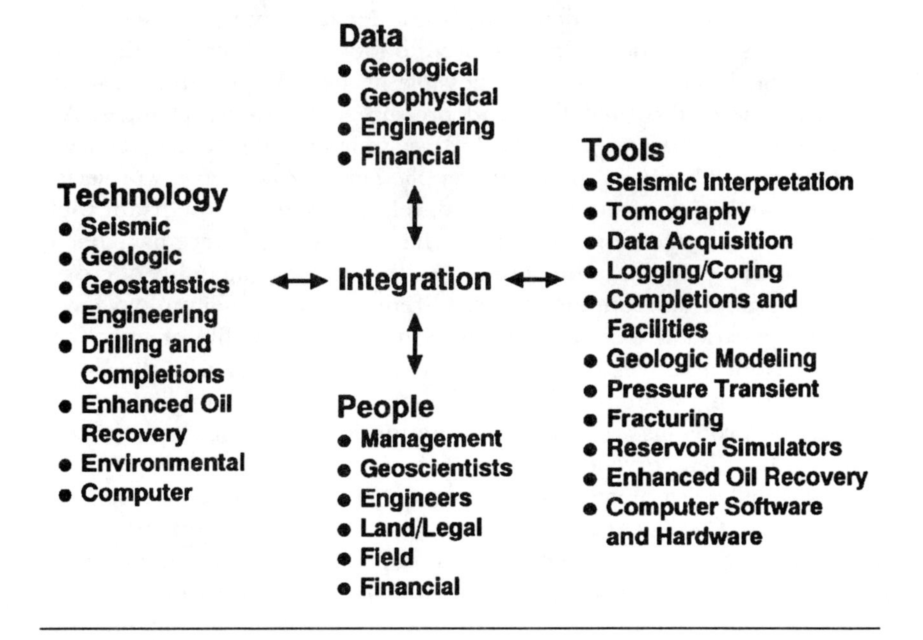

For this reason, geological and engineering data should be reviewed and coordinated with the geophysicists to determine whether or not an extension is possible for the drilling of an exploratory well. Most of the difficulties encountered in incorporating geological and engineering knowledge into seismic results and vice versa may be averted by an exchange of these ideas between the three disciplines.

Robertson of Arco points out that the geologic detail needed to properly develop most hydrocarbon reservoirs substantially exceeds the detail required to find them.[21] This perception has accelerated the application of 3D-seismic analysis to reservoir management. A 3D-seismic analysis can lead to identification of reserves that may not be produced optimally (or perhaps not produced at all) by the existing reservoir management plan. In addition, it can save costs by minimizing dry holes and poor producers.

The initial interpretation of a 3D-seismic survey affects the original development plan. With the development of the field, additional information is collected and is used to revise and refine the original interpretation. Note that the usefulness of a 3D-seismic survey lasts for the life of the reservoir.

The geophysicists' interpretation of the 3D-seismic data may be combined with the other relevant information regarding the reservoir (i.e., trap, fault, fracture pattern, shapes of the deposits). The 3D-seismic data guide interwell interpolations of reservoir properties. The reservoir engineer can use the seismic volume to understand lateral changes.

The 3D-seismic analysis can be used to look at the flow of fluids in a reservoir. Such flow surveillance is possible by acquiring baseline 3D-data before and after the fluid flow and pressure/temperature changes. Although flow surveillance with multiple 3D-seismic surveys is at an early stage of application, it has been successfully applied in thermal recovery projects.

Cross-well seismic tomography is developing into an important tool for reservoir management, and within the last few years there have been notable advances in the understanding of the imaging capability of cross-well tomograms. The fundamental requirements for the technology have been demonstrated. High-frequency seismic waves capable of traveling long interwell distances can be generated without damaging the borehole, and tomographic inversion techniques can give reliable images as long as the problems associated with nonuniform and incomplete sampling are handled correctly.

Cross-well seismology is becoming an important tool in reservoir management. Current applications focus on the monitoring of enhanced oil recovery processes, but perhaps most important is the potential of the method to improve our geological knowledge of the reservoir.

So far, most cross-well seismic surveys have been done for the purpose of mapping steam zones in steamflood operations. Seismology is well-suited for this application, since the presence of live steam in the reservoir sharply reduces the P-wave velocity. Moreover, for high-gravity oils in unconsolidated sand reservoirs the seismic P-wave velocity decreases significantly with increased temperature. Consequently, seismic velocities can be used as a measure of reservoir temperature and/or an indicator of live steam within the reservoir.

A number of oil fields in the San Joaquin Valley of California are characterized by high-gravity oil and shallow, unconsolidated reservoir rocks (e.g., Midway Sunset). Many of these fields, including Midway Sunset, have undergone years of steam drive to enhance oil recovery. Although most of these steam drive programs have been successful, there are common problems of poor sweep efficiency, gravity override, and steam channeling through zones of high permeability. It may be possible to reduce these problems by injecting foam along with the steam in order to partially plug these high-permeable zones.

It is important to establish the location of the steam front during the course of the steam-drive and to determine the effectiveness of the foam injection in altering any undesirable steam movement. Traditionally, temperature-monitoring wells are drilled. However, they are expensive and measure temperature at one location only; consequently, an alternate method is needed. Since seismic velocities are a sensitive function of temperature and phase of the reservoir fluids, and since surface seismic data is of very poor quality at Midway Sunset, cross-well seismic tomography surveys have been conducted in order to map the steam front.

The role of geology in reservoir simulation studies was well described by Harris in 1975. He described the geological activities required for constructing realistic mathematical reservoir models. These models are used increasingly to evaluate both new and mature fields and to determine the most efficient management scheme. Part of the information contained in the model is provided by the geologist, based on studies of the physical framework of the reservoir. However, for the studies to be useful the geologist must develop quantitative data. It is important that the geologist and the engineer understand each other's data.[22]

As described by Harris, both engineering and geological judgment must guide the development and use of the simulation model. The geologist usually concentrates on the rock attributes in four stages: (1) rock studies establish lithology and determine depositional environment, and reservoir rock is distinguished from nonreservoir rock; (2) framework studies establish the structural style and determine the three-dimensional continuity character and gross-thickness trends of the reservoir rock; (3) reservoir-quality studies determine the framework variability of the reservoir rock in terms of porosity, permeability, and capillary properties (the aquifer surrounding the field is similarly studied); and

(4) integration studies develop the hydrocarbon pore volume and fluid transmissibility patterns in three dimensions.

Throughout his work, the geologist requires input and feedback from the engineer. Examples of this "interplay of effort" are indicated in Figure 2–6.[22] Core-analysis measurements of samples selected by the geologist provide data for the preliminary identification of reservoir rock types. Well-test studies aid in recognizing flow barriers, fractures, and variations in permeability. Various simulation studies can be used to test the physical model against pressure-production performance; adjustments are made to the model until a match is achieved.

Many companies have initiated the development of a three-dimensional geological modeling program to automate the generation of geologic maps and cross-sections from exploration data. A good example of putting geology into reservoir simulations is described by Johnson and Jones.[23] The models are directly interfaced to the reservoir simulator; thus, the reservoir engineer utilizes the complex reservoir description provided by the geologist for field development planning. The reservoir engineer routinely and readily updates the model with new data or interpretations and quickly provides consistent maps and cross-section.

According to Johnson and Jones, the geologist can input structural and stratigraphic concepts as a series of computer grids honoring the geologic tops. Interpolations of logged porosity and other data from wells are controlled by this stratigraphic framework and fill a 3-D matrix of

FIGURE 2–6. General Geological Activities in Reservoir Description and Input from Engineering Studies *(Copyright © 1975, SPE, from* JPT, *May 1975*[22]*)*

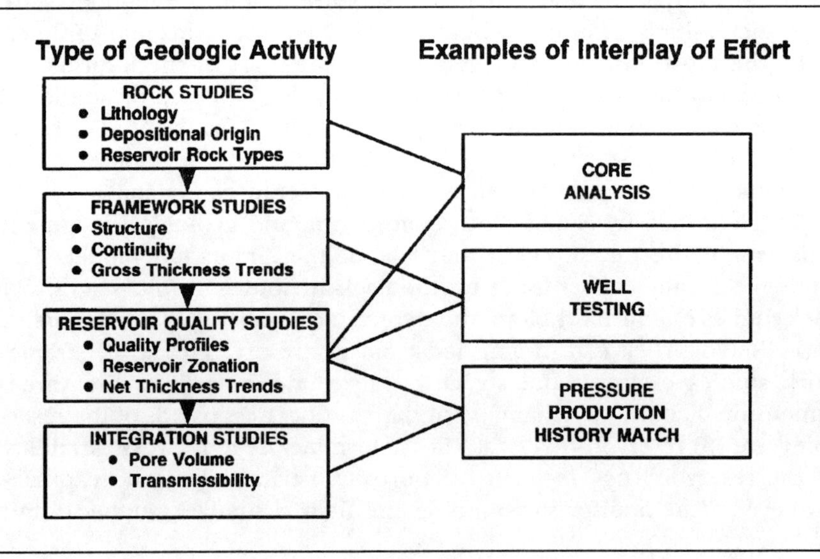

cells. Additional geological features critical to reservoir performance can be added to complete the geologist's picture of the reservoir.

Recently, Frank, Van Reet, and Jackson presented an excellent example of synergistic combination of geostatistics, 3-D seismic data, and well log data which contributed to the success in pinpointing infill drilling targets in Kingdom Abo field in Terry County, Texas.[24] In this project, geostatistics is a powerful tool for reservoir characterization, since it utilizes all well data in a manner that adheres to a model based on statistical and user-defined spatial correlations. When coupled with 3-D seismic, the end product is an interpretation of the reservoir that can be used to pinpoint additional development drilling locations.

INTEGRATING EXPLORATION AND DEVELOPMENT TECHNOLOGY

New developments in computer hardware, technology, and software are enhancing integration of multidisciplinary skills and activities. The mainframe supercomputers, more powerful personal computers, and workstations have revolutionized interdiscipline technical activities and industry business practices, making them more responsive and effective.

Recently, *Oil and Gas Journal* published a special report on "Integrating Exploration and Development Technology" using state-of-the-art computing and communications.[24–27] The *OGJ* special report states that integration is changing the way oil companies work. However, integration also creates challenges, from managing computer systems to designing organizations, to making best use of interdiscipline teams.

Neff and Thrasher captured the significant impacts that the late 1970s and 1980s new technologies made on the petroleum industry.[25] Major computer technologies include the supercomputers, interactive workstations, networking, rapid access mass-storage devices, and 3-D visualization hardware. These resources are enhancing the integration of the activities of the multidisciplinary groups by utilizing their own professional technologies, tools, and data.

Advancements in 3-D seismic acquisition and processing are credited to the massive number-crunching supercomputers such as Cray computers. 3-D seismic data along with computer-processed logs and core analyses characterize or describe more realistically and accurately the reservoir providing the 3-D computer maps. The reservoir engineers use these maps along with rock and fluid properties and production/injection data to simulate reservoir performance and to design depletion and development strategies for new and old fields. The supercomputers made reservoir simulators work faster and more accurately. The integration process from reservoir characterization to reservoir simulation, which requires interdisciplinary teamwork has been made practical and efficient by utilization of computers.

Interactive workstations interface several machines together locally in a physical cluster or using networks and software to link central processing units (CPUs) from various sites into a virtual cluster. The machines include high-end PCs, Suns, DECs, IBMs, MicroVAXes, Hewlett-Packard (HPs), and Silicon Graphics hardware. Contrary to the workings of the supercomputers and mainframe computers, the interactive workstations allow data migration, analysis, and interpretation on truly interactive domain rather than batch mode. The workstations are also capable of utilizing many geoscience and engineering software interactively. The demands for workstations of various kinds are ever increasing in the industry because they are becoming the workhorse of the integrated geoscience and engineering teams.

The computer networks that link the IBM mainframe computers, Cray supercomputers, Unix workstations, and PC token ring networks together provide the mechanism for effective communication and coordination from various geographical office locations. Major oil companies have worldwide computer links between all divisions and regional offices. The office-to-office communication has been made very quick (almost instantaneous), productive, and cost-effective by computer networking. The IBM mainframe-based PROFS/Office Vision electronic mail facilities, videoconference centers in various geographical locations, and workstations' images of maps, graphs, and reports via network communications are excellent examples of networking. The networks have made the tasks of the integrated teams easier, faster, and immensely productive.

While networks provide an efficient means to move digital data, retrieval and storing of data pose a major challenge in the petroleum industry today. The problems are:

- Incompatibility of the software and data sets from the different disciplines.
- Databases usually do not communicate with each other.

Many oil companies are staging an integrated approach to solving these problems.[28] In late 1990, several major domestic and foreign oil companies formed Petrotechnical Open Software Corporation (POSC) to establish industry standards and a common set of rules for applications and data systems within the industry. POSC's technical objective is to provide a common set of specifications for computing systems, which will allow data to flow smoothly between products from different organizations and will allow users to move smoothly from one application to another. POSC members are counting on POSC and its major software vendors to provide a long-term solution to database-related issues.

3-D computer visualization via a video monitor of a reservoir at a micro- or macro-scale is the latest major breakthrough in computer technology. The awesome power of visualization lies in its ability to synthesize diverse data types viz., geology, land, geophysics, petrophysics, drilling, and reservoir engineering, and attributes for better understanding and capturing by human senses. Figure 2–7 is an example of visualization of a Gulf Coast Mexico salt dome, blending many types of information. Figure 2–8 shows computer visualization of electron microscope pictures of rock samples alongside classic rock displays. 3-D visualization technique will enhance our understanding of the reservoir, providing better reservoir description and simulation of reservoir performance. It may very well be the most powerful and persuasive communication tool of the integrated teams for decades to come.

FIGURE 2–7. Integrating Exploration and Development Technology *(courtesy* OGJ, *May 1993[25], from Wyatt et al., "Ergonomics in 3-D depth migration," 62nd SEG Int. Mtg. and Exp., October 1992)*

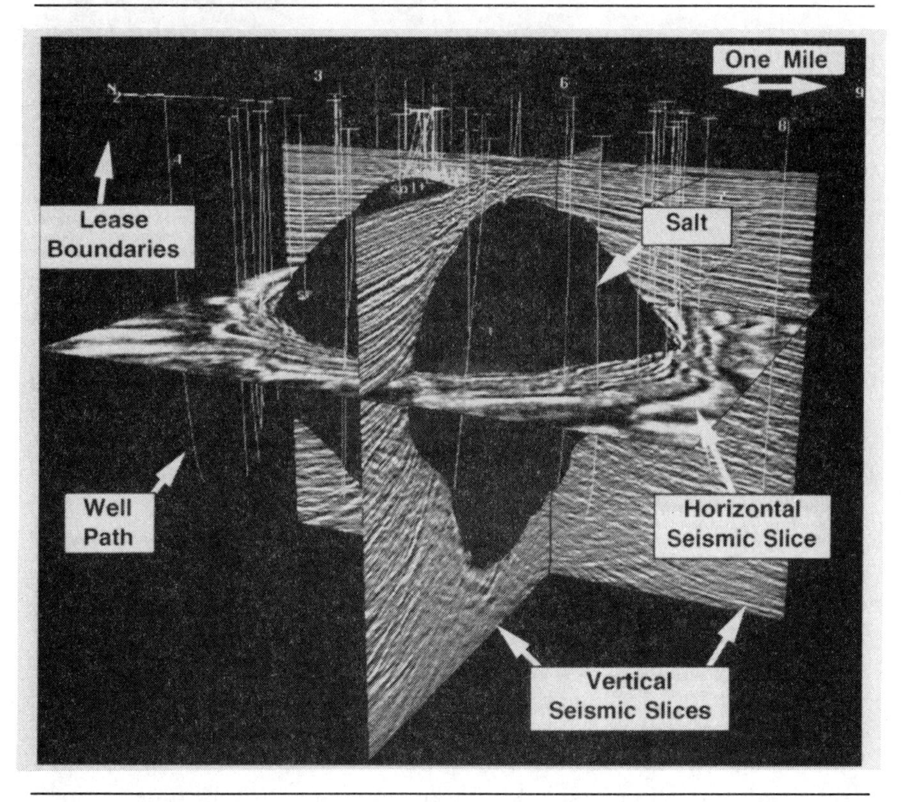

FIGURE 2–8. Integrating Exploration and Development Technology *(courtesy* OGJ, *May 1993*[25]*)*

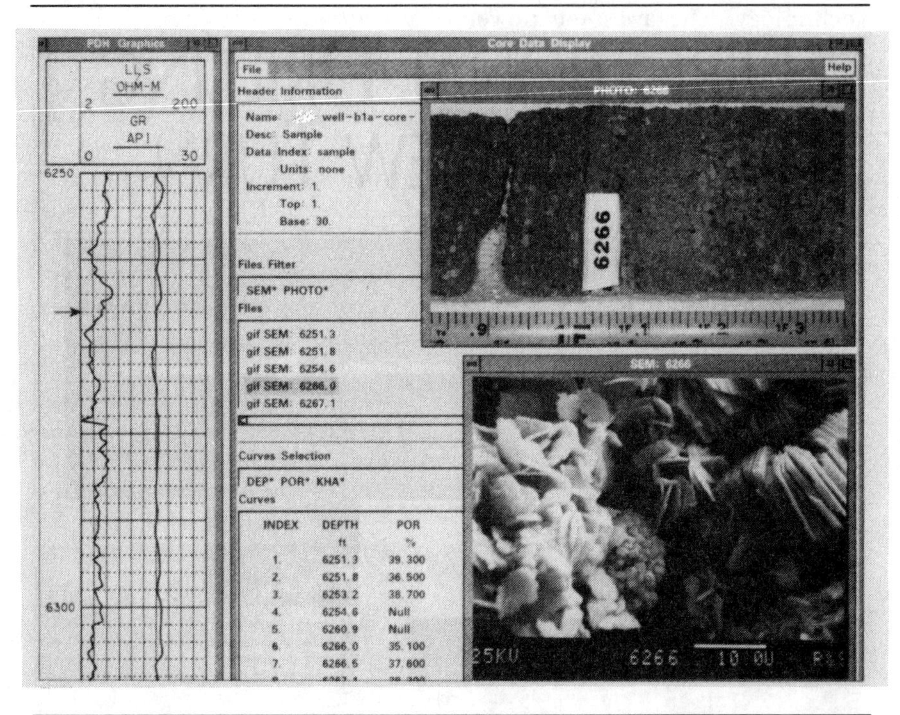

Now, a time and cost-effective way to integrate exploration and production activities using existing hardware and software is available.[26] A fully open-data exchange system, which was jointly created by Finder Graphics Systems Inc., GeoQuest, and Schlumberger, is being distributed as the Geoshare standards. Members of the Geoshare user's group, which consists of many geoscientific software developers and oil and gas operators, will soon be able to transfer data and interpretations among their various data bases in support of E&P techniques.

Guthery, Landgren, and Breedlove concluded that the published Geoshare standard provides means for exchanging data and results between any petroleum applications, regardless of their formats, configurations or hardware platforms.[26] It is a completely open and expandable standard whose future lies with the Geoshare user's group.

Traditionally, finding and producing hydrocarbons were considered the essence of success in the upstream end of the petroleum industry. Now, companies are viewing their options as far more flexible, and a diversified portfolio of skills within an integrated and flexible business framework is emerging (see Figure 2–9).[27] Patterson and Altieri discussed

FIGURE 2–9. Two Types of Organization *(Copyright © 1993, Gemini Consulting, courtesy* OGJ, *May 1993*[27]*)*

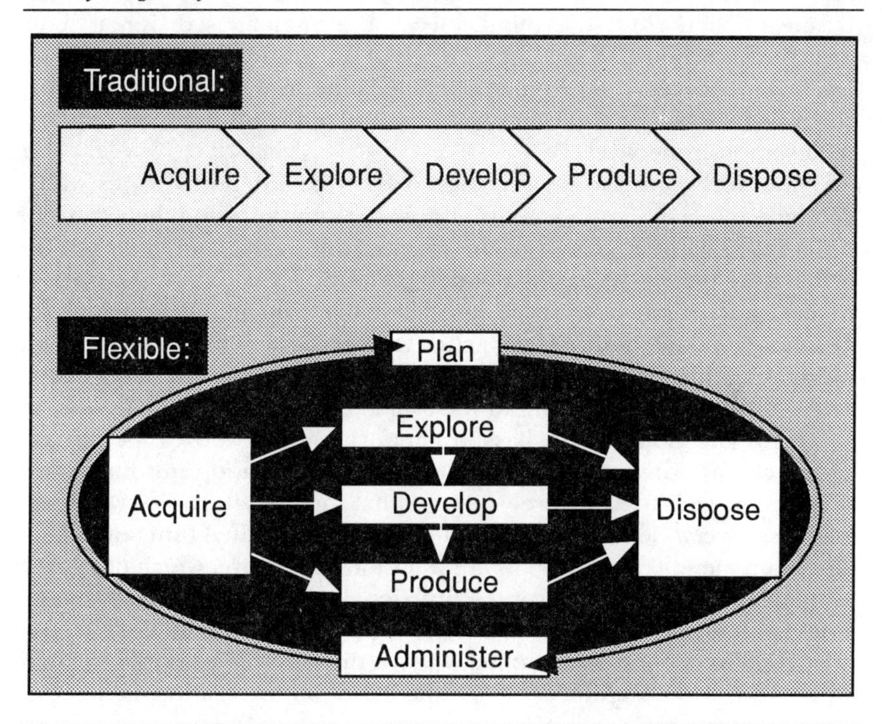

two dominant approaches to convert the "stovepipe" organization of separate functional units into an integrated organization.

In the "top-down" first approach, changes in strategy and management style made at the executive level are expected to filter down and throughout the organization over time. The use of asset management teams is now standard practice in many companies. Even though this kind of teamwork and flexibility is a step in the right direction, it does not really address broader organization and information technology issues. In the emerging second approach, which works from the bottom up, changes in information technology alone (both hardware and software) are intended to eliminate problems of knowledge transfer and communication.

While significant business improvements may be produced from the best of both approaches, neither provides the blueprints necessary for achieving the linkage between a company's strategic direction and its organization, operations, and information systems. Patterson and Altieri found that by modeling the "what" (the work of the business) and "why" (the purpose of the work), it is possible to build a stable blueprint that can be used to redesign and align the entire business. Once the fundamental purpose of business is defined, the modeling of the work that supports the purpose is approached first at the middle levels of the business. Then from the middle level, the analysis is driven in both directions—upward, to get higher levels of work, and downward, to lower levels. This process will produce a detailed blueprint of the essential work required to achieve the business purpose and of the information required to accomplish the work.

The work needs to be carried out by cross-functional teams with a common objective and smooth line of communication between the different functional groups of the organization. The team maps out the actual work flows, comparing them to the work defined in the completed model, which serves as the zero base. Since the team members analyze the business from the standpoint of actual work requirements (not functional assignments), they are able to see that individuals from different functional units are contributing to the same essential work.

Using the model as the basic framework, the old "stove-pipe" organization can be converted into an integrated organization in which cross-functional teams focus entirely on work that supports the objectives of the business. Everyone involved gains, not just an individual functional area. The team members share a common business objective, which effectively eliminates most political disputes, functional rivalries, and fear of change. The final result is a new, seamless organization that is flexible and adaptable to change. It can quickly move the focus of its work anywhere within the business life cycle to maximize value creation in an ever changing marketplace.

REFERENCES

1. Thakur, G. C. "Reservoir Management: A Synergistic Approach." Paper SPE 20138 presented at the SPE Permian Basin Oil and Gas Recovery Conference, Midland, Texas, March 8–9, 1990.
2. Satter, A., et al. "Reservoir Management: Technical Perspective." SPE Paper 22350 presented at the SPE International Meeting on Petroleum Engineering held in Beijing, China, March 24–27, 1992.
3. Haldorsen, H. H. and T. Van Golf-Racht. "Reservoir Management into the Next Century." Paper NMT 890023 presented at the Centennial Symposium at New Mexico Tech., Socorro, NM, October 16–19, 1989.
4. Wiggins, M. L. and R. A. Startzman. "An Approach to Reservoir Management." SPE Paper 20747, Reservoir Management Panel Discussion, SPE 65th Ann. Tech. Conf. & Exhib., New Orleans, LA, September 23–26, 1990.
5. Robertson, J. D. "Reservoir Management Using 3D Seismic Data," *JPT* (July 1989): 663–667.

6. Wyllie, M. R. J. "Reservoir Mechanics—Stylized Myth or Potential Science?" *JPT* (June 1962): 583–588.
7. Essley, P. L. "What is Reservoir Engineering?" *JPT* (January 1965): 19–25.
8. Craig, F. F., et al. "Optimized Recovery Through Continuing Interdisciplinary Cooperation," *JPT* (July, 1977): 755–760.
9. Harris, D. G. and C. H. Hewitt. "Synergism in Reservoir Management—The Geologic Perspective," *JPT* (July 1977): 761–770.
10. Calhoun, J. C. "A Definition of Petroleum Engineering," *JPT* (July 1963): 725–727.
11. Amyx, Bass and Whiting. *Petroleum Reservoir Engineering.* New York: McGraw-Hill Book Company, 1960.
12. Goolsby, J. L. "The Relation of Geology to Fluid Injection in Permian Carbonate Reservoirs in West Texas," *S.W. Pet.* Short Course, Lubbock, TX, 1965.
13. Jordan, J. K. "Reliable Interpretation of Waterflood Production Data," *JPT* (August 1955): 18–24.
14. Halbouty, M. T. "Synergy is Essential to Maximum Recovery," *JPT* (July 1977): 750–754.
15. Sneider, R. M. "The Economic Value of a Synergistic Organization," Archie Conference, Houston, TX, October 22–25, 1990.
16. Talash, A. W. "An Overview of Waterflood Surveillance and Monitoring," *JPT* (December 1988): 1539–1543.
17. "Teamwork, New Technology, and Mature Reservoirs," *JPT* (January 1992): 38–40.
18. Lantz, J. R. and N. Ali. "Development of a Mature, Giant Offshore Oil Field," *JPT* (April 1991): 392–397.
19. Sessions, K. P. and D. H. Lehman. "Nurturing the Geology—Reservoir Engineering Team: Vital for Efficient Oil and Gas Recovery." SPE Paper 19780 presented at the Annual Technical Conference and Exhibition," San Antonio, TX, October 8–11, 1989.
20. Satter, A. "Reservoir Management Training—An Integrated Approach." SPE Paper 20752, Reservoir Management Panel Discussion, SPE 65th Annual Technical Conference & Exhibition, New Orleans, LA, September 23–26, 1990.
21. Robertson, J. D. "Reservoir Management Using 3-D Seismic Data," *Geophysics: The Leading Edge of Exploration* (February 1989): 25–31.
22. Harris, D. G. "The Role of Geology in Reservoir Simulation Studies," *JPT* (May 1975): 625–632.
23. Johnson, C. R. and T. A. Jones. "Putting Geology Into Reservoir Simulations: A Three-Dimensional Modeling Approach." SPE Paper 18321, presented at the Annual Technical Conference and Exhibition, Houston, TX, October 2–5, 1988: 585–594.
24. Frank, Jr., J. R., E. Van Reet and W. D. Jackson. "Combining Data Helps Pinpoint Infill Drilling Targets in Texas Field," *Oil & Gas J.* (May 31, 1993): 48–53.

25. Neff, D. B. and T. S. Thrasher. "Technology Enhances Integrated Teams' Use of Physical Resources," *Oil & Gas J.* (May 31, 1993): 29–35.
26. Guthery, S., K. Landgren and J. Breedlove. "Data Exchange Standard Smooths E&P Integration," *Oil & Gas J.* (May 31, 1993): 36–42.
27. Patterson, S. and J. Altieri. "Business Modeling Provides Focus for Upstream Integration," *Oil & Gas J.* (May 31, 1993): 43–47.
28. Johnson, J. P. "POSC Seeking Industry Software Standards, Smooth Data Exchange," *Oil & Gas J.* (October 26, 1992): 64–68.

CHAPTER 3

▼ ▼ ▼

Reservoir Management Process

The modern reservoir management process involves establishing a purpose or strategy and developing a plan, implementing and monitoring the plan, and evaluating the results (Figure 3–1).[1] None of the components of reservoir management is independent of the others. Integration of all these are essential for successful reservoir management. It is dynamic and ongoing. As additional data become available, the reservoir

FIGURE 3–1. *(Copyright © 1992, SPE, from paper 22350[1])*

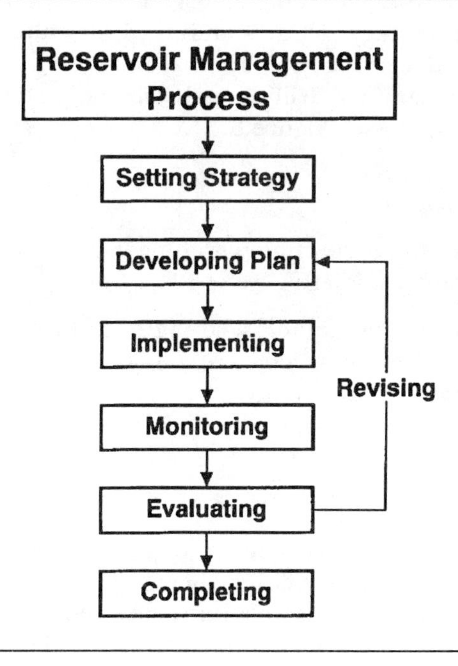

management plan is refined and implemented with appropriate changes. While a comprehensive plan for reservoir management is highly desirable, every reservoir may not warrant such a detailed plan because of cost effectiveness. However, the key to success is to have a management plan (whether so comprehensive or not) and implement it from day one.

SETTING GOALS[1]

Recognizing the specific need and setting a realistic and achievable purpose is the first step in reservoir management. The key elements for setting a reservoir management goal are:

- Reservoir characteristics.
- Total environment.
- Available technology.

Understanding of each of these elements is the prerequisite to establishing short- and long-term strategies for managing reservoirs.

Reservoir Characteristics

The nature of the reservoir being managed is vitally important in setting its management strategy. Understanding the nature of the reservoir requires a knowledge of the geology, rock and fluid properties, fluid flow and recovery mechanisms, drilling and well completions, and past production performance (see Figure 3–2).

Total Environment

Understanding of the following environments is essential in developing management strategy and effectiveness:

- Corporate—goal, financial strength, culture, and attitude.
- Economic—business climate, oil/gas price, inflation, capital, and personnel availability.
- Social—conservation, safety, and environmental regulations.

Technology and Technological Toolbox

The success of reservoir management depends upon the reliability and proper utilization of the technology being applied concerning exploration, drilling and completions, recovery processes, and production. Many technological advances have been made in all of these areas (see Table 3–1). However, they offer opportunities that may or may not be appropriate for every reservoir.

FIGURE 3–2. Reservoir Knowledge *(Copyright © 1992, SPE, from paper 22350[1])*

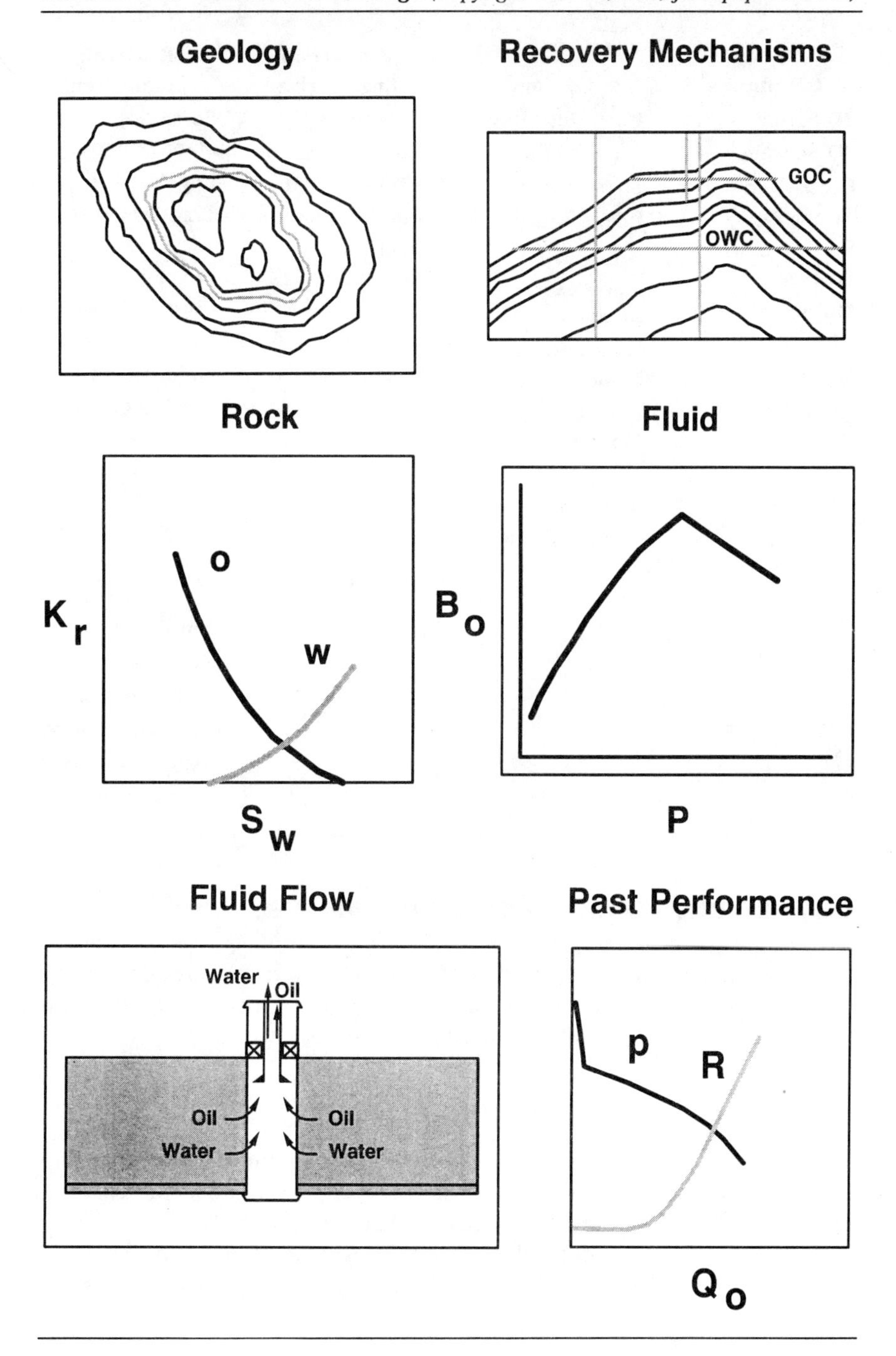

TABLE 3–1. Technology *(Copyright © 1992, SPE, from paper 22350[1])*

Geophysics	Geology	Production Engineering	Reservoir Engineering
2D Seismic	Core Description	Economics	Portfolio Management
3D Seismic	Thin Sections	Data Acquisition & Management	Log Analysis
Cross-Hole Tomography	Microscopes	Well Stimulation	Transient Well Tests
Vertical Seismic Profile	Image Analysis	Pipeflow Simulation	Conventional Core Analysis
Multicomponent Seismic	X-Ray	Wellbore Simulation	CT Scan, NMR
Shear Wave Logging	Stable Isotope Analysis	Nodal Analysis	Fluid Analysis
	Depositional Models		Decline Curve Analysis
	Diagenetic Models		Material Balance
	Maps, Cross-Sections		Waterflood
	Remote Sensing		Streamtube Models
			Reservoir Simulation
			Geostatistics
			EOR Screening
			EOR Technology
			Expert Systems
			Neural Networks

DEVELOPING PLAN AND ECONOMICS[1]

Formulating a comprehensive reservoir management plan is essential for the success of a project. It needs to be carefully worked out involving many time-consuming development steps (see Figure 3–3).

Development and Depletion Strategy

The most important aspect of reservoir management deals with the strategies for depleting the reservoir to recover petroleum by primary and applicable secondary and enhanced oil recovery methods.

Development and depletion strategies will depend upon the reservoir's life stage. In case of a new discovery, we need to address the

FIGURE 3–3. Developing Plan *(Copyright © 1992, SPE, from paper 22350[1])*

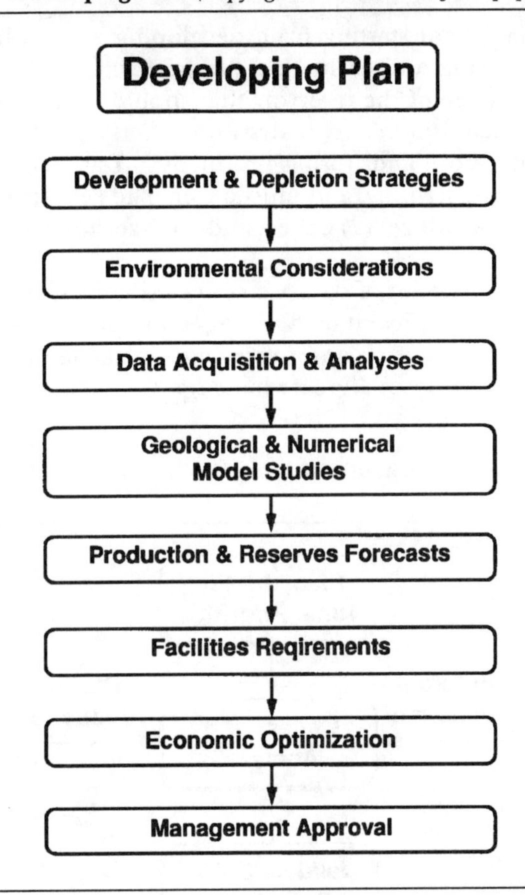

question of how to best develop the field (i.e., well spacing, number of wells, recovery schemes, primary, and subsequently secondary and tertiary). If the reservoir has been depleted by primary means, secondary and even tertiary recovery schemes need to be investigated.

Environmental Considerations

In developing and subsequently operating a field, environmental and ecological considerations have to be included. Regulatory agency constraints will also have to be satisfied. These are very sensitive and important aspects of the reservoir management process.

Data Acquisition & Analysis

Reservoir management starting from developing a plan, implementing the plan, monitoring and evaluating the performance of the reservoir requires a knowledge of the reservoir that should be gained through an integrated data acquisition and analysis program. Figure 3–4 shows a list of data needed before and during production. Data analyses require a great deal of effort, scrutiny, and innovation. The key steps are (1) plan, justify, time, and prioritize, (2) collect and analyze, and (3) validate/store (data base).

An enormous amount of data are collected and analyzed during the life of a reservoir. An efficient data management program—consisting of collecting, analyzing, storing and retrieving—is needed for sound reservoir management. It poses a great challenge.

FIGURE 3–4. Data Acquisition and Analysis *(Copyright © 1992, SPE, from paper 22350[1])*

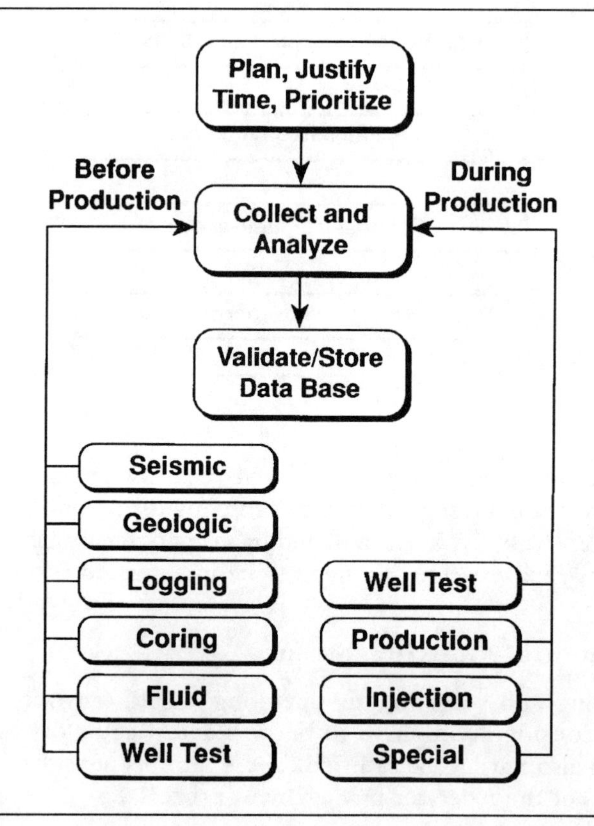

Geological and Numerical Model Studies

The geological model is derived by extending localized core and log measurements to the full reservoir using many technologies, such as geophysics, mineralogy, depositional environment and diagenesis. The geological model, particularly the definition of geological units and their continuity and compartmentalization, is an integral part of geostatistical and ultimately reservoir simulation models.

Production and Reserves Forecasts

The economic viability of a petroleum recovery project is greatly influenced by the reservoir production performance under the current and future operating conditions. Therefore, the evaluation of the past and present reservoir performance and forecast of its future behavior is an essential aspect of the reservoir management process (see Figure 3–5). Classical

FIGURE 3–5. Production and Reserves Forecasts *(Copyright © 1992, SPE, from paper 22350[1])*

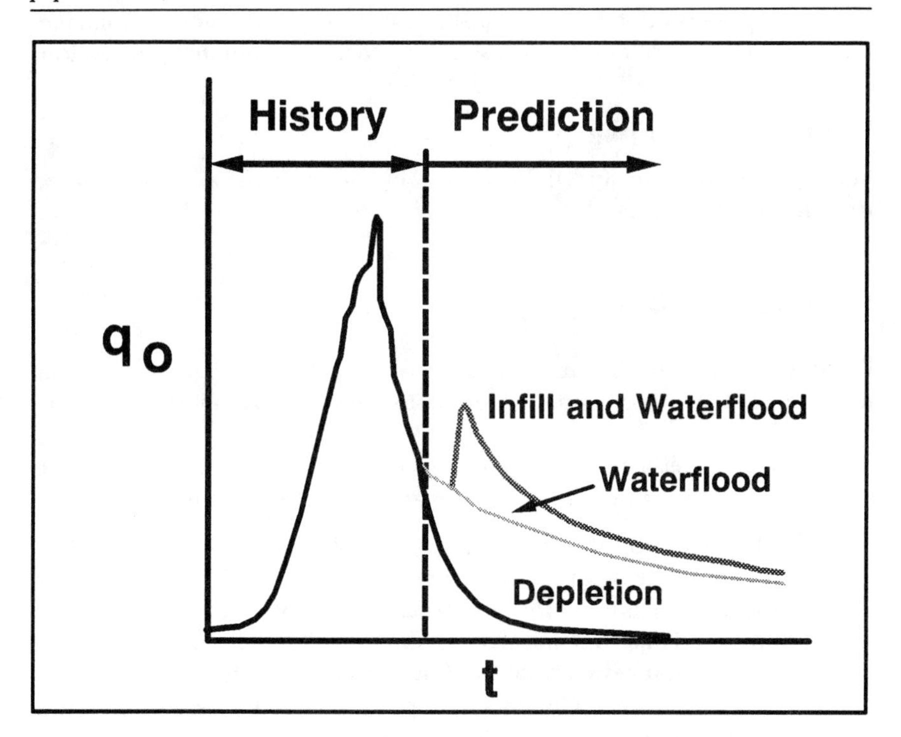

volumetric, material-balance, and decline-curve analysis methods, and high-technology black oil, compositional and enhanced oil recovery numerical simulators are used for analyzing reservoir performance and estimating reserves. Reservoir simulators play a very important role in formulating initial development plans, history matching and optimizing future production, and in planning and designing enhanced oil recovery projects.

Facilities Requirements

Facilities are the physical link to the reservoir. Everything we do to the reservoir, we do through the facilities. These include drilling, completion, pumping, injecting, processing, and storing. Proper design and maintenance of facilities has a profound effect on profitability. The facilities must be capable of carrying out the reservoir management plan, but they cannot be wastefully designed.

Economic Optimization

Economic optimization is the ultimate goal selected for reservoir management. Figure 3–6 presents the key steps involved in economic optimization.

Management Approval

Management support and field personnel commitment are essential for the success of a project.

IMPLEMENTATION

Once the goals and objectives have been set and an integrated reservoir management plan has been developed, the next step is to implement the plan.

Table 3–2 describes a step-by-step procedure on how to improve success in implementing a reservoir management program.

- The first step involves starting with a plan of action, including all functions. It is common for many reservoir management efforts to devise a plan, but this plan usually does not involve all functional groups. Thus, not all groups buy into these programs, and the cooperation between various functions is below the desired level. If a plan is to be developed and implemented in the best way, it must have commitment from all disciplines, including management.

FIGURE 3–6. Economic Optimization *(Copyright © 1992, SPE, from paper 22350[1])*

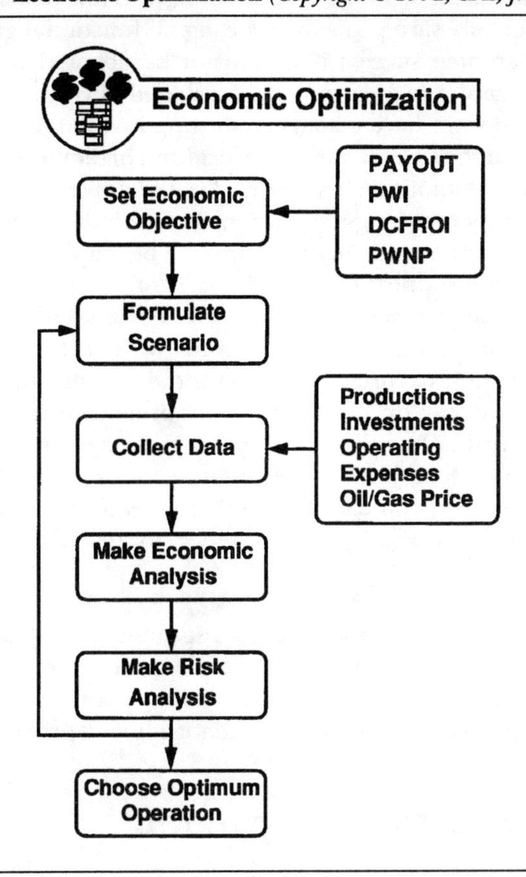

TABLE 3–2. How to Improve Success in Implementing a Reservoir Management Program *(Copyright © 1990, SPE, from paper 20748[2])*

- Start with a plan of action, involving all functions.
- Flexible plan.
- Management support.
- Commitment of field personnel.
- Periodic review meetings, involving all team members (inter-disciplinary cooperation in teaching each other's functional objectives).

- The plan must be flexible. Even if the reservoir management team members prepare plans by involving all functional groups, it does not guarantee success if it can not be adapted to surrounding circumstances (e.g., economic, legal, and environmental).
- The plan must have management support. No matter how technically good the plan, it must have local and higher level management blessings. Without their support, it would not be approved. Thus, it is necessary that we get management involved from "day one."
- No reservoir management plan can be implemented properly without the support of the field personnel. Time and time again we have seen reservoir management plans fail because either they are imposed on field personnel without thorough explanations or they are prepared without their involvement. Thus, the field personnel do not have a commitment to these plans.
- It is critical to have periodic review meetings, involving all team members. Most, if not all, of these meetings should be held in the field offices. The success of these meetings will depend upon the ability of each team member to teach their functional objectives.[2]

The important reasons for failure to successfully implement a plan are: (1) lack of overall knowledge of the project on the part of all team members, (2) failure to interact and coordinate the various functional groups, and (3) delay in initiating the management process.

SURVEILLANCE AND MONITORING[1]

Sound reservoir management requires constant monitoring and surveillance of the reservoir performance as a whole in order to determine if the reservoir performance is conforming to the management plan. In order to carry out the monitoring and surveillance program successfully, coordinated efforts of the various functional groups working on the project are needed.

An integrated and comprehensive program needs to be developed for successful monitoring and surveillance of the management project. The engineers, geologists, and operations personnel should work together on the program with management support. The program will depend upon the nature of the project. Ordinarily, the major areas of monitoring and surveillance involving data acquisition and management include: (1) oil, water and gas production, (2) gas and water injection, (3) static and flowing bottom hole pressures, (4) production and injection tests, (5) injection and production profiles, and any others aiding surveillance. In case of enhanced oil recovery projects, the monitoring and surveillance program is particularly critical because of the inherent uncertainties.

A detailed waterflooding and surveillance technique is described in a recent SPE Distinguished Author Series paper[3] which is included in appendix A.

EVALUATION[1]

The plan must be reviewed periodically to ensure that it is being followed, that it is working, and that it is still the best plan. The success of the plan needs to be evaluated by checking the actual reservoir performance against the anticipated behavior.

It would be unrealistic to expect the actual project performance to match exactly the planned behavior. Therefore, certain technical and economic criteria need to be established by the functional groups working on the project to determine the success of the project. The criteria will depend upon the nature of the project. A project may be a technical success but an economic failure.

How well is the reservoir management plan working? The answer lies in the careful evaluation of the project performance. The actual performance (e.g., reservoir pressure, gas-oil-ratio, water-oil-ratio, and production) needs to be compared routinely with the expected (see Figure 3–7).

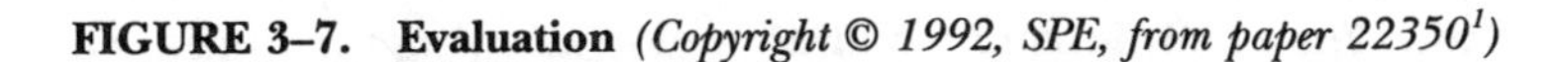

FIGURE 3–7. Evaluation *(Copyright © 1992, SPE, from paper 22350[1])*

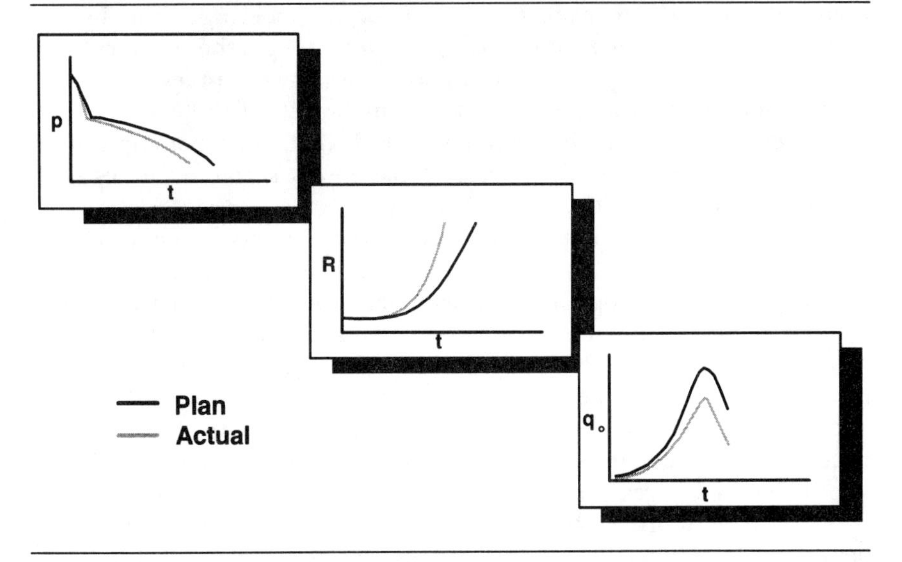

In the final analysis, the economic yardsticks will determine the success or failure of the project.

REVISION OF PLANS & STRATEGIES[1]

Revision of plans and strategies is needed when the reservoir performance does not conform to the management plan or when conditions change. The answers to questions such as is it working, what needs to be done to make it work, what would work better, must be asked and answered on an ongoing basis in order for us to say we are practicing sound reservoir management.

REASONS FOR FAILURE OF RESERVOIR MANAGEMENT PROGRAMS

There are numerous reasons why reservoir management programs have failed. Some of the reasons are listed below:

Unintegrated System

It was not considered as a part of a coupled system consisting of wells, surface facilities, and the reservoir. Not all of these were emphasized in a balanced way. For example, one could do well in studying the fluids and their interaction with rock (i.e., reservoir engineering); but, by not considering the well and/or the surface system design, the recovery of oil and/or gas was not optimized. Most people can cite examples of mistakes made where we thoroughly studied various aspects of the reservoir and made decisions resulting in too many wells drilled, improper application of well completion technology, and/or inadequate surface facilities available for future expansion.

Perhaps the most important reason why a reservoir management program is developed and implemented poorly is an unintegrated group effort. Sometimes the operating decisions are made by people who do not recognize the dependence of one system on the other. Also, the people may not have the required background knowledge in critical areas (e.g., reservoir engineering, geology and geophysics, production and drilling engineering, and surface facilities). Although it may not be absolutely necessary for reservoir-management decision makers to have a working knowledge in all areas, they must have an intuitive feel for them.

The team approach to reservoir management involving interaction between various functions has been recently emphasized (see Figure 3–8).[2]

FIGURE 3–8. Reservoir Management Approach *(Copyright © 1990, SPE, from paper 20748[2])*

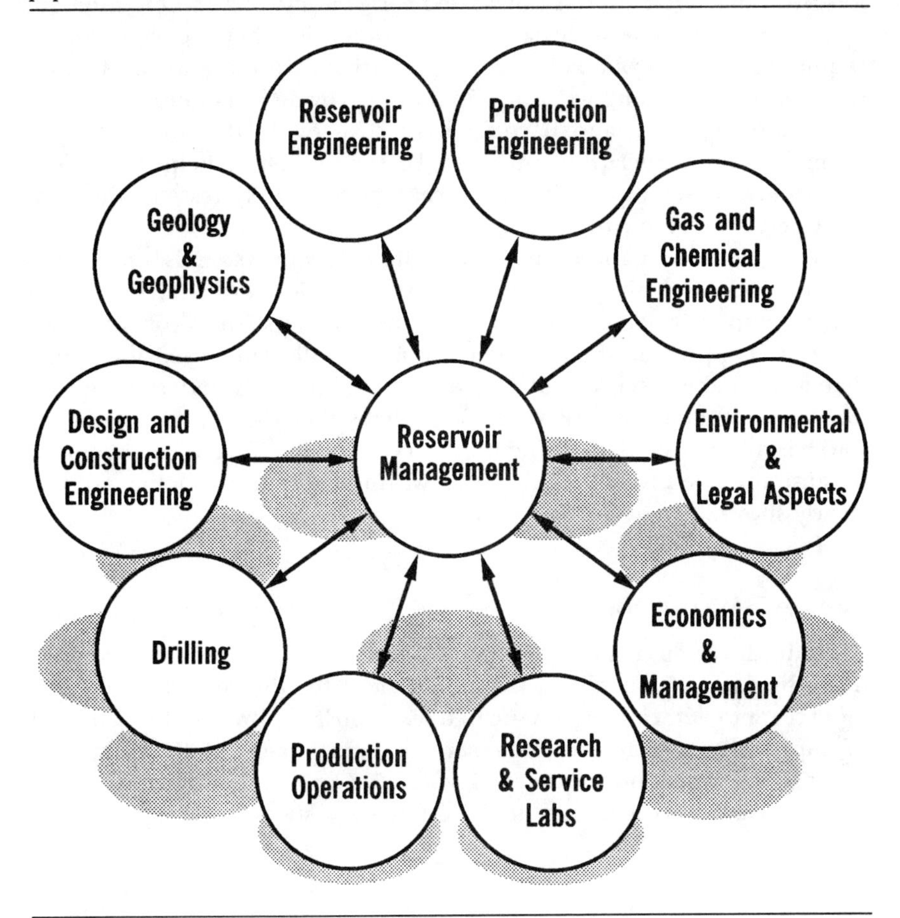

It is suggested that the team members work as a well-coordinated "basketball team" rather than a "relay team." Constant interaction between various functions is required in the team effort. Note that the synergism of the team approach can yield a "whole is greater than the sum of its parts" effect. Thus, interaction between various engineering functions, production operations, geology and geophysics; and their interaction with management, economics, proration, legal, and environmental groups are both critical to a successful reservoir management program. This statement is basically an extension of the idea advocated by Talash that "Teamwork between reservoir engineers and production/operations engineers is essential to waterflood project management."[5]

Starting Too Late

Reservoir management was not started early enough; and when initiated, management became necessary because of a crisis that occurred, and it required a major problem to be solved. Early initiation of a coordinated reservoir management program could have provided a better monitoring and evaluating tool, and it could have cost less in the long run. For example, a few early Drill Stem Tests (DSTs) could have helped decide if and where to set pipe. Also, performing some early tests could have indicated the size of the reservoir.

Early definition and evaluation of the reservoir system is a prerequisite to good reservoir management. The collection and analysis of data play an important role in the evaluation of the system. Most often, an integrated approach of data collection is not followed, especially immediately after the discovery of a reservoir. Also, in this endeavor not all functions are generally involved. Sometimes the reservoir management staff has difficulties in justifying the data collection effort to management because the need for the data, along with its costs and benefits, are not clearly shown.

Lack of Maintenance

Calhoun draws an analogy between reservoir and health management.[6] According to his concept, it is not sufficient for the reservoir management team to determine the state of a reservoir's health and then attempt to improve it. One reason for reservoir management ineffectiveness is that the reservoir and its attached system's (wells and surface facilities) health (condition) is not maintained from the start.

RESERVOIR MANAGEMENT CASE STUDIES

A comprehensive plan for reservoir management, including a team approach, is highly desirable. However, every reservoir may not warrant such a detailed plan because of cost-benefit considerations. With this in mind, two approaches utilizing case studies are described as follows.

The first case study, North Ward Estes field, illustrates the application of a comprehensive approach; whereas the second, Columbus Gray lease, discusses a problem-solving approach to reservoir management. Both approaches have shown positive results. Although they are philosophically quite different, each has its own merits.

The problem-solving approach is based upon the following:

- An action plan for evaluating and increasing the net worth of reservoirs is prepared by involving a selected group of personnel, and it is based upon the best available data.
- In the problem-solving sessions, an informal exchange of ideas takes place, and problems associated with current operating practices are defined. Next, specific recommendations aimed at enhancing reservoir performance are suggested, and pros and cons for each recommendation are evaluated. If the required relevant data are not available, then either they are assumed or collected in the field, keeping the cost-benefit analysis in mind.

North Ward Estes Field[2]

The North Ward Estes (NWE) field, located in Ward and Winkler counties, Texas (see Figure 3–9), was discovered in 1929. It is an 18 mile × 4 mile anticlinorium. Cumulative oil production from primary and secondary recovery has been in excess of 320 million barrels, or about 25% OOIP, from more than 3,000 wells. The field has been waterflooded since 1955. Geologically, the field resides on the western edge of the

FIGURE 3–9. North Ward Estes Field Location Map *(Copyright © 1990, SPE, from paper 20748[2])*

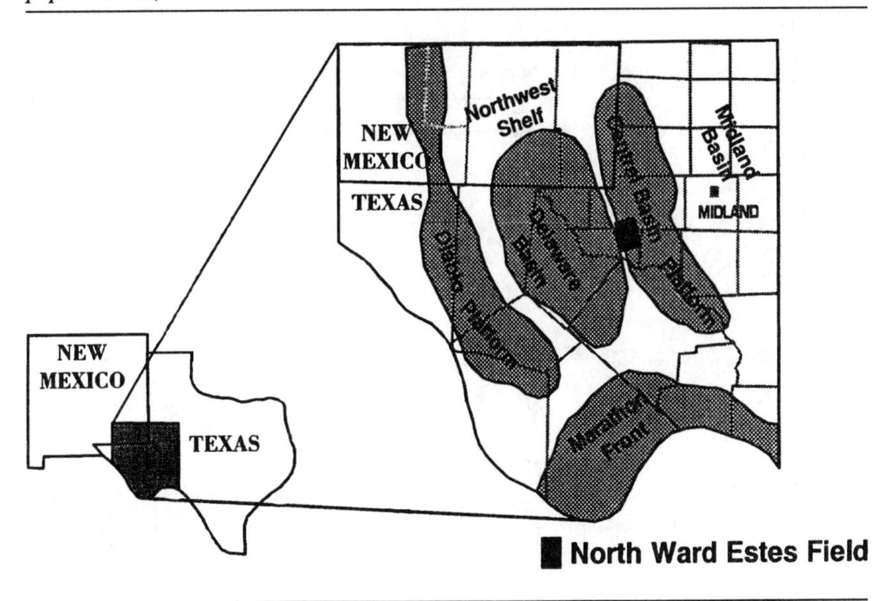

Central Basin Platform. The field is part of an Upper Guadalupian productive trend that extends uninterrupted for 90 miles on the edge of the platform (see Figure 3–10).

The average reservoir depth is 2,600 feet; porosity and permeability average 19% and 19 MD, respectively. The reservoir temperature is 83°F. The flood patterns are generally 20-acre, five spots and line drives.

Field Information and Geology

The field was initially developed on 20-acre spacing. Later, however, the most productive parts of the field were drilled on 10-acre spacing. Until the 1950s, the wells were mostly completed open-hole and shot with nitroglycerine. Perforated liners were then hung from the casing, which was set above the productive formation in the gas sands.

After 1950, the wells were completed cased-hole, hydraulically fractured, and acid stimulated. About one-half of the current producers and injectors are cased-hole. Table 3–3 provides additional information on the field history, structure, and stratigraphy.

The producing formations are Yates and Queen sands, but most of the production has been from the Yates sands (see Figure 3–11). They consist of very fine-grained sandstones to siltstones, separated by dense dolomite beds. These sands, as shown in Figure 3–11, are: A, BC, D, E, F, stray sands, J_1, J_2, and J_3.

FIGURE 3–10. *(Copyright © 1990, SPE, from paper 20748[2])*

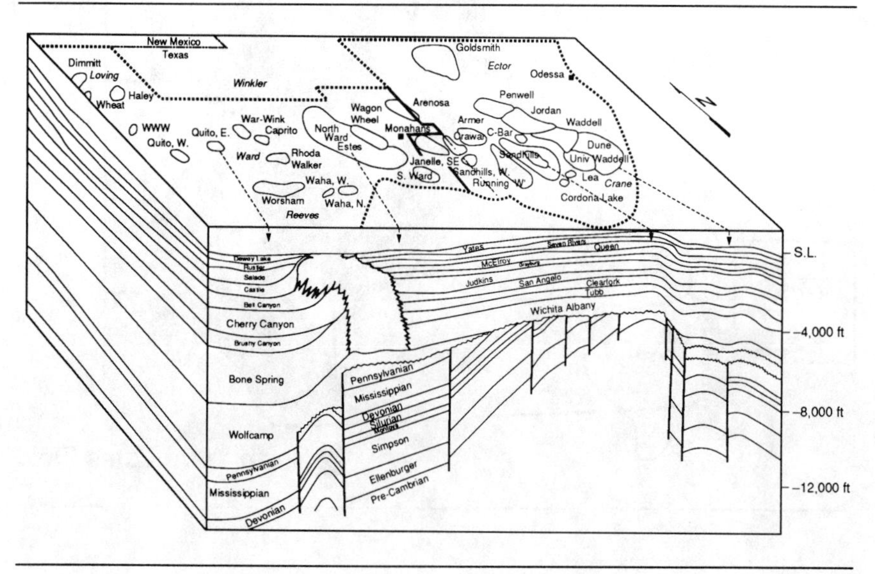

TABLE 3–3. North Ward Estes Field *(Copyright © 1990, SPE, from paper 20748[2])*

History
- 1929—North Ward Field discovered—G. W. O'Brien #4, Section 19
- 1936—Estes Field discovered—E. W. Estes #1 Section 38
- 1944—Fields combined
- 1955—Waterflood began
- 1981—Polymer projects began
- 3,000 + wells drilled
- Active wells—1,301 producers, 982 injectors, 10-acre spacing

Structure
- Low relief antiform
- Central basin platform homocline

Stratigraphy
- Primary production—Yates—Average depth—2,600 ft
- Secondary production—Queen—Average depth—3,100 ft
- Age—Permian (Late Guadalupian)
- Lithology—Very fine grain sand and siltstones dolomite/anhydrite interbedded
- Average porosity—19%
- Average permeability—19 md.
- Environment—Tidal Flat

FIGURE 3–11. Type Log for North Ward Estes Field *(Copyright © 1990, SPE, from paper 20748[2])*

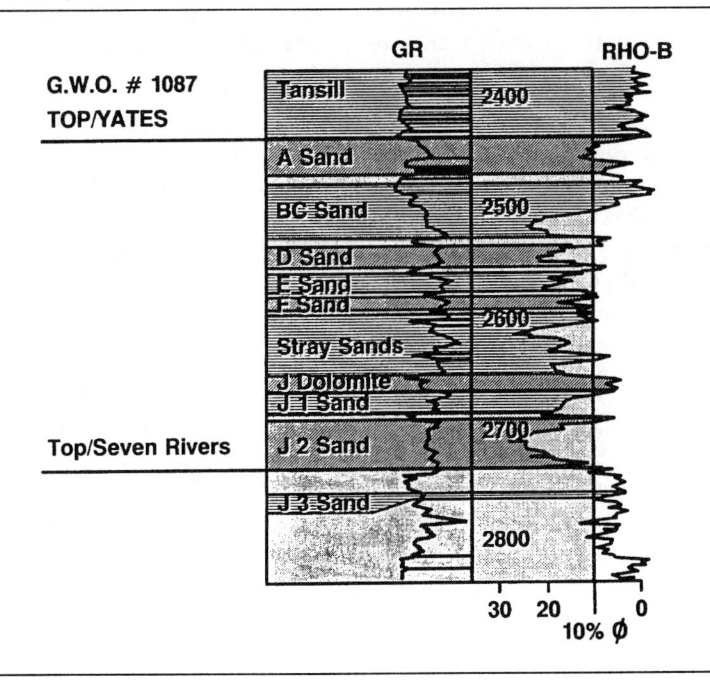

Most of the BC was in the original gas cap and consists of siltstones to fine-grained sandstone with clay. The D and E sands are similar to BC. The stray is composed of thin-bedded, lenticular, siltstones, and fine-grained sandstones, with high clays. The J_1 and J_2 sands are composed of coarser sands with much less clay content; therefore, they have higher porosities and permeabilities. Generally, the J_3 is not well developed and is wet in most areas.

The Queen formation, which lies below the Yates sands, is composed of intervals of fine-grained sandstones to siltstones and numerous thin, lenticular sands with poor lateral continuity. Thus, the Queen sand has been difficult to waterflood.

Reservoir Management Team

A team including all functional groups, as shown in Figure 3–8, was formed to investigate all pertinent options for optimizing recovery from the field. The following describes the results of the team effort.

Geological Characterization

A correlation scheme was developed for the field based upon laterally continuous key dolomites that bracket the productive sands and segment the reservoir into discrete mappable units. A computer database was built by our geologists to facilitate the processing and integrating of large volumes of data to aid in the geological characterization study. The database components were:

- Wireline log data from 3,300 wells, which included about 15 million curve feet.
- Core data from 538 wells, which totaled about 30,000 feet of analyses and lithology description.
- Marker data for more than 60,000 correlation markers.
- Fluid contact data (i.e., original gas-oil and oil-water).
- Production data, consisting of historical and wellbore data, including diagrams.

Core analyses were depth-corrected. Logs were normalized using a 60-foot interval of laterally continuous anhydritic dolomite. Core porosity data were cross-plotted versus bulk density log values to develop transforms for derivation of porosity.[7] Corrections for hole rugosity, overburden pressure, and lithologic complications were applied to refine the porosity transform.[8] The final transforms are shown in Figure 3–12.

As seen in Figure 3–12, the correlation between porosity and permeability is poor. However, when the correlation based upon lithofacies was made, increased correlation coefficients were obtained.[8] Structure and

FIGURE 3–12. *(Copyright © 1990, SPE, from paper 20748[2])*

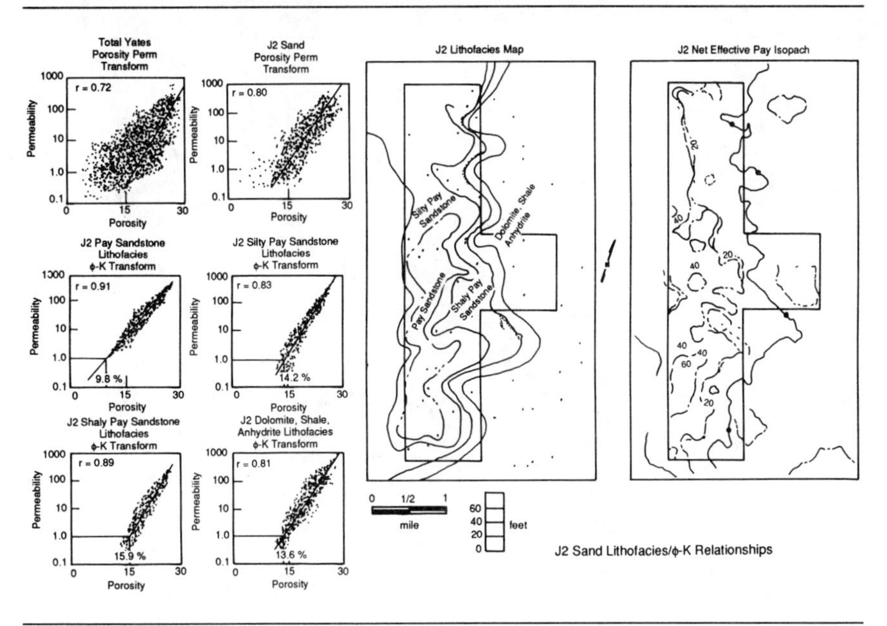

porosity-feet maps were merged with fluid contact and water saturation data to calculate volumetrics. Facies relationships and actual-to-apparent pay ratios were applied to determine effective hydrocarbon pore volume. Computer-generated net isopach maps of the sands display a north-south strike. The sands pinch out into an evaporite facies updip and a carbonate facies downdip.

About 11 man-years and $1.6 million were spent to achieve the above results.[4] Figure 3–13 summarizes computer-aided characterization study steps. Normalized log and core data, markers, fluid contacts, and production data were quality checked and corrected for any errors. The output included maps (e.g., structure, isopach and porosity-thickness), porosity vs. permeability plots, water saturation and volumetric data, production plots, and cross-sections, including wellbore diagrams. An example of a sand trend cross-section is shown in Figure 3–14. It is based upon basic geologic data and is supported by production data.

One of the results of the characterization study has been identification of well workovers. In addition, several waterflood projects were designed and implemented. A waterflood project that did not prove as successful as others was later analyzed in terms of the characterization study. If the project had been considered after the study, it probably would not have been implemented, and considerable savings could have been attained.

FIGURE 3–13. North Ward Estes Computer-Aided Characterization Study *(Copyright © 1990, SPE, from paper 20748[2])*

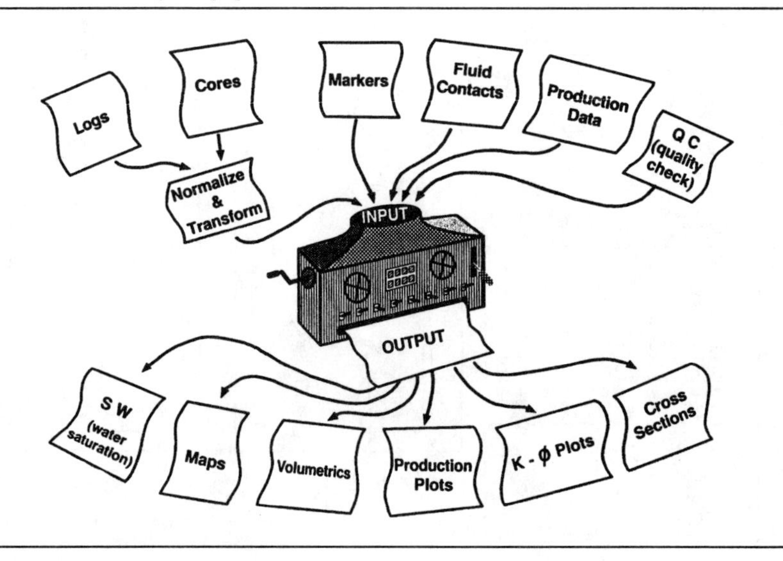

FIGURE 3–14. Sand Trends for North Ward Estes Field *(Copyright © 1990, SPE, from paper 20748[2])*

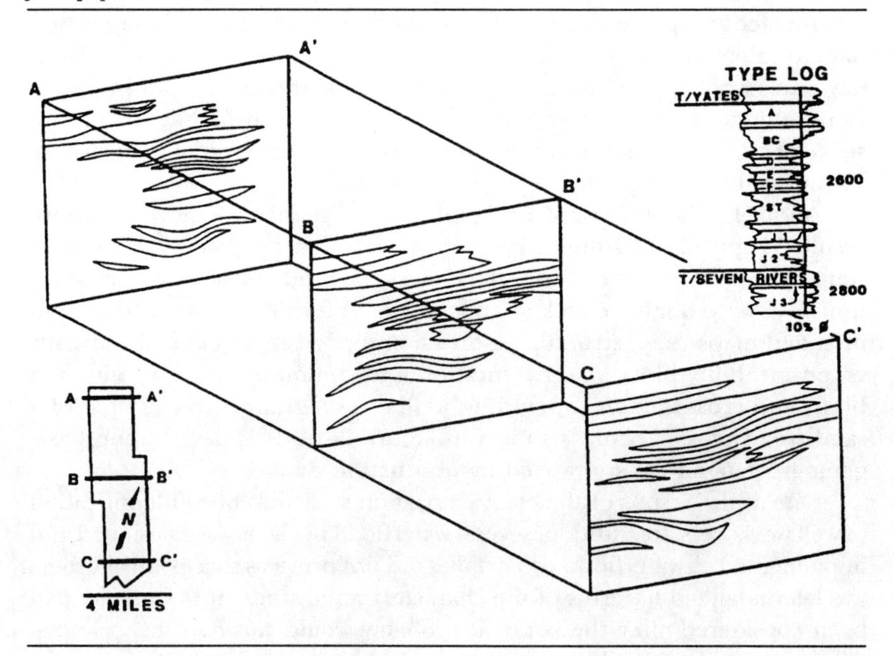

CO_2 Injectivity Test

A CO_2 injectivity test was conducted to investigate any injectivity reductions during CO_2 and water injection cycles. An injector in good mechanical condition and with no hydraulic fracturing was selected. Geological cross-sections through this well showed well-developed sands. The injectivity test provided the following valuable information:

- No reduction in injection rates was observed during or after CO_2 injection.
- The CO_2 injection rate was about 20% higher than the water injection rate.
- No significant change in injection profile was observed during or after CO_2 injection.

In addition to the previously mentioned results, the CO_2 injectivity test implanted a valuable "seed" of team effort that led to fruitful results during the design and implementation of the CO_2 project.

CO_2 Project Design and Implementation

The CO_2 flood design was based upon a history match of the waterflood performance of the six-section project area, the selection of typical patterns including a detailed reservoir characterization, a prediction for continuation of the waterflood, predictions for CO_2 flooding, and the scale up of the pattern predictions to the entire project area.[9] Predictions were made for the continuation of the waterflood and for CO_2 flooding. Additional reservoir simulation was conducted to determine the optimum CO_2 slug size.

Management approval of this project was obtained in December 1987. In January 1988, a task force was formed, and the CO_2 injection was initiated in March 1989.

The CO_2 plant compresses, desulfurizes, and dehydrates all CO_2-rich gas produced from the project. The plant is designed to process 65 MMcf/D of produced gas. In addition to reinjection gas, the plant will also produce 4 tons per day of marketable sulfur from moderate concentrations of H_2S (2%) in the hydrocarbon gas.

Team Effort

Why a Team Effort?

The North Ward Estes field is one of Chevron U.S.A.'s largest fields, and it has significant EOR potential. CO_2 flooding was the only economic option available to recover significant reserves from this field. For about

1,300 producing wells, the average production rate is only 7 BOPD at 95% water cut, and about 700 wells make 5 BOPD or less. Also, 300 wells are now capable of producing only at or below the present economic limit. Thus, if CO_2 flooding was not implemented right away, economics would have dictated the plugging and abandoning of uneconomic wells.

Keeping the previous points in mind and considering the average age of wells in the field at 35 years, a "window of opportunity" became quite obvious. If the wells were abandoned, it was unlikely that the project would have been undertaken because economics would not have justified redrills. Thus, it became urgent to start an EOR project (i.e., either move quickly or risk losing the chance). To design and implement an EOR project and to improve the performance of the existing waterfloods, a study team (see Figure 3–8) was formed.

What Did the Team Achieve?

During the design phase, as many as 25 to 30 members of various functional groups worked together on a comprehensive design of a six-section CO_2 project, reviewed hundreds of workover candidates, and evaluated several waterflood modification projects.

CO_2 injection was started in the six-section area within 15 months of project initiation. In addition, many workovers and waterflood modification projects were implemented during this time period. Moreover, within 1½ years, a gas processing plant was built and started. The team's goal for every aspect of the project—from well workovers, reservoir studies, CO_2 injection, and gathering system construction to start up—was accomplished in a short time without sacrificing quality.

In summary, the teamwork across the function lines has resulted in design and implementation of many successful projects in the North Ward Estes field.

Columbus Gray Lease[2]

The Columbus Gray lease in the Fuhrman Mascho field is located six miles southwest of Andrews, Texas (see Figure 3–15). The field was discovered in 1930, and the first well was completed on the lease in 1937. It was developed on 40-acre spacing, and waterflood began in 1965 on an 80-acre, 5-spot pattern. The flood reached its peak in 1967 at 720 BOPD, and it declined at a rate of 15%, thereafter. Injection was suspended in 1975 in all but five leaseline wells. In 1979, an infill program was started on 20-acre spacing, and injection was restarted to support the new wells. The infill program was completed in mid-1980s, and it resulted in an

FIGURE 3–15. Location Plat, Columbus Gray Lease *(Copyright © 1990, SPE, from paper 20748[2])*

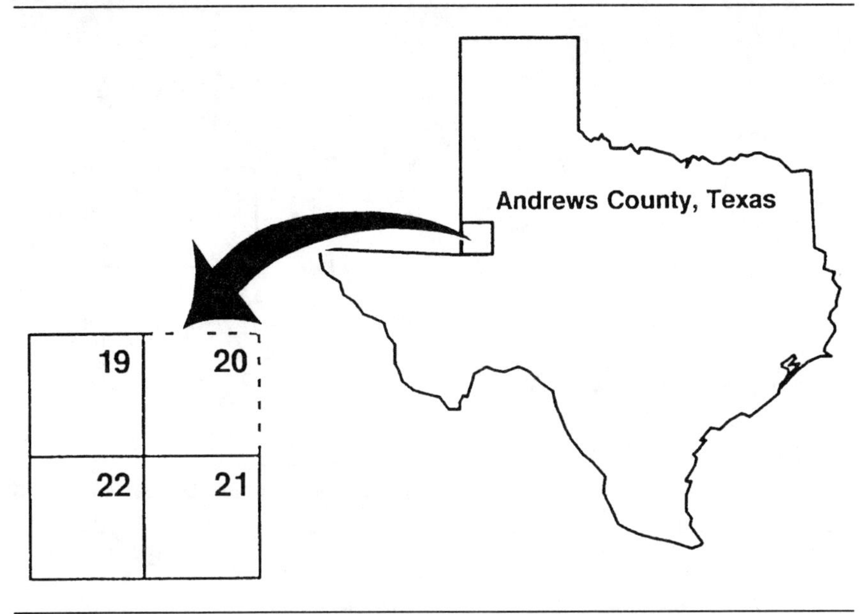

injection pattern of 80-acre, inverted nine-spot. The peak response occurred in 1984 at 1,000 BOPD, and then the production declined at 20% per year (Figure 3–16). The current flood pattern is shown in Figure 3–17.

Geology

The field is located on the eastern margin of the Central Basin Platform, 13 miles east of its edge. The San Andres formation, Guadalupian in age (Middle Permian), was deposited in an open marine, shallow-water shelf environment. Other San Andres fields on the Northern Central Basin Platform similar to Fuhrman Mascho include the Means, Shafter Lake, Seminole, and Emma fields.

The San Andres formation has a gross productive thickness of 300 ft in the lease area. It can be divided into two intervals based on porosity development and vertical continuity (see Figure 3–18). The Upper San Andres (USA) has an average thickness of 225 feet, and it is comprised of light-colored dolomites that are finely crystalline and probably vuggy or sucrosic. Also, it is anhydritic and contains scattered interbeds of gray or

FIGURE 3–16. Performance of Columbus Gray *(Copyright © 1990, SPE, from paper 20748[2])*

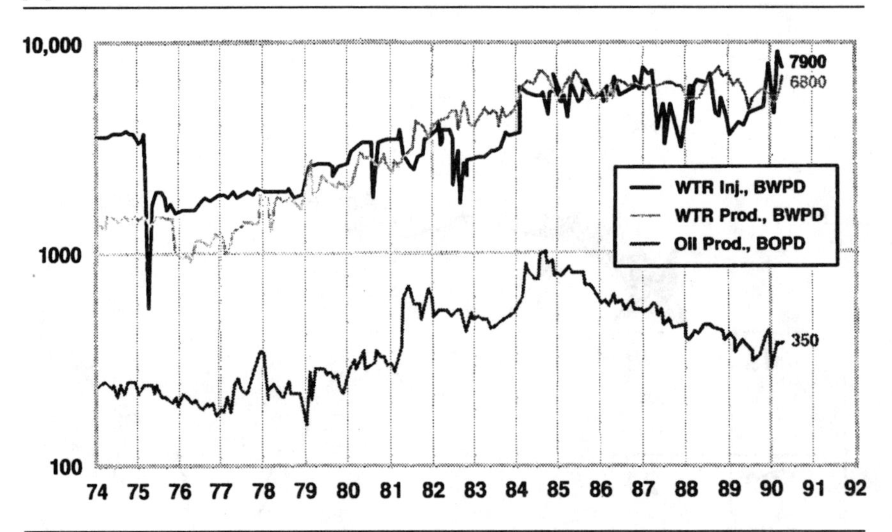

green shales. Porosity and permeability are both horizontally and vertically discontinuous. Thickness and reservoir quality change over short distances. The trapping mechanism in this zone is both structurally and stratigraphically controlled. The lateral variability of porosity development makes well-to-well correlations difficult.

The average depth to the top of the USA is 4,250 feet. The average porosity, permeability, net pay, and water saturation are 4.8%, 2.6 md, 80 ft, and 35%, respectively. Although no discrete oil-water contact (0WC) is present within this interval, some water is produced.

The Lower San Andres (LSA) is about 60 feet thick, and it comprises the lowest part of the producing interval. This zone is vertically and laterally continuous relative to the USA in the lease area, and it is generally capped by a dense dolomite. Core data indicate that the LSA lacks porosity occlusion by anhydrites as compared to the USA. The main difference between these zones is that the moldic and vuggy porosity is preserved in the LSA, while it is generally plugged by anhydrite in the USA.

The LSA has an average porosity, permeability, net pay, and water saturation of 11.8%, 29 MD, 25 ft, and 30% respectively. The oil-water contact defines the productive limits of this zone.

Results

Studies performed on the lease indicate that the majority of the remaining oil is in the USA because the LSA is 10 times more permeable. As a result,

FIGURE 3–17. Current Pattern, Columbus Gray *(Copyright © 1990, SPE, from paper 20748[2])*

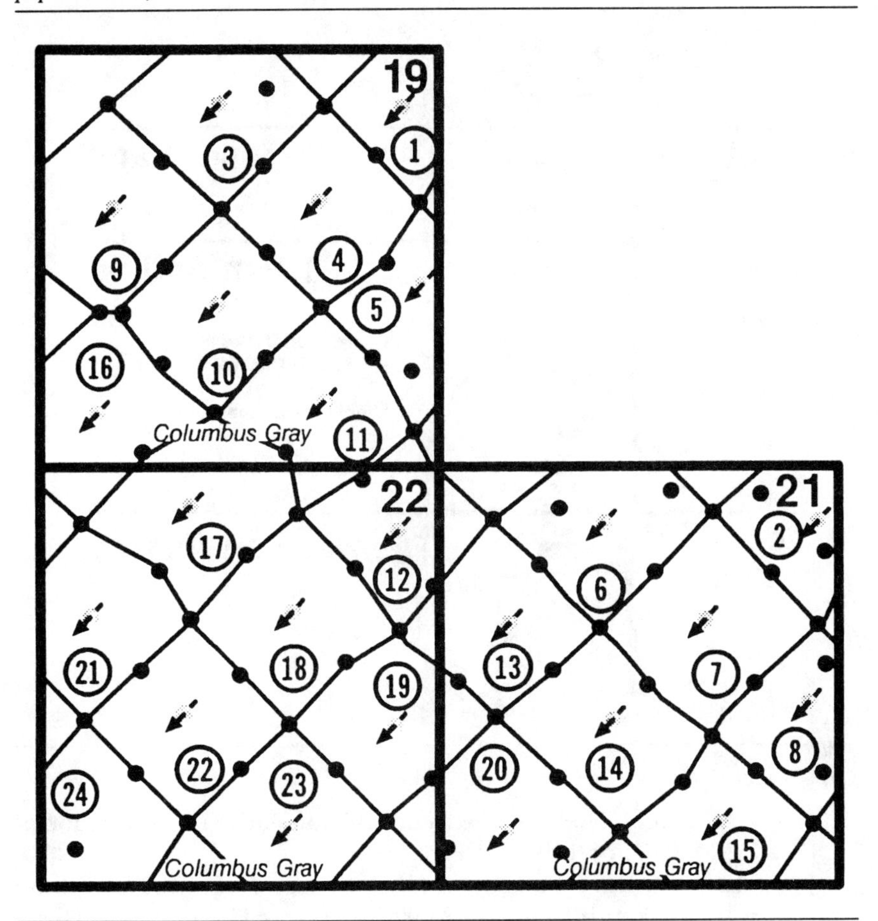

very little water has been injected in the upper zone. A study conducted in June 1989 made recommendations to increase recovery from the upper zone, and it estimated that an additional 500,000 BO could be obtained.

The June 1989 study was the result of a problem-solving session involving two (production and reservoir) engineers and a geologist over a period of two weeks. The results of the study were also discussed in a half-day session with the field foreman and surface facilities engineer. The following describes the results of the study:

- The original oil-in-place (OOIP), determined by volumetric analysis in 1984, was reviewed and modified. The values of OOIP for the upper and lower zones are estimated at 29.4 and 15.5 MMBO,

FIGURE 3–18. Type Log, Columbus Gray *(Copyright © 1990, SPE, from paper 20748[2])*

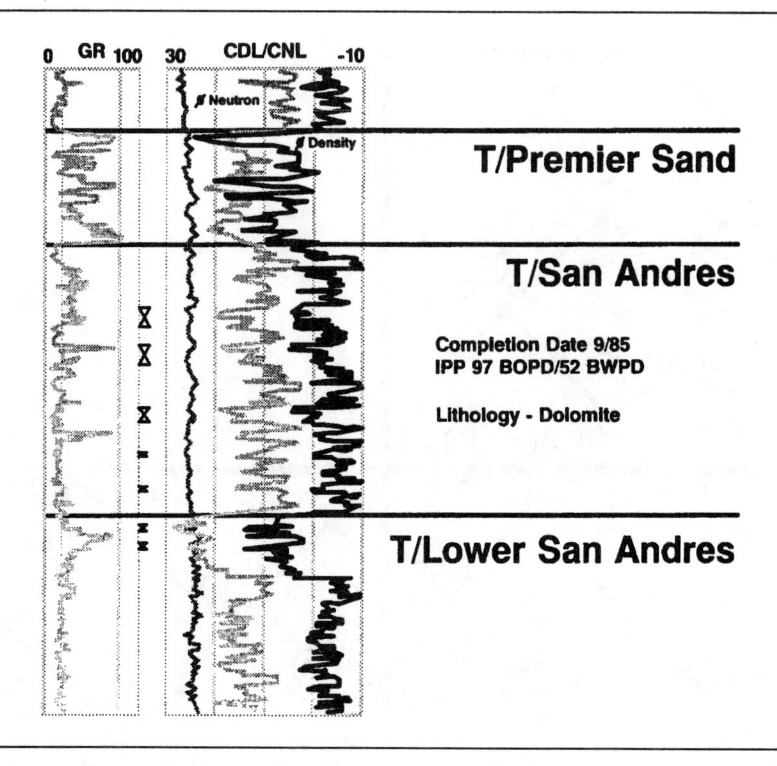

respectively. The lease has a current cumulative oil production of 7.2 MMBO, or 16% OOIP.

- The lease has a current producing rate of 350 BOPD with an average water injection of 8,000 BWPD (see Figure 3–16). This is an improvement in production of 50 BOPD over the June 1989 rate, or it is about 100 BOPD more than the rate based upon the expected decline.

  The improvement is attributed to the workovers performed on the lease as recommended by the study. For example, Well 1928, which was producing 17 BOPD before, is averaging 65 BOPD after the installation of a larger pump. Another example is Well 2126, once producing 5 BOPD, is now averaging 17 BOPD after workover (i.e., cleaning out, adding perforations, and acidizing).

- Although by recompleting in the LSA some response has already been observed, it is believed that additional production increase will be seen in the southern portion of Section 21. The wells in this area were only completed in the USA and, as previously described, this interval has a much lower porosity and permeability.

- Since the parting pressure in Section 19 increased, the injection pressure and rate were also raised.
- The producing water-cut is too high for the relatively low volume of water injected (about 10% and 60% HCPV in the USA and LSA, respectively). This high water-cut indicates: (1) channeling from the injectors to producers through fractures and/or high permeability zones, and (2) poor volumetric sweep in the LSA.
- Several recommendations involving pumping down the high fluid level wells, increasing injection pressure due to the increased frac pressure, running profiles before and after the increase in pressure, and diverting injection to match production profile.
- Several of the previous recommendations have already been incorporated. These cost about $750,000, and they will result in an estimated total production increase of 250,000 BO.

These results show that identifying and methodically solving reservoir problems has increased the performance of the Columbus Gray lease.

Reservoir Management

The reservoir management approach followed was very simple in this case because the lease production rate was only about 300 BOPD at the time of the study. Based upon reservoir heterogeneity and past performance, the expected increase in production was not considered high. Thus, a decision was made to design and implement a cost-effective reservoir management program that the lease could support. Depending upon the success of the implemented program, additional work could be recommended.

Conclusions

- Reservoir management has been described as the judicious use of various means available to maximize benefits from a reservoir.
- There are numerous reasons why some reservoir management programs fail. Perhaps the most important reason why a reservoir management program is developed and implemented poorly is unintegrated group effort. A procedure to improve success in implementing such a program has been employed.

- Both the comprehensive approach and problem-solving approach to reservoir management have resulted in positive results. Although they are philosophically quite different, each has shown its merits.
- The North Ward Estes field illustrates an application of comprehensive reservoir management, whereas the Columbus Gray lease depicts a problem-solving approach to reservoir management.

REFERENCES

1. Satter, A., J. E. Varnon, and M. T. Hoang. "Reservoir Management: Technical Perspective." SPE Paper 22350, SPE International Meeting on Petroleum Engineering, Beijing, China, March 24–27, 1992.
2. Thakur, G. C. "Implementation of a Reservoir Management Program," SPE Paper 20748 presented at the SPE Annual Technical Conference and Exhibition, New Orleans, Sept. 23–26, 1990.
3. Thakur, G. C. "Waterflood Surveillance Techniques—A Reservoir Management Approach," *J. Pet. Tech.* (Oct. 1991): 1180–1188.
4. Thakur, G. C. "Reservoir Management: A Synergistic Approach." SPE Paper 20138, presented at the Permian Basin Oil and Gas Conference, Midland, Texas, March 8–9, 1990.
5. Talash, A. W. "An Overview of Waterflood Surveillance and Monitoring," *J. Pet. Tech.* (December 1988): 1539–1543.
6. Calhoun, J. C. "A Definition of Petroleum Engineering," *J. Pet. Tech.* (July 1963): 725–727.
7. Havlena, D. "Interpretation, Averaging and Use of the Basic Geological Engineering Data," *J. Canadian Pet. Tech.*, part 1, vol. 5, no. 4 (October–December 1966): 153–164; part 2, vol. 7, no. 3 (July–September 1968): 128–144.
8. Stanley, R. G., et al. "North Ward Estes Geological Characterization," *AAPG Bulletin*, 1990.
9. Winzinger, R., et al. "Design of a Major $C0_2$ Flood—North Ward Estes Field, Ward County, Texas." SPE Paper 19654, presented at the SPE Annual Technical Conference, San Antonio, Texas, October 8–11, 1989.
10. Harris, D. G. "The Role of Geology in Reservoir Simulation Studies," *J. Pet. Tech.* (May 1975): 625–632.

CHAPTER 4

▼ ▼ ▼

Data Acquisition, Analysis and Management

Throughout the life of a reservoir, from exploration to abandonment (see Figure 1–1), an enormous amount of data are collected. An efficient data management program consisting of acquisition, analysis, validating, storing, and retrieving plays a key role in reservoir management. It requires planning, justifying, prioritizing, and timing. As emphasized in Chapter 3, an integrated approach involving all functions is necessary to lay the foundation of reservoir management.

DATA TYPES[1,2,3]

The types of data collected before and after production are shown in Figure 3–4. Table 4–1 lists the data under the various broad classification including the timing of acquisition and the professionals responsible for acquisition and analyses. It is emphasized that the multidisciplinary professionals need to work as an integrated team to develop and implement an efficient data management program.

DATA ACQUISITION AND ANALYSIS

Multidisciplinary groups (i.e., geophysicists, geologists, petrophysicists, drilling, reservoir, production and facilities engineers) are involved in collecting various types of data throughout the life of a reservoir (see Table 4–1). Land and legal professionals also contribute to the data

TABLE 4–1. Reservoir Data

Classification	Data	Acquisition Timing	Responsibility
Seismic	Structure, stratigraphy, faults, bed thickness, fluids, interwell heterogeneity	Exploration	Seismologists, Geophysicists
Geological	Depositional environment diagenesis, lithology, structure, faults, and fractures	Exploration, discovery & development	Exploration & development geologists
Logging	Depth, lithology, thickness, porosity, fluid saturation, gas/oil, water/oil and gas/water contacts, and well-to-well correlations	Drilling	Geologists, petrophysicists and engineers
Coring		Drilling	Geologists, drilling and reservoir engineers, and laboratory analysts
Basic	Depth, lithology, thickness, porosity, permeability, and residual fluid saturation		
Special	Relative permeability, capillary pressure, pore compressibility, grain size, and pore size distribution		
Fluid	Formation volume factors, compressibilities, viscosities, gas solubilities, chemical compositions, phase behavior, and specific gravities	Discovery, delineation, development, and production	Reservoir engineers and laboratory analysts

TABLE 4–1. Reservoir Data (continued)

Classification	Data	Acquisition Timing	Responsibility
Well Test	Reservoir pressure, effective permeability-thickness, stratification, reservoir continuity, presence of fractures or faults, productivity and injectivity indices, and residual oil saturation	Discovery, delineation, development, production and injection	Reservoir and production engineers
Production & Injection	Oil, water, and gas production rates, and cumulative productions, gas and water injection rates and cumulative injections, and injection and production profiles	Production & injection	Production & reservoir engineers

collection process. Most of the data, except for the production and injection data, are collected during delineation and development of the fields.

An effective data acquisition and analysis program requires careful planning and well-coordinated team efforts of interdisciplinary geoscientists and engineers throughout the life of the reservoir. On one hand, there may be the temptation to collect lots of data; and on the other hand, there may be the temptation to short-cut data acquisition to reduce costs. Justification, priority, timeliness, quality, and cost-effectiveness should be the guiding factors in data acquisition and analysis. It will be more effective to justify to management data collection if the need for the data, the costs, and the benefits are clearly defined.

Dandona reminds that certain types of data such as core derived information, initial fluid properties, fluid contacts, and initial reservoir pressures can only be obtained at an early development stage.[3] Coring, logging, and initial reservoir fluid sampling should be made at appropriate times using the proper procedure and analyses. Normally, all wells are logged; however, an adequate number of wells should be cored to validate the log data. Initial bottom-hole pressure measurements should be made, preferably at each well and at selected "key wells" periodically. According to Woods and Abib, key wells represent 25% of the total wells.[2] Also, they found it beneficial to measure pressures in all wells at least every two to three years to aid in calibrating reservoir models.

It is essential to establish the specification of what and how much data need to be gathered and the procedure and frequency to be followed. A logical, methodical, and sequential data acquisition and analysis program suggested by Raza is shown in Table 4–2.[4]

DATA VALIDATION[2,4]

Field data are subjected to many errors (i.e., sampling, systematic, random, etc.). Therefore, the collected data need to be carefully reviewed and checked for accuracy as well as for consistency.

In order to assess validity, core and log analyses data should be carefully correlated and their frequency distributions made to identify different geologic facies. Log data should be carefully calibrated using core data for porosity and saturation distributions, net sand determination, and geological zonation of the reservoir. The reservoir fluid properties can be validated by using the equation of state calculations and by empirical correlations. The reasonableness of geological maps should be

TABLE 4–2. An Efficient Data Flow Diagram *(Copyright © 1992, SPE, from* JPT, *April 1992*[4]*)*

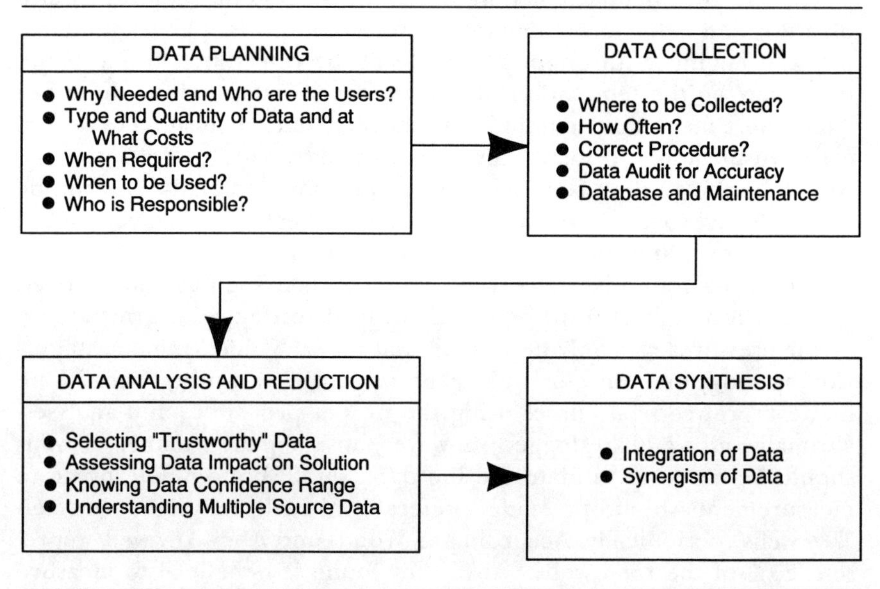

established by using the knowledge of depositional environment. The presence of faults and flow discontinuities as evidenced in a geological study can be investigated and validated by pressure interference and pulse and tracer tests.

The reservoir performance should be closely monitored while collecting routine production and injection data including reservoir pressures. If past production and pressure data are available, classical material-balance techniques and reservoir modeling can be very useful to validate the volumetric original hydrocarbons-in-place and aquifer size and strength.

Laboratory rock properties, such as oil-water and gas-oil relative permeabilities, and fluid properties, such as PVT data, are not always available. Empirical correlations can be used to generate these data.[5–15]

DATA STORING AND RETRIEVAL

The reconciled and validated data from the various sources need to be stored in a common computer database accessible to all interdisciplinary end users. As new geoscience and engineering data are available, the database will require updating. The stored data are used to carry out multipurpose reservoir management functions including monitoring and evaluating the reservoir performance.

As discussed in Chapter 2, storing and retrieval of data during reservoir life cycle poses a major challenge in the petroleum industry today because of noncommunicating nature and data bases and incompatibility of the software and data sets from the different disciplines. The Petrotechnical Open Software Corporation (POSC) is working on establishing industry standards and a common set of rules for applications and data systems within the industry.[16]

DATA APPLICATION

A better representation of the reservoir is made from 3-D seismic information. The cross-well tomography provides interwell heterogeneity.

Geological maps such as gross and net pay thicknesses, porosity, permeability, saturation, structure, and cross-section are prepared from seismic, core and log analysis data. These maps, which also include faults, oil-water, gas-water and gas-oil contacts, are used for reservoir delineation, reservoir characterization, well locations, and estimates of original oil and gas-in-place.

The more commonly used logging systems are:

- Open Hole Logs:
  — Resistivity, Induction, Spontaneous Potential, Gamma ray,
  — Density, Sonic Compensated Neutron, Sidewall neutron
  — Porosity, Dielectric, and Caliper.
- Cased Hole Logs:
  — Gamma Ray, Neutron (except SNP), Carbon/Oxygen, Chlorine, Pulsed Neutron and caliper.

The well log data that provide the basic information needed for reservoir characterization are used for mapping, perforations, estimates of original oil and gas in place, and evaluation of reservoir perforation. Production logs can be used to identify remaining oil saturation in undeveloped zones in existing production and injection wells. Time-lapse logs in observation wells can detect saturation changes and fluid contact movement. Also, log-inject-log can be useful for measuring residual oil saturation.

Core analysis is classified into conventional, whole-core, and sidewall analyses. The most commonly used conventional or plug analysis involves the use of a plug or a relatively small sample of the core to represent an interval of the formation to be tested. Whole core analysis involves the use of most of the core containing fractures, vugs, or erratic porosity development. Sidewall core analysis employs cores recovered by sidewall coring techniques.

Unlike log analysis, core analysis gives direct measurement of the formation properties, and the core data are used for calibrating well log data. These data can have a major impact on the estimates of hydrocarbon-in-place, production rates, and ultimate recovery.

The fluid properties are determined in the laboratories using equilibrium flash or differential liberation tests. The fluid samples can be either sub-surface sample or a recombination of surface samples from separators and stock tanks. Fluid properties can be also estimates by using correlations.[10–15]

Fluid data are used for volumetric estimates of reservoir oil and gas in place, reservoir type, (i.e., oil, gas or gas condensate), and reservoir performance analysis. Fluid properties are also needed for estimating reservoir performance, wellbore hydraulics, and flowline pressure losses.

The well test data are very useful for reservoir characterization and reservoir performance evaluation. Pressure build-up or falloff tests provide the best estimate of the effective permeability-thickness of the reservoir in addition to reservoir pressure, stratification, and presence of faults and fractures. Pressure interference and pulse tests provide reservoir continuity and barrier information. Multiwell tracer tests used in waterflood and in enhanced oil recovery projects give the preferred flow paths between the injectors and producers. Single-well tracer tests are used to determine residual oil saturation in waterflood reservoirs. Repeat formation tests can measure pressures in stratified reservoirs indicating a varying degree of depletion in the various zones.

Production and injection data are needed for reservoir performance evaluation.

Table 4–3 lists required data for the techniques used for analyzing reservoir performance.

EXAMPLE DATA

Engineering studies of G-1, G-2, and G-3 reservoirs of Meren Field offshore Nigeria included geologic evaluation, material balance calculations, and 3-phase, 3-D reservoir performance simulations.[17] Geological, reservoir, and production data used for the studies are presented in Figures 4–1 through 4–8 and Tables 4–4 and 4–5.

TABLE 4–3. Data Required for Reservoir Performance Analyses

Data Group	Volumetric	Decline Curve	Material Balance	Mathematical Models
Geometry	Area, thickness	No	Area, thickness Homogeneous	Area, thickness Heterogeneous
Rock	Porosity, saturation	No	Porosity, saturation, relative permeability, compressibility Homogeneous	Porosity, saturation, relative permeability, compressibility, capillary pressure Heterogeneous
Fluid	Form. vol. factors	No	PVT Homogeneous	PVT Heterogeneous
Well	No	No	PI for rate vs. time	Locations Perforations PI
Production & Injection	No	Production	Yes	Yes
Pressure	No	No	Yes	Yes

FIGURE 4–1. Structure Map, Sand G-1 *(Copyright © 1982, SPE, from* JTP, *April 1982*[17]*)*

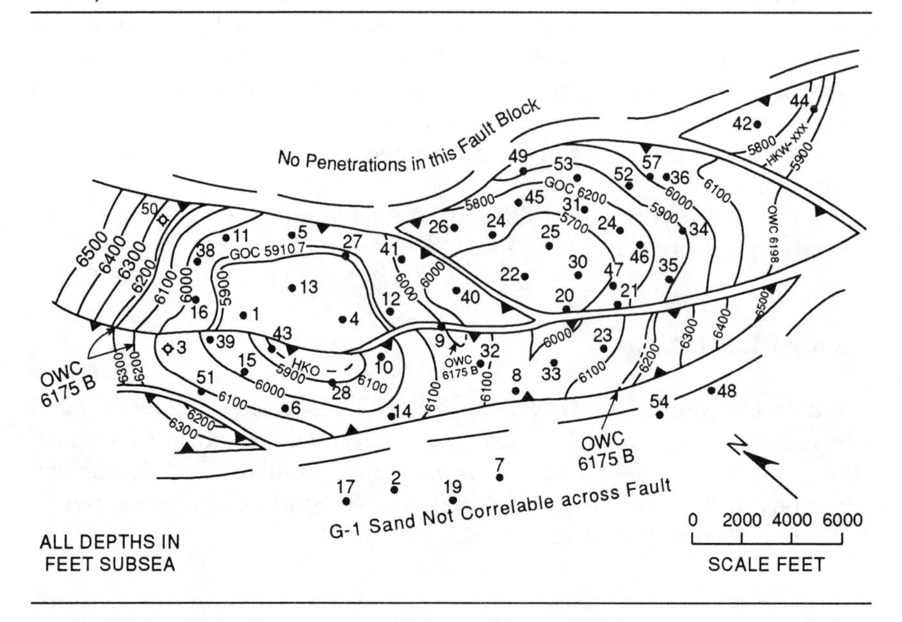

FIGURE 4–2. Structure Map, Sand G-2 *(Copyright © 1982, SPE, from* JPT, *April 1982*[17]*)*

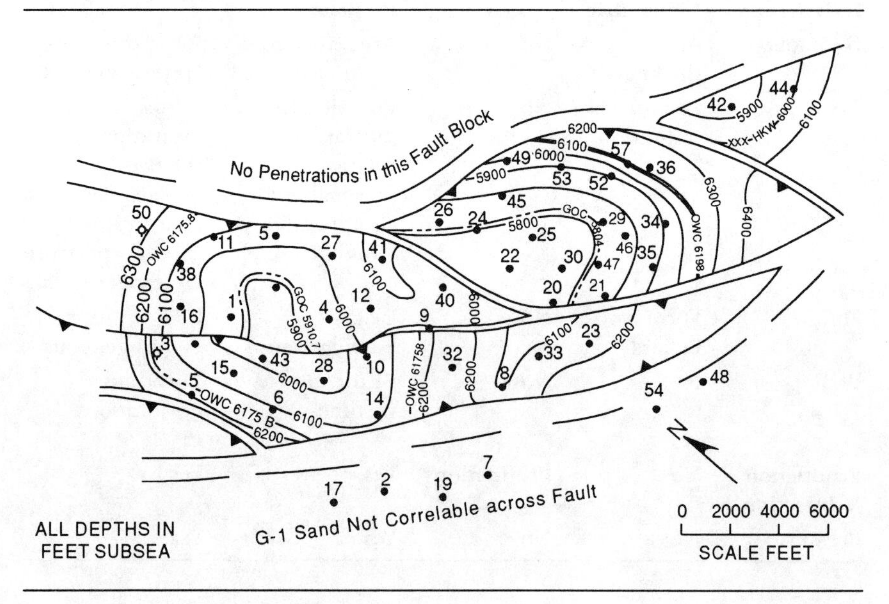

FIGURE 4–3. Meren Field Cross Section Showing Sand G Juxtaposition *(Copyright © 1982, SPE, from* JPT, *April 19982[17])*

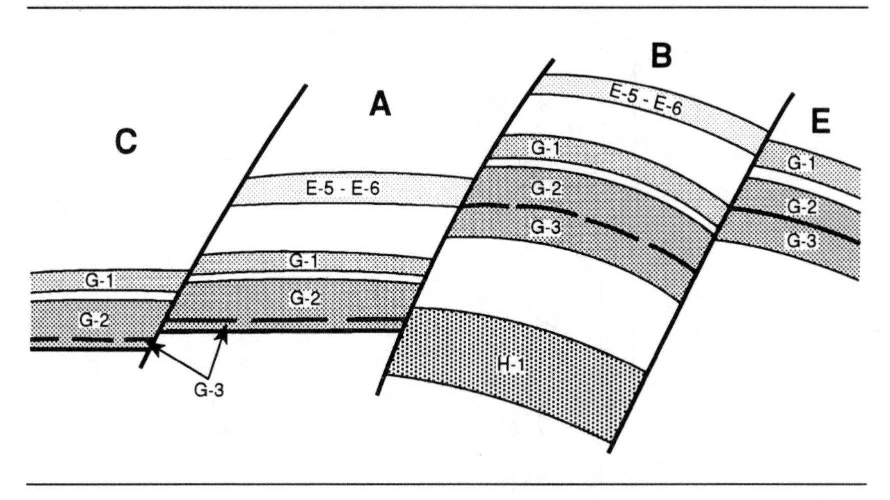

FIGURE 4–4. Cross Section *(Copyright © 1982, SPE, from* JPT, *April 1982[17])*

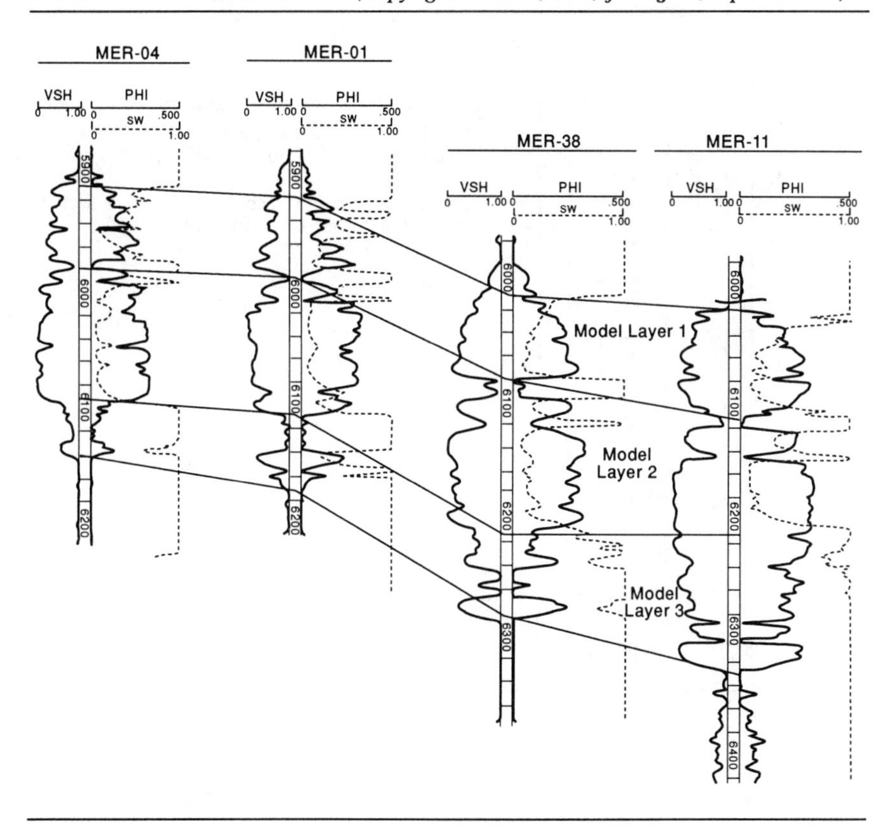

FIGURE 4–5. Type Log of Meren Field *(Copyright © 1982, SPE, from* JPT, *April 1982[17])*

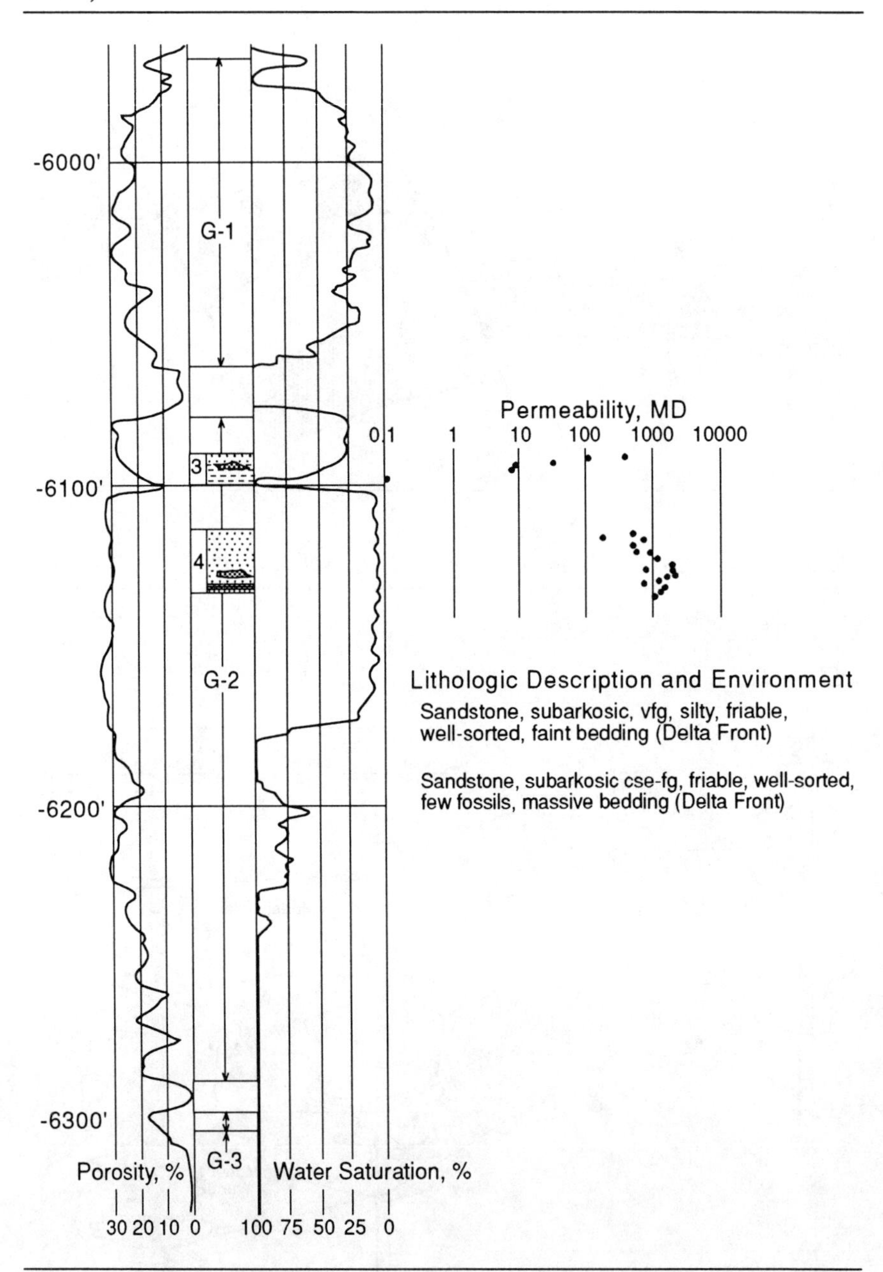

FIGURE 4–6. Porosity-permeability Relationships, Merem Field Sand G *(Copyright © 1982, SPE, from* JPT, *April 1982[17])*

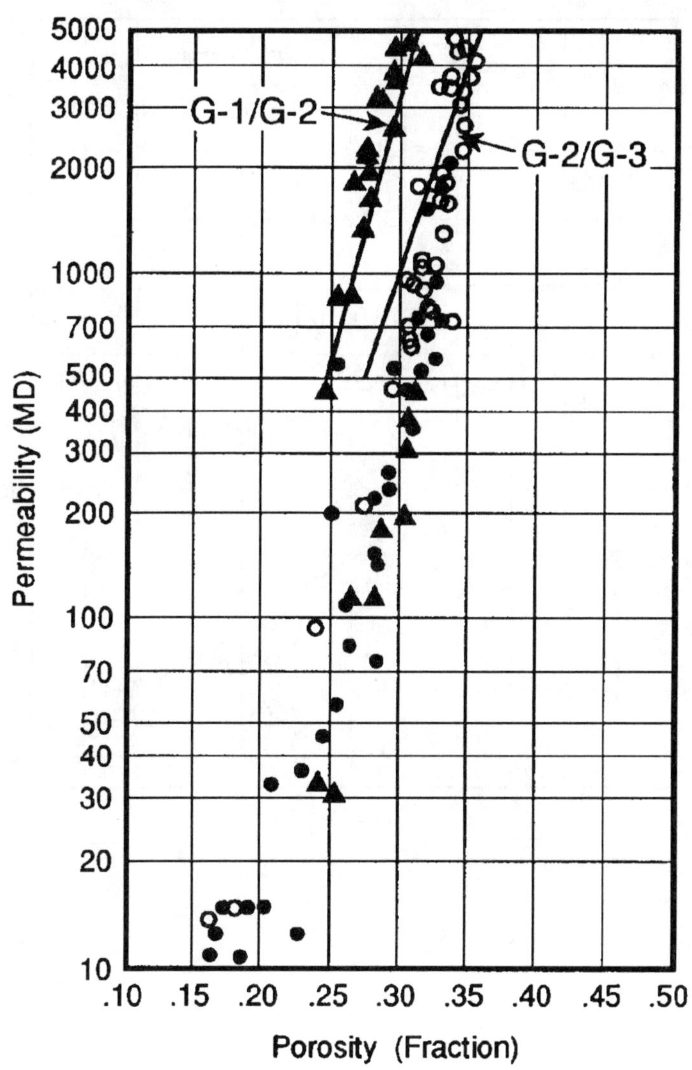

FIGURE 4–7. Reservoir Performance, Sands G-1/G-2 *(Copyright © 1982, SPE, from* JPT, *April 1982[17])*

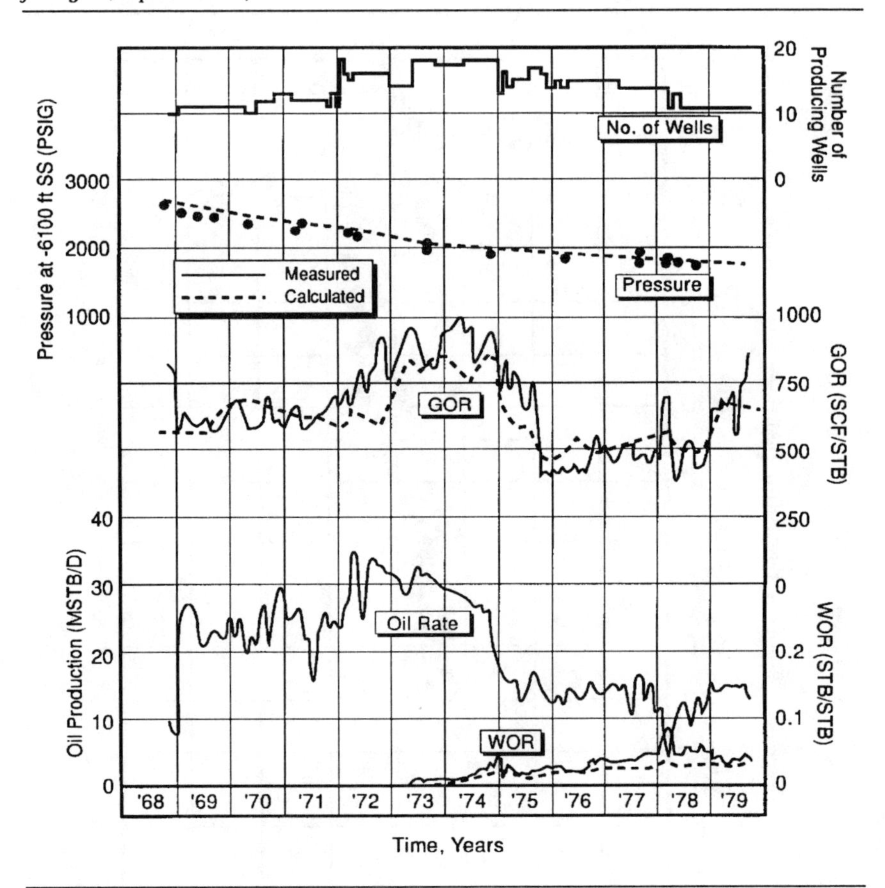

FIGURE 4–8. Well 11 Performance, Sands G-1/G-2 *(Copyright © 1982, SPE, from* JPT, *April 1982[17])*

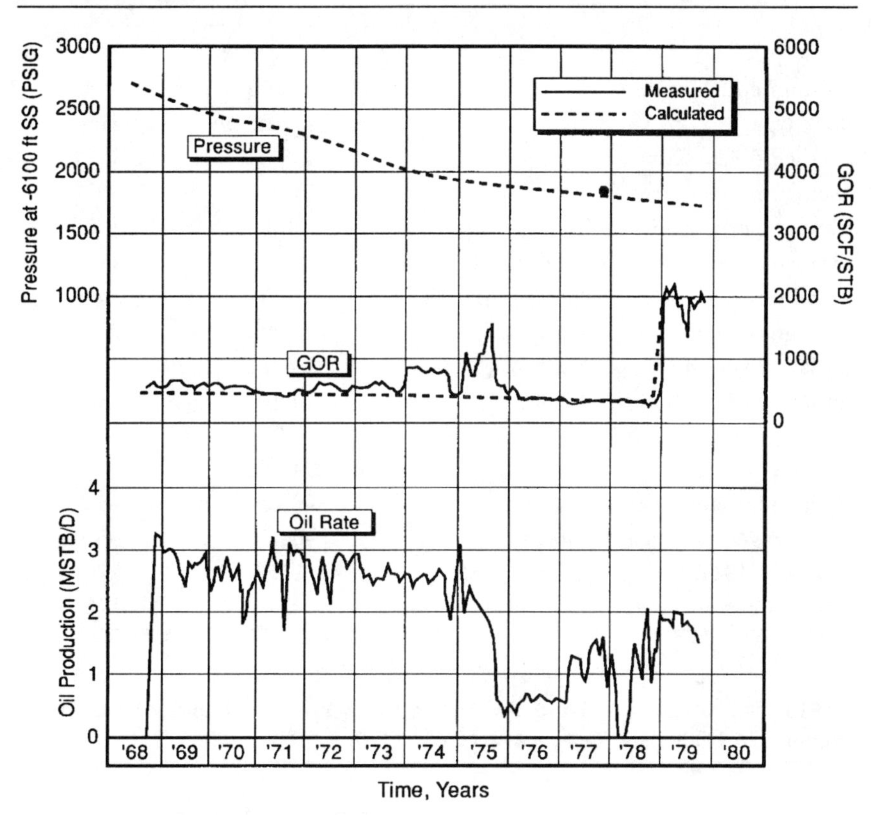

TABLE 4–4. Basic Reservoir Data *(Copyright © 1992, SPE, from* JPT, *April 1982[17])*

	Reservoir	
	G-1/G-2	**G-2/G-3**
Datum depth, ft subsea	–6,100	–6,000
Rock type	sandstone	sandstone
Average thickness, ft	138	126
Average porosity, %	27	32
Average permeability, md	1,150	1,775
Average connate water saturation, %	24	14
Initial reservoir pressure (at datum), psig	2,660	2,560
Average bubble-point pressure, psig	2,629	2,500
Initial oil volume factor, RB/STB	1,327	1,312
Initial solution GOR, scf/STB	566	588
Initial oil viscosity, cp	0.575	0.460
Oil gravity, °API	34	33
Initial oil/water contact, ft subsea	–6,175	–6,197
Initial gas/oil contact, ft subsea	–6,000	–5,804
OOIP, MMSTB	281.5	276.8
Original gas in place, Bscf	205.6	176.7

TABLE 4–5. Reservoir Fluid Properties for a Pressure Range of Saturation Pressure to 1,500 psig *(Copyright © 1992, SPE, from* JPT, *April 1982[17])*

	Reservoir	
	G-1/G-2	**G-2/G-3**
Oil volume factor, RB/STB	.327 to 1.245	1.312 to 1.214
Reciprocal gas volume factor, scf/RB	910 to 475	833 to 490
Water volume factor, RB/STB	1.048 to 1.056	1.049 to 1.056
Solution GOR, scf/STB	566 to 368	588 to 360
Oil phase viscosity, cp	0.575 to 0.751	0.460 to 0.633
Oil phase pressure gradient, psi/ft	0.3068 to 0.3211	0.3201 to 0.3340
Gas viscosity, cp	0.0184 to 0.0146	0.0195 to 0.0142
Gas gradient, psi/ft	0.0616 to 0.0319	0.0536 to 0.0323

Additional rock and fluid properties of Meren G-1/G-2 reservoirs can be calculated using correlations. The calculated data are presented in Figures 4–9 through 4–11.

FIGURE 4–9. Oil Properties

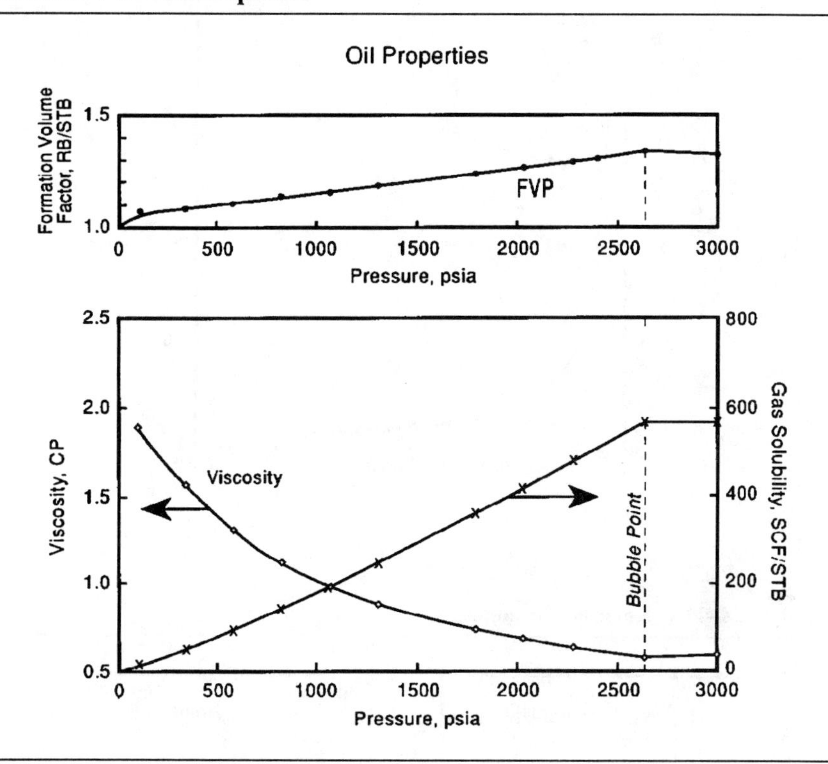

FIGURE 4–10. Gas Properties

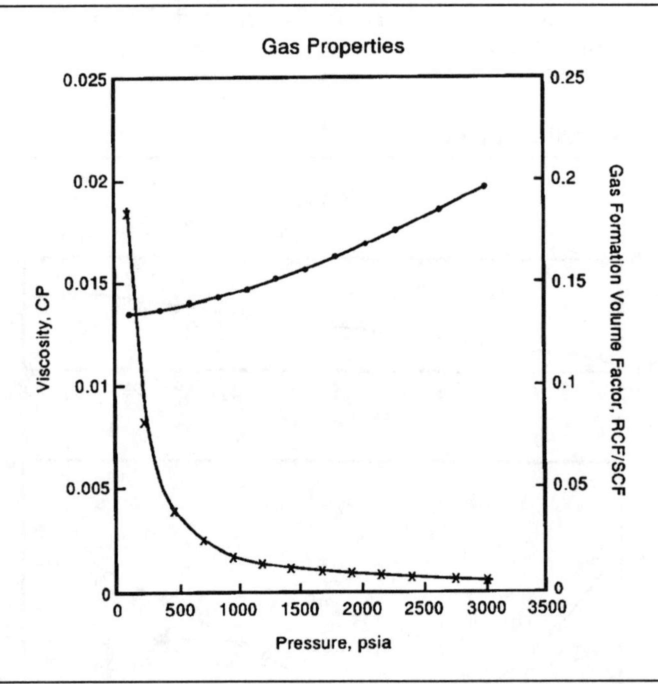

FIGURE 4–11. Relative Permeabilities

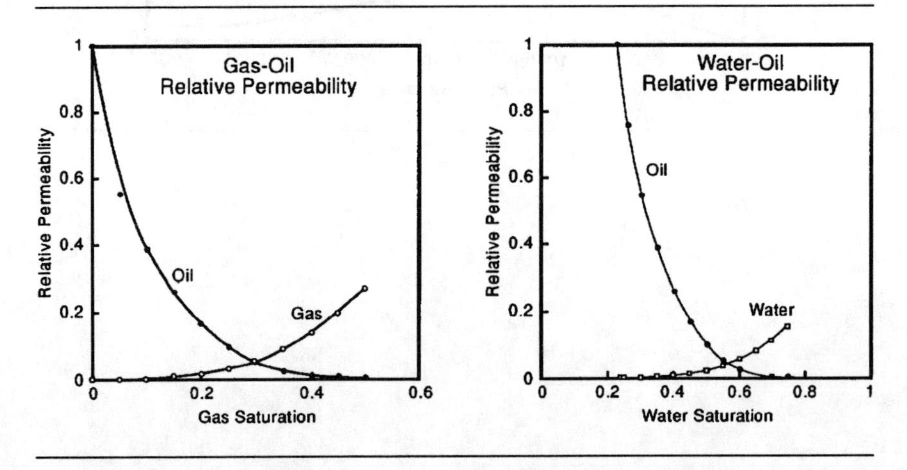

REFERENCES

1. Satter, A., J. E. Varnon and M. T. Hoang. "Reservoir Management: Technical Perspective." SPE Paper 22350, SPE International Meeting on Petroleum Engineering, Beijing, China, March 24–27, 1992.
2. Woods, E. G. and Osmar Abib. "Integrated Reservoir Management Concepts." Reservoir Management Practices Seminar, SPE Gulf Coast Section, Houston, Texas, May 29, 1992.
3. Dandona, A. K., R. B. Alston and R. W. Braun. "Defining Data Requirements for a Simulation Study." SPE Paper 22357, SPE International Meeting on Petroleum Engineering, Beijing, China, March 24–27, 1992.
4. Raza, S. H. "Data Acquisition and Analysis: Foundational to Efficient Reservoir Management," *JPT* (April 1992): 466–468.
5. Corey, A. T. *Prod. Monthly* **19**, no. 1, (1954): 38.
6. Wyllie, M. R. J. and G. H. F. Gardner. "Generalized Kozeny-Carmen Equation, parts 1 and 2," *World Oil*, (March 1958): 121–126, (April 1958):210–228.
7. Stone, H. L. "Estimation of Three-Phase Oil Relative Permeability," *JPT* (1970): 214–218.
8. Dietrich, J. K. and P. L. Bondor. "Three-Phase Oil Relative Permeability Models." SPE 6044 paper presented in the SPE Annual Meeting in New Orleans, October, 1976.
9. Honarpour, M., L. F. Koederitz and H. A. Harvey. "Empirical Equations for Estimating Two-Phase Relative Permeability in Consolidated Rock," *JPT* (December 1982): 2905–2908.
10. Beal, C. "The Viscosity of Air, Water, Natural Gas, Crude Oil and its Associated Gases at Oil Field Temperature and Pressure," *Trans. AIME* (1946): 94.
11. Standing M. B. "A Generalized Pressure-Volume-Temperature Correlation For Mixture of California Oils and Gases," *Drilling and Production Practice API* (1947): 275.
12. Lasater, J. A. "Bubble Point Pressure Correlation," *Trans. AIME* (1958): 379.
13. Chew, J. and C. A. Connally. "A Viscosity Correlation for Gas Saturated Crude Oils," *Trans. AIME* (1959): 20.
14. McCain, W. D. *The Properties of Petroleum Fluids.* Tulsa, Oklahoma: Petroleum Publishing Co., (1973): page number.
15. Beggs, H. D. and J. R. Robinson. "Estimating the Viscosity of Crude Oil Systems," *JPT* (September 1975): 1140.
16. Johnson, J. P. "POSC Seeking Industry Software Standards, Smooth Data Exchange," *Oil & Gas J.* (October 26, 1992): 64–68.
17. Thakur, G. C., et al. "Reservoir Studies of G-1, G-2, and G-3 Reservoirs, Meren Field, Nigeria," *JPT* (April 1982): 721–732.

CHAPTER 5

▼ ▼ ▼

Reservoir Model

This chapter presents an approach to building an integrated reservoir model based upon geological, geophysical, petrophysical, and engineering data.

The reservoir model is not just an engineering or a geoscience model; rather it is an integrated model, prepared jointly by geoscientists and engineers. The integrated reservoir model requires a thorough knowledge of the geology, rock and fluid properties, fluid flow and recovery mechanisms, drilling and well completions, and past production performance (see Figure 3–2).[1]

The geological model is derived by extending localized core and log measurements to the full reservoir using many technologies such as geophysics, mineralogy, depositional environment, and diagenesis. The geological model (particularly the definition of geological units and their continuity and compartmentalization) is an integral part of geostatistical and ultimately reservoir simulation models.

ROLE OF RESERVOIR MODEL

The economic viability of a petroleum recovery project is greatly influenced by the reservoir production performance under the current and future operating conditions. Therefore, the evaluation of the past and present reservoir performance and the forecast of its future behavior is an essential aspect of the reservoir management process (see Figure 3–5).[1] Classical volumetric, material balance and decline curve analysis methods, and high-technology black oil, compositional and enhanced oil-recovery numerical simulators are used for analyzing reservoir performance and estimating reserves. The accuracy of the results is dictated by the quality of the reservoir model used to make reservoir performance analysis.

As opposed to one reservoir life, the simulator can simulate many lives for the reservoir under different scenarios and thus provide a very powerful tool to optimize the reservoir operation. Thomas discusses with examples the role reservoir simulators play in formulating initial development plans, history matching and optimizing future production, and in planning and designing enhanced oil recovery projects.[2] Figure 5–1 presents key steps involved in reservoir simulation.[3]

Historically, reservoir simulators have been used for studying large fields and those undergoing complex recovery processes. They have not been suitable for small reservoirs because the simulation studies are costly, require highly trained professionals, and are too time consuming for an operating company work environment. This is changing. A proposed minisimulation technique using personal computers can play an important role in managing small reservoirs.[4]

The process of developing a sound reservoir model plays a very important part in reservoir management because:

- It requires integration among geoscientists and engineers.
- It allows geoscientists' interpretations and assumptions to be compared to actual reservoir performance as documented by production history and pressure tests.

FIGURE 5–1. Reservoir Simulation *(Copyright © 1988, SPE, from Saleri, N. G. and R. M. Toronyi, "Engineering Control in Reservoir Simulation," SPE paper 18305, SPE Annual Tech. Conf. and Exh., October 1988[3])*

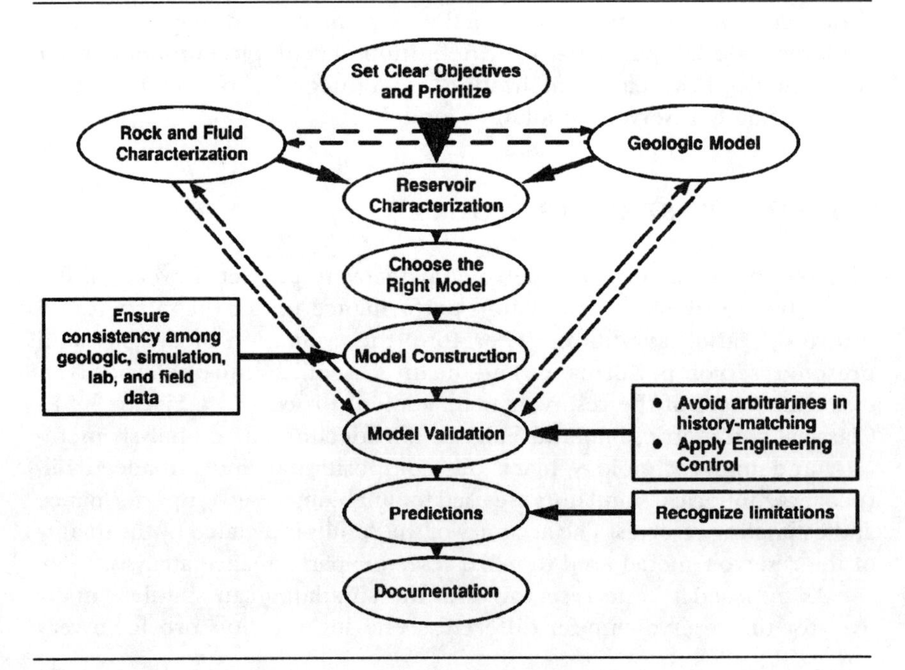

- It provides a means of understanding the current performance and predicts the future performance of a reservoir under various "what if" conditions so that better reservoir management decisions can be made.[5]

In addition, the reservoir model should be developed jointly by geoscientists and engineers because:

- An interplay of effort results in better description of the reservoir and minimizes the uncertainties of a model. The geoscientists' data assist in engineering interpretations, whereas the engineering data sheds new light on geoscientists' assumptions.
- The geoscientist-engineer team can correct contradictions as they arise, preventing costly errors later in the field's life.
- It has been well documented that a cross-functional team can gain valuable insight in describing a reservoir. For example, as described in Chapter 2, while working in a complex fault system, Amoco International Oil Company's reservoir engineers worked with geologists to "produce an accurate *a priori* reservoir description." The team tested the description against field performance using a simulation model and gained invaluable information related to fault configuration and the relationships of gas in place, permeability, and reserves. The model led to confident planning of future platform and compression requirements, and it provided much needed lead time to install equipment.
- In a fragmented effort (i.e., when engineers and geoscientists do not communicate), each discipline may study only a fraction of the available data; thus, the quality of the reservoir management can suffer and adversely affect drilling decisions and depletion plans throughout the life of the reservoir.
- Multidisciplinary teams using the latest technology provide opportunities to tap unidentified reserves. For example, improved 3-D seismic data can aid the surveillance of production operations in mature projects and can identify the presence or lack of continuity between wells, and thus improve the description of the reservoir model.
- Utilizing reservoir models developed by multidisciplinary teams can provide practical techniques of accurate field descriptions to achieve optimal production.

Thus, it is important that we prepare a simulation model that takes into account realistic geology and other rock-fluid characteristics. With a realistic simulation model, we can do or obtain guidance on the following:

- Determine the performance of an oil field under water injection or gas injection, or under natural depletion.

- Judge the advisability of flank waterflooding as opposed to pattern waterflooding.
- Determine the effects of well location and spacing.
- Estimate the effect of producing rate on recovery.
- Calculate the total gas field deliverability for a given number of wells at certain specified locations.
- Estimate the lease-line drainage in heterogeneous oil or gas fields.[5,6]

Breitenbach presents a comprehensive history of reservoir simulation, starting in the 1940s.[7] In the 1940s analog models played an important role. During the 1950s, with the advent of 2D and 3D finite-difference equations, it became possible to solve problems related to multiphase flow in heterogeneous porous media. In the 1960s, reservoir simulation was devoted largely to three-phase, black oil reservoir problems. During the 1970s, the efforts extended to enhanced oil recovery processes. In the 1980s, the reservoir simulation application continued to expand and included the area of reservoir description. The use of geostatistics to describe reservoir heterogeneities and technology to model naturally fractured reservoirs were developed.[8,9,10]

Today, desktop computers and a wide variety of reservoir simulation systems provide engineers and geoscientists an economical means to solve complex reservoir problems in a reasonable time frame.

The simulation process consists of describing the reservoir (i.e., preparing a reservoir model), matching historical performance, and predicting the future performance of the reservoir under a variety of scenarios. The reservoir description step results in a reservoir model that includes overall geometry (i.e., permeability, porosity, and height of each grid block).[11,12,13] After constructing the reservoir model, it is generally validated by determining whether it can duplicate past field behavior. Often the reservoir description information is changed within geological and engineering bounds to history-match the past production performance.

There are many educational values of simulation models:

- Too often we tend to demand accurate determination of all types of input data before we accept the computed results as meaningful or reliable. On the other hand, interest in accuracy of input data should be proportional to the sensitivity of computed results to variations in those data.
- Sensitivity to errors in reservoir description data can be determined by performing simulation runs with variations in those data covering a reasonable range of uncertainty.[5,6,14]

Thus, efforts should be made on obtaining those data that have the greatest effect on calculated performance. Also, a general guide for developing a model is to “select the least complicated model and grossest reservoir description that will allow the desired estimation of reservoir performance."[6]

Many authors have documented applications of 3-D seismic in reservoir management and modeling.[15,16] Geophysicists are playing a more important role than ever in identifying key reservoir features in a simulation model. Also, they are assisting in validating the geologic model.

GEOSCIENCE

Geoscientists probably play the most important role in developing a reservoir model. The purpose of this section is to provide the geoscientist's activities in developing a reservoir model. The model requires variations in porosity, permeability, and capillary properties.

The distributions of the reservoir and nonreservoir rock types and of the reservoir fluids determine the geometry of the model and influence the type of model to be used.[17] For example, the number and scale of the shale (or dense carbonate) breaks in the physical framework determine the continuity of the reservoir facies and influence the vertical and horizontal dimensions of each cell. Real variations in reservoir parameters may require several cross-sections or a three-dimensional model. Other influences on cell dimension include computing cost, well spacing, fluid-phase distribution, and the purpose of the study.

Incorporation of geologic model into a simulation model requires recognition and capture of detailed reservoir heterogeneities. With the advent of advanced simulation technology and the understanding of complex subsurface structures, these heterogeneities can be recognized early in the life of a field and incorporated into the simulation model.

Both engineering and geological judgment must guide the development and use of the simulation model. The geologist usually concentrates on the rock attributes in four stages:

1. Rock studies establish lithology and determine depositional environment, and reservoir rock is distinguished from nonreservoir rock.
2. Framework studies establish the structural style and determine the three-dimensional continuity character and gross-thickness trends of the reservoir rock.
3. Reservoir-quality studies determine the framework variability of the reservoir rock in terms of porosity, permeability, and capillary properties (the aquifer surrounding the field is similarly studied).
4. Integration studies develop the hydrocarbon pore volume and fluid transmissibility patterns in three dimensions.[17,18]

Geoscientists and engineers need feedback from each other throughout their work, and an example of this interaction is shown in Figure 2–6. Core analyses provide data for identifying reservoir rock types, whereas well test studies assist in recognizing flow barriers and fractures. Simulation studies can be utilized to validate the geologic model against pressure-production performance. Often adjustments are required in the model to history match the actual performance.

Loudon Field Surfactant Pilot

Loudon field surfactant flooding pilot study details information required for simulation, and this work provides a comprehensive example to follow as a case study. The pilot, located in central Illinois (see Figure 5–2), contains five wells drilled on a 0.625-acre, unconfined five-spot pattern and an observation well midway between the producer and the northernmost injector. Typical well logs and core description data for the pilot are shown in Figure 5–3. The sandstone interval is the Weiler sandstone, which is the major reservoir unit at Loudon. The core-description graph records depositional environment and key rock units, oil staining, texture, lithology, sedimentary structures, and calcite cement of geologic origin. The logs and core-description data show that the cored interval consists of six lithologic units. Representative photos of these intervals are shown in Figure 5–4.

FIGURE 5–2. Location and Plan of the Pilot-Test Site in the Loudon Field, Illinois *(Copyright © 1975, SPE, from* JPT, *May 1975*[12]*)*

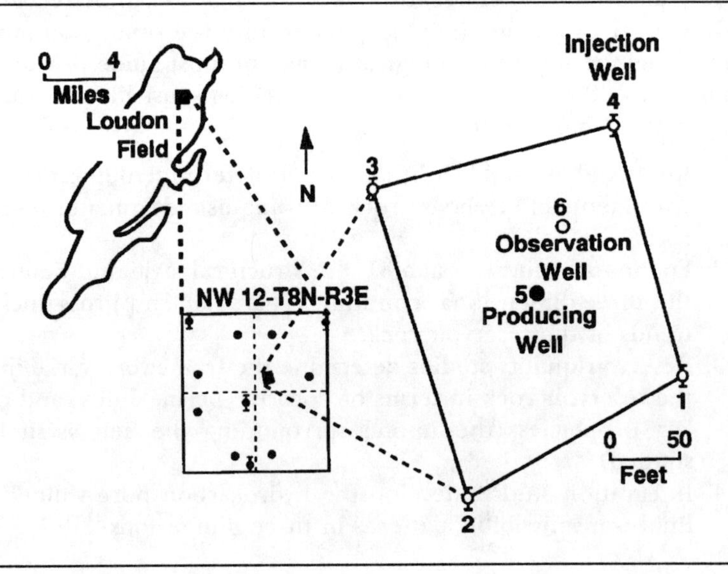

FIGURE 5–3. Example Reservoir-Quality Profile Showing Porosity and Permeability Relationships to Lithology *(Copyright © 1975, SPE, from* JPT*, May 1975*[12]*)*

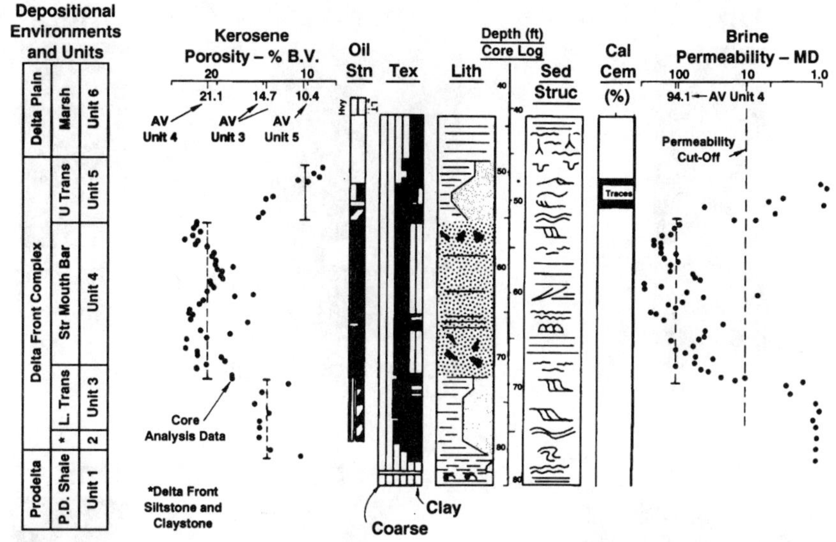

The environment of deposition provides the basis for rock framework studies, and it assists in understanding quality and three-dimensional distribution of the reservoir rock. The depositional conditions are recognized by examining slabbed cores (see Figure 5–4). The subdivisions of the depositional framework in the pilot are indicated in Figure 5–3. The three-dimensional relationships of the subdivisions illustrated in Figure 5–5 indicate documented information from studies of modern deltas.

Core analysis data and well logs provide more definitive methods of reservoir characterization. It is helpful to obtain core analysis data on representative lithologies and prepare graphs of porosity vs. permeability as shown in Figure 5–6. The rock type for each plug value is indicated to aid identification.

Regional subsurface and surface studies demonstrated that the Loudon field is located on the flank of a large, ancient river system feeding a coastal deltaic complex (see Figure 5–7). Most of the Loudon field, situated in the thinner, discontinuous delta front, was produced initially by solution gas drive. Later, waterflooding operations were begun to increase oil recovery.

To ascertain the reservoir continuity in a field, well logs and seismic sec-tion are analyzed. The objective of correlation studies is to determine the vertical and areal limits of reservoir. When cores are lacking, the

FIGURE 5–4. Typical Photographs of the Six Lithologic Units (The photo for unit 3 is about 1 ft. long) *(Copyright © 1975, SPE, from* JPT, *May 1975*[12]*)*

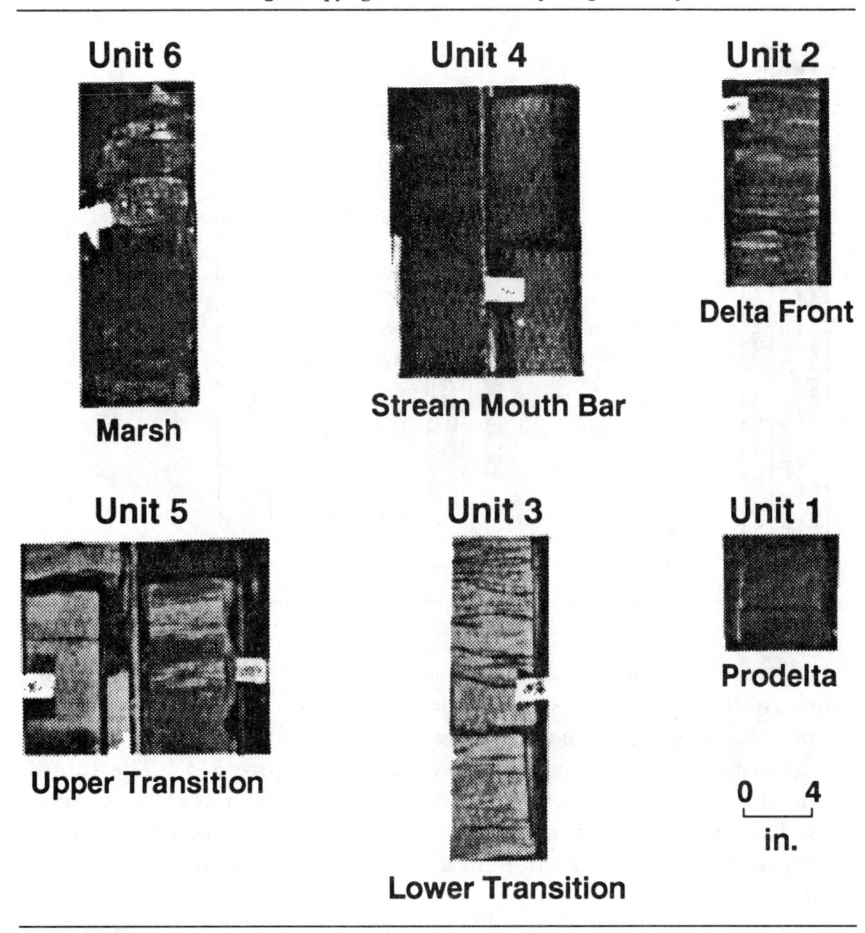

depositional units often can be estimated by comparing well-log curves in cored and uncored sections.

The method used in the case study is shown in Figure 5–8. Wells in the pilot and in the 160-acre area surrounding the pilot are correlated. The system used recognized points of depositional association in various well locations that formed a series of "loops" on a map of a 40-acre area. Key curve points are numbered and, on cross-sections, lines were drawn to connect these points.

The pattern of lines in the interval above and below the Weiler, together with the depositional units (see Figure 5–3), aided in determining the vertical and lateral limits of reservoir units in various wells.

FIGURE 5–5. Block Diagram Showing the Vertical and Areal Distribution of Units in a Typical Modern Delta *(Copyright © 1975, SPE, from* JPT, *May 1975*[12]*)*

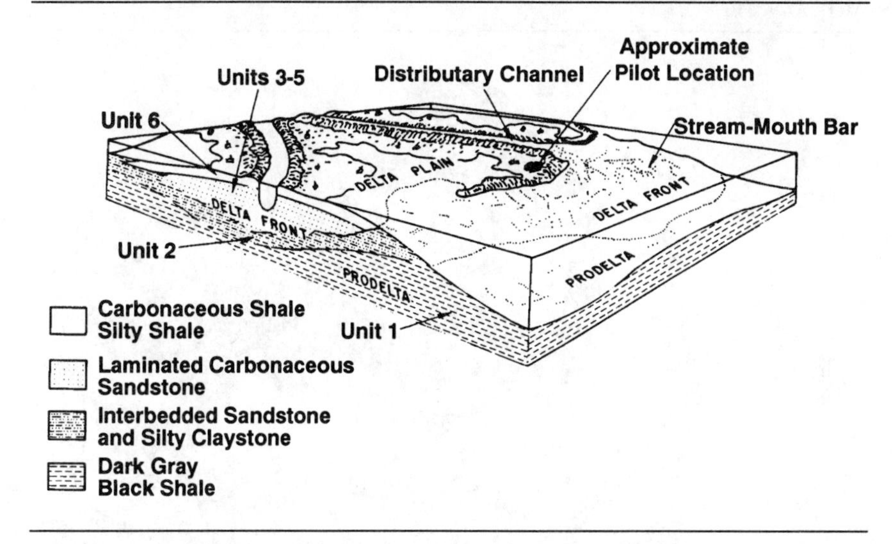

FIGURE 5–6. Porosity Versus Permeability Cross-Plot for Various Rock Types Aids Identification of Reservoir Rocks *(Copyright © 1975, SPE, from* JPT, *May 1975*[12]*)*

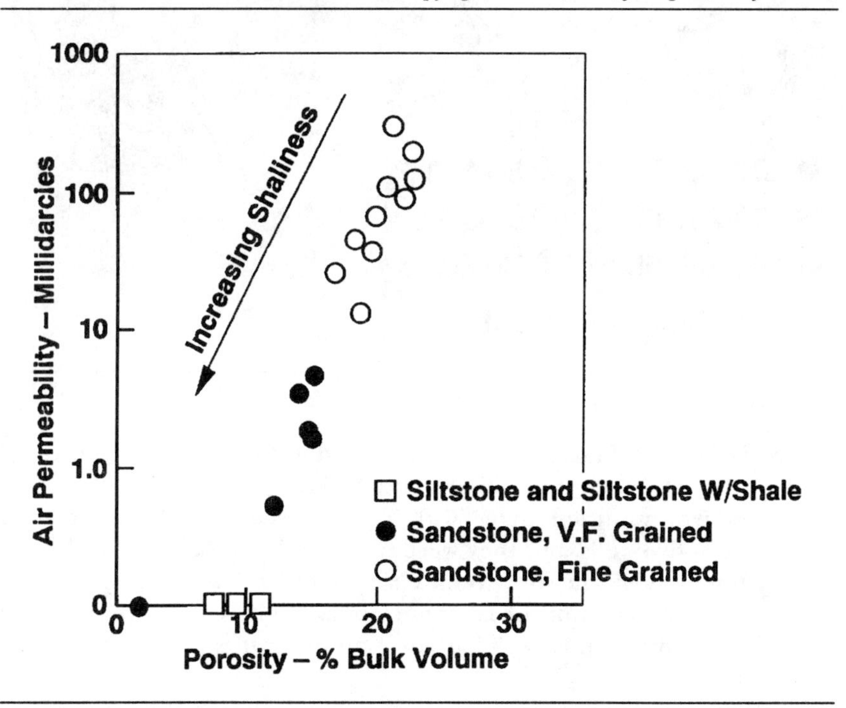

FIGURE 5–7. Regional Thickness Map of the Chester (Weiler) Sandstone Reflects Depositional Controls on Sand Distribution *(Copyright © 1975, SPE, from* JPT, *May 1975*[12]*)*

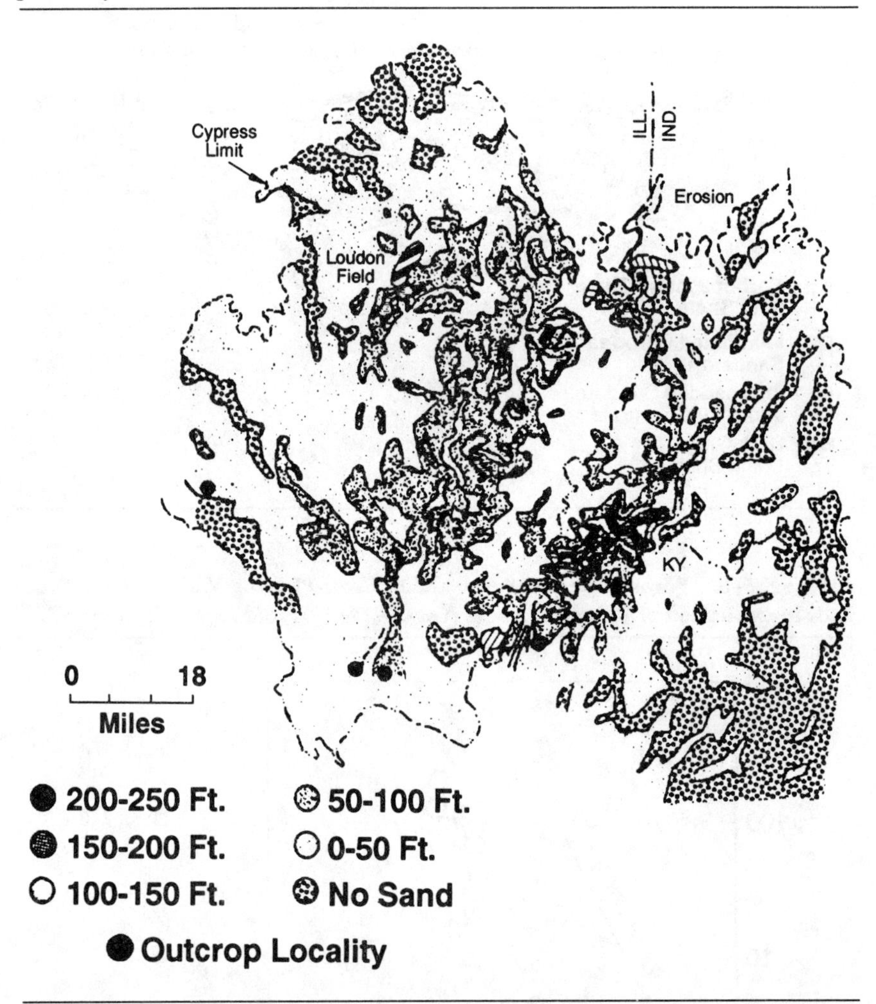

Shale breaks of two scales were identified in the pilot (see Figures 5–9 and 5–10): (1) those with radii of a least 200 ft. but less than 1,000 ft. and (2) those with radii less than 50 ft. Tight streaks of carbonate cement occurred in five wells, but they were not correlative between wells.

In summary, framework studies are aimed at determining the number and the distribution of reservoir zones. This is accomplished with core-description graphs, well-log correlations, and structure and gross-thickness maps.

FIGURE 5–8. Example of Well-Log Correlation in the Pilot Indicating Detail Required for Sand Continuity Studies *(Copyright © 1975, SPE, from* JPT, *May 1975*[12]*)*

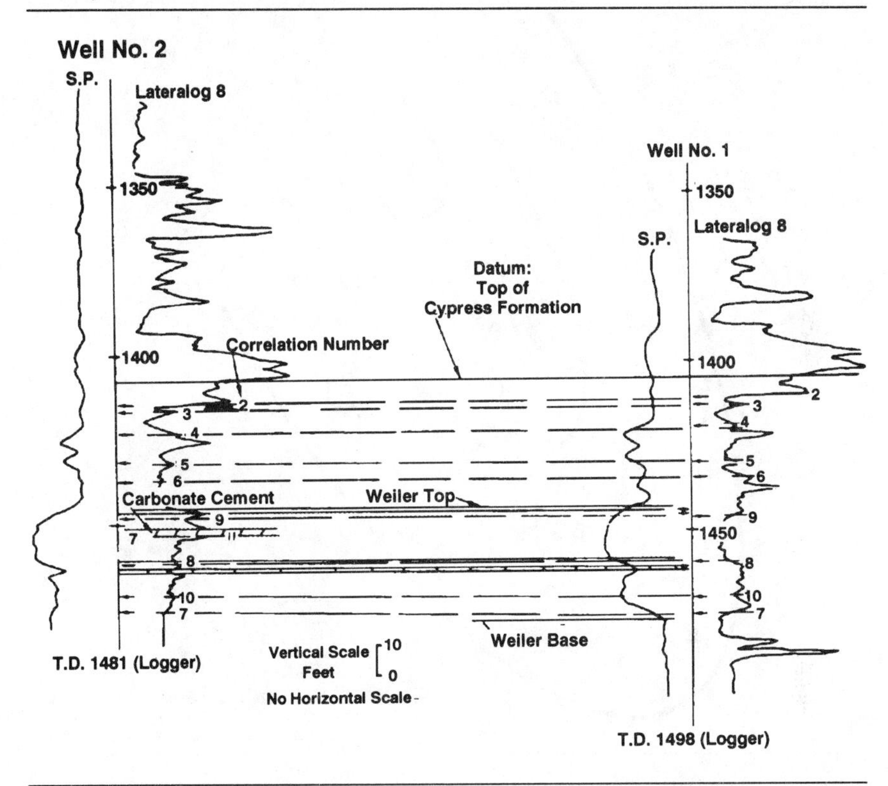

For each reservoir zone, it is necessary to establish the vertical and lateral distribution of pay, porosity, and permeability values. Laboratory analysis suggested that 10 md was the practical lower limit for effective surfactant flooding, and it was used as a cutoff value for determining net sand.

Net sand counts and preliminary evaluation of variations in porosity and permeability were developed, as shown in Figure 5–10. Also shown in this figure is the permeability cutoff for determining pay.

If core plugs are widely spaced, then the rock interval between plugs can be estimated by spreading the plug values to include intervals with a similar lithology. Conventional log-analysis data, such as porosity or water saturation, can be used. The results from either of these methods can be compared with well-test data. In the pilot, porosity was fairly constant, but permeability showed some variability.

Next, reservoir maps of pore volume and permeability capacity are developed. Such maps are important because they represent the three-

FIGURE 5–9. Thickness Map of the Reservoir and Nonreservoir Rocks in Unit 4 *(Copyright © 1975, SPE, from* JPT, *May 1975[12])*

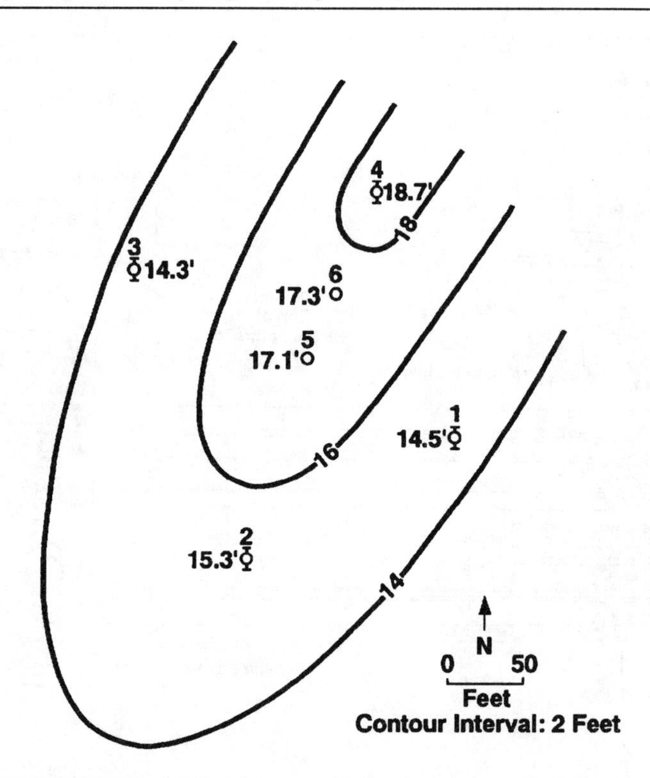

dimensional attributes of certain parameters defining the physical model. Thus, these maps point up critical areas that may require careful evaluation during the construction of a simulation model. Pore volume and other maps can be constructed by cross-contouring a net-sand map with a porosity map (or a map of another parameter).

One of the most important steps in the integration study is verifying the geologic model against the actual performance, also known as "history matching." This step often requires adjusting some of the parameters of the reservoir model. However, any adjustments should be made cautiously and within reason by involving geoscientists and engineers. In the pilot case, a good match of tracer response was achieved. The simulated pore volume was 16,500 bbls, whereas the geologic model calculated value was 16,250 bbls.

FIGURE 5–10. Example Reservoir-quality Profile Showing Porosity and Permeability Relationships to Lithology *(Copyright © 1975, SPE, from* JPT, *May 1975*[12]*)*

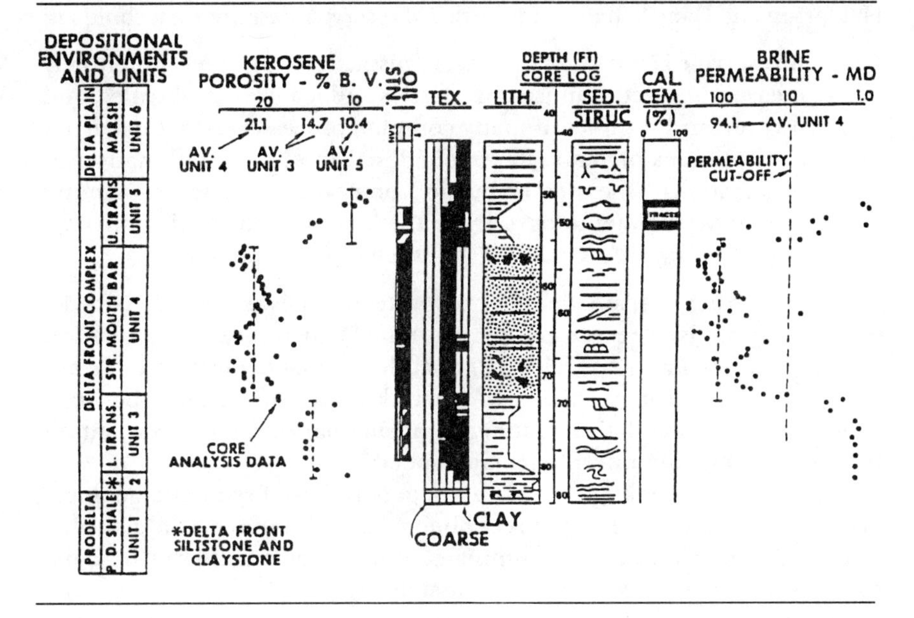

SEISMIC DATA

Three-dimensional seismic surveys help identify reserves that may not be produced optimally. The analysis can save costs by minimizing dry holes and poor producers. A 3-D survey shot during the evaluation phase is used to assist in the design of the development plan. With the development and production, data are constantly being evaluated to form the basis for locating production and injection wells, managing pressure maintenance, and performing workovers. These activities generate new information (logs, cores, DST's, etc.) that change maps, revise structure, and alter the reservoir stratigraphic model.

A 3-D seismic survey impacts the original development plan. With the drilling of development wells, the added information is used to refine the original interpretation. As time passes and the data builds, elements of the 3-D data that were initially ambiguous begin to make sense. The usefulness of a 3-D seismic survey lasts for the life of a reservoir.

A 3-D seismic data can be used to assist in (1) defining the geometric framework, (2) qualitative and quantitative definition of rock and fluid properties, and (3) flow surveillance.

GEOSTATISTICS

Haldorsen and Damsleth explain the use of reservoir-description techniques:

> A reservoir is intrinsically deterministic. It exists; it has potentially measurable, deterministic properties and features at all scales; and it is the end product of many complex processes ... that occurred over millions of years. Reservoir description is a combination of observations (the deterministic component), educated aiming (geology, sedimentology, [and the depositional environment]) and formalized "guessing" (the stochastic component).[19]

They further explain that stochastic techniques are applied to describe deterministic reservoirs because of (1) incomplete information about the reservoir on all scales, (2) complex spatial deposition of facies, (3) variability of rock properties, (4) unknown relationships between properties, (5) the relative abundance of singular pieces of information from wells, and (6) convenience and speed.

"Stochastic modeling" refers to the generation of synthetic geologic properties in one, two, or three dimensions. A number of plausible solutions can be created and simulated, and those results can be compared to see their effect on history matching.

Recent interest has focused on the use of fractal geostatistics and other stochastic techniques to map the variability of uncertainty associated with heterogeneous reservoirs.[20,21,22] Fractal geostatistics is based on the assumption that fractal statistics can be used to represent reservoir heterogeneity between wells as a random fractal variation superimposed on a smooth interpolation of correlated well-log values.[10] Standard fractal statistics are used to determine the characteristics of the random fractal variation from analyses of well-log and core data. This approach is successful because the variation of properties of many natural systems is fractal in nature.

Geostatistics/stochastic modeling of reservoir heterogeneities is playing an important role in generating more accurate reservoir models. It provides a set of spatial data analysis tools as probabilistic language to be shared by geologists, geophysicists, and reservoir engineers, and it's a vehicle for integrating various sources of uncertain information.[22]

Geostatistics is useful in modeling the spatial variability of reservoir properties and the correlation between related properties such as porosity and seismic velocity. These models can then be used in the construction of numerical models for interpolating a property whose average is critically important and stochastic simulations for a property whose extremes are critically important. Geostatistics enable geologists to put their valuable information in a format that can be used by reservoir engineers. In the absence of sufficient data in a particular reservoir, statistical characteristics from other more mature fields in similar geologic environments and/or outcrop studies are utilized. By capturing the critical heterogeneities in a quantitative form, a more realistic geologic description is created.

Variograms (and correlograms) are used to study spatial continuity of a particular variable. Also, they can be applied to study cross-continuity of different variables at different locations. For example, one could compare porosity at a particular location to travel time at a nearby location. Once modeled, this spatial cross-correlation can be used in a multivariate regression procedure known as "cokriging" for building a porosity map not only from the available porosity data, but also from the seismic information.

Any unsampled value (e.g., porosity) can be estimated by generalized regression from surrounding measurements for the same value once the statistical relationship between the unknown being estimated and the available sample values has been defined. This prior model of the statistical similarity between data values is defined by correlogram. This generalized regression can also include nearby measurements of some different variable—seismic travel time or a facies code. When using such secondary information, one also needs a prior model of the cross-correlogram, which provides information on the statistical similarity between different variables at different locations. These generalized regression algorithms are collectively known as "krigimg" and "co-riging."

Integrating widely different types of data, such as qualitative geological information and direct laboratory measurements, is not easy to achieve by multivariate regression or co-kriging. At a particular location where a porosity measurement does not yet exist, a consideration of the lithofacies may provide a reasonable range within which the unknown value should fall. In addition, if we have a porosity histogram developed for the lithofacies in question, we then can tell if the unknown porosity is more likely to be on the low end or the high end of the range.

The result of this updating is a probability distribution that provides the probability for the unsampled value to belong within any given class of values. For such a distribution, any optimal estimate for the unsampled value can be derived once an optimality criterion (e.g., least square error criterion) has been specified. Geostatistics is used in building probability distributions that characterize the present uncertainty about a reservoir's rock and fluid properties. If there are variable probability distributions representing the uncertainties in rock and fluid properties at different locations, each drawing from such a multivariate probability distribution represents a stochastic image of the reservoir. Such stochastic images honor all of the available information entered through the correlograms and cross-correlograms, yet they are different one from the other.

The differences between these stochastic images provide a direct usable visualization of the uncertainty about the reservoir rock and fluid properties. Where all the different outcomes agree, there is little or no uncertainty; where they differ most, there is maximum uncertainty.

Stochastic images represent alternative, equiprobable numerical models of the reservoir rock and fluid properties, and they can be used as input to flow simulators for sensitivity analysis.

To illustrate an application of the geostatistical concepts, porosity and permeability distributions for a reservoir were generated with fractal geostatistics based on the work of Hewett and Behrens.[23] For this field, the interwell permeabilities were determined with a porosity vs. permeability correlation, and the porosity distributions were obtained from the geostatistics.[24] Four different porosity/permeability representations were created by honoring any permeability barriers that existed. The resulting permeability distributions for this model are shown in Figure 5–11. The result of this procedure was to generate four different geologic models of varying complexities for this reservoir.

FIGURE 5–11. Permeability representations: N/T Model B. *(Copyright © 1992, SPE, from SPERE, Sept. 1992[23] and copyright © 1992, SPE, from* JPT, *Dec. 1992[24])*

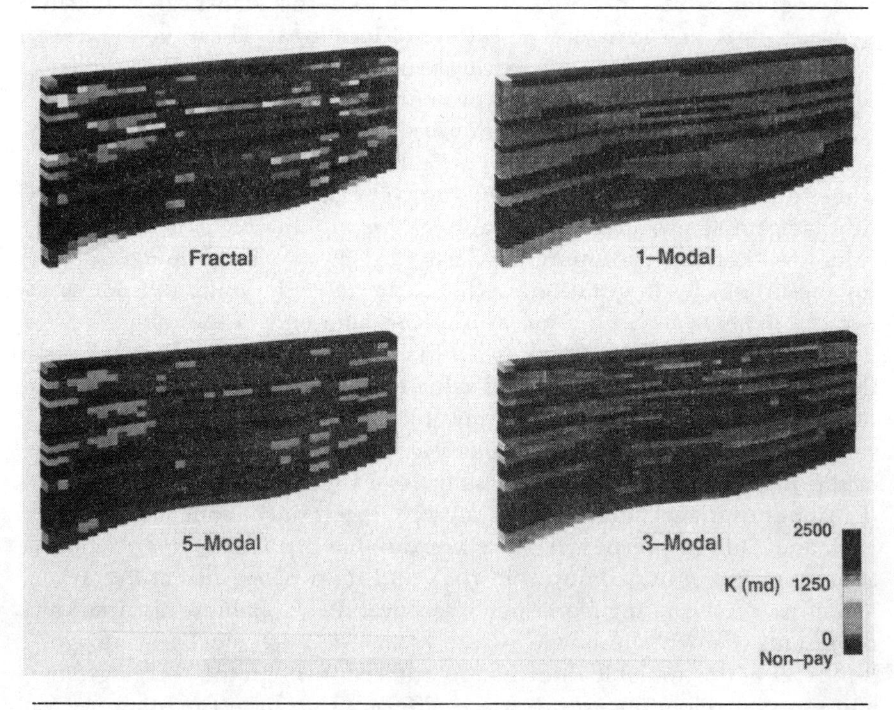

ENGINEERING

Optimizing oil and gas recovery requires the following six steps:

1. Identify and define all individual reservoirs in a given field and their physical properties.
2. Deduce past and predict reservoir performance.
3. Minimize drilling of unnecessary wells.
4. Define and modify (if necessary) wellbore and surface systems.
5. Initiate operating controls at the proper time.
6. Consider all pertinent economic and legal factors.

After identifying the geologic model, additional engineering/production information are input in the model. These include reservoir fluid and rock properties, well location and completion, well-test pressures and pulse-test responses to determine well continuity and effective permeability. The use of material balance method permits the calculation of original hydrocarbon in place, aquifer influx, and areal extent of the reservoir. The use of injection/production profiles provide vertical fluid distribution.

INTEGRATION

Traditionally, data of different types have been processed separately, leading to several different models—a geologic model, a geophysical model, and a production/engineering model. Indicator geostatistics provide an approach to merge all of the relevant information and then produce reservoir models consistent with that information.

The importance of geology to the prediction of reservoir performance is recognized by reservoir engineers. However, it is important that the geologic (including the qualitative inferences) picture be transferred into a simulation model and not be time-consuming and frustrating.

Three-dimensional geological modeling programs have been developed to automate the generation of geological maps and cross-sections from exploration data.[11] Because these models are directly interfaced to the reservoir simulator, the reservoir engineer can easily utilize the complex reservoir description provided by the geologist for field development planning. In addition, the reservoir engineer can routinely and readily update their model with new data or interpretations and quickly provide consistent maps and sections for assessing results of activities like infill drilling. Three-dimensional geological modeling programs can provide maps and cross-sections in large numbers. This permits the engineer to become thoroughly familiar with the geology prior to designing the simulation model.

A revolution in simulation techniques has come with the advent of numerical simulation models. Today, the reservoir engineer seeks more data, both in quantity and detail, from the geologist and production engineer. On the other hand, the history matching of the reservoir can lead to a feedback of geological information to the geologist.

The degree of interaction between geoscientists and engineers has been well documented in the literature. Craig, et al. (1977)[25] and Harris and Hewitt (1977)[17] explained the value of synergism between engineering and geology. Craig, et al. emphasized the value of detailed reservoir description, utilizing geological, geophysical and reservoir simulation with the knowledge of geophysical tools, to provide a more accurate reservoir description for use in engineering calculations. Harris and Hewitt presented a geologic perspective of the synergism in reservoir management. They explained the reservoir inhomogeneity due to complex variations of reservoir continuity, thickness patterns, and pore-space properties (e.g., porosity permeability and capillary pressure).

A major breakthrough in reservoir modeling has occurred with the advent of integrated geoscience (reservoir description) and engineering (reservoir production performance) software designed to manage reservoirs more effectively and efficiently (see Figures 5–12 and 5–13). Several service, software and consulting companies are now developing and marketing integrated software installed in a common platform. These interactive and user-friendly software provides more realistic reservoir models. The users from different disciplines can work with the software cooperatively as a basketball team rather than passing their own results like batons in a relay race.

CASE STUDIES

North Sea Lemen Field

As one early example, Craig, et al. used a multidisciplinary approach on the East Unit of the North Sea Lemen field from the time the field came on stream in 1968.[23] The field contained more than 10 Tcf gas—then the world's largest producing offshore gas field.

Working in a complex fault system, the company's reservoir engineers worked closely with geologists to "produce an accurate *a priori* reservoir description." The team tested the description against field performance in a 2D fine-grid, single-phase model and refined it with measured pressures from the first six years of production.

Geologists reviewed the location of faults and reservoir boundaries of the historical map. The resulting model successfully predicted pressure for an additional two years. The proven accuracy of the model led to confident planning of future platform and compression requirements, providing more than three years' lead time to install equipment.

FIGURE 5–12. Integrated Geoscience-Engineering Software System Concept *(courtesy of Western Atlas Software)*

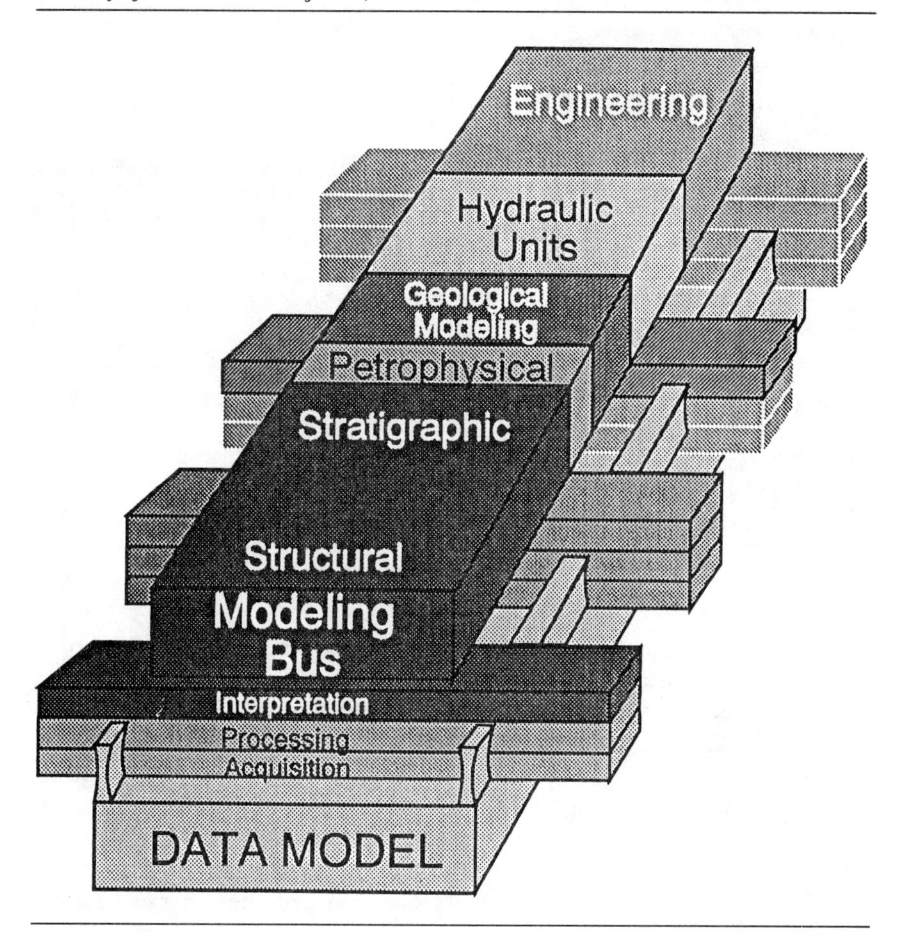

Brassey Oil Field

Recently, Woofter and MacGillivray[26] presented a case study of the development of an aeolian sand, Brassey oil field, British Columbia. The case study presented an aggressive engineering/geoscience team approach to prepare development plans. Production and miscible injection commenced simultaneously in 1989, only two years after discovery. The integration of the reservoir description, volumetrics, seismic delineation, and well-test data provided the basis for determining reservoir continuity and size. (See Appendix B for a complete description of this case study.)

A geologic model was developed before the miscible flood was designed and concurrent with seismic delineation. Excellent core control

FIGURE 5–13. Integrated Geoscience-Engineering Software System *(courtesy of Western Atlas Software)*

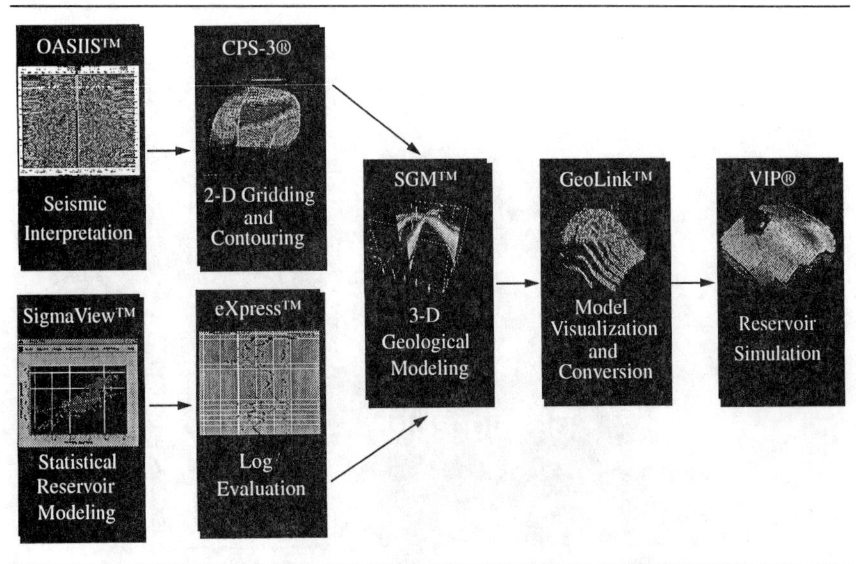

throughout the field aided detailed sedimentologic and petrographic interpretations, but the wide well spacing of 0.5 mile or more limited interwell correlations. A geologic model was developed with the aid of geophysical and engineering information.

Reservoir simulations were constructed with the geologic model. Although the well spacing and thinness of the reservoir limited interwell correlations to two layers, additional model sensitivities were run with more layers on the basis of case permeability variations.

Teamwork was critical to the development of the reservoir model. The geophysicist defined the pool edges; the engineers described pools by using well-test pressures and pulse-test responses to determine well continuity; and the geologist mapped the reservoir properties. The team's goal was to continue to improve the model as production history and tracer information became available.

Original oil in place was calculated from material-balance analysis and derived from a hand-contoured volume. Material-balance calculations show 22 to 27 MMSTB, compared with 24 MMSTB of oil from mapped volumes.

Agreement of independently calculated volumes gave the confidence to proceed with field development, and it was the basis for the predicted field recovery and producing allowable.

Brookeland Well and Teal Prospect

In an *Oil & Gas Journal* article, Texaco's achievement of considerable cost savings by using horizontal well technology on its Brookeland and Teal acreage has been documented.[27] A cross-discipline team of Texaco geologists, geophysicists, engineers, and field technicians contributed to the success of both projects.

Currently, a cross-discipline team is conducting an in-depth field study of the Teal prospect using production history, sequence stratigraphy models, 3-D seismic interpretation, and horizontal well technology.

REFERENCES

1. Satter, A., Varnon, J. E. and Huang, M. T. "Reservoir Management: Technical Perspective." SPE Paper 22350, SPE International Meeting on Petroleum Engineering, Beijing, China, March 24–27, 1992.
2. Thomas, G. W. "The Role of Reservoir Simulation in Optimal Reservoir Management." SPE Paper 14129, International Meeting on Petroleum Engineering, Beijing, China, March 17–20, 1986.
3. Saleri, N. G. and R. M. Toronyi. "Engineering Control in Reservoir Simulation." SPE Paper 18305, SPE 63rd Annual Technical Conf. & Exhib., Houston, TX, October 2–5, 1988.
4. Satter, A., D. F. Frizzell and J. E. Varnon. "The Role of Mini-Simulation in Reservoir Management." SPE Paper 22370, International Meeting on Petroleum Engineering, Beijing, China, March 24–27, 1992.
5. Coats, K. H. "Reservoir Simulation: State of the Art," *JPT* (August 1982): 1633–1642.
6. Coats, K. H. "Use and Misuse of Reservoir Simulation Models." SPE Reprint Series no. 11, Numerical Simulation, (1973): 183–190.
7. Breitenbach, E. A. "Reservoir Simulation: State of the Art," *JPT* (September 1991): 1033–1036.
8. Tang, R. W. et al. "Reservoir Studies With Geostatistics To Forecast Performance," SPE Reservoir Engineering (May 1991): 253–258.
9. Harris, D. E. and R. L. Perkins "A Case Study of Scaling Up 2D Geostatistical Models to a 3D Simulation Model." SPE Paper 22760 presented at the 1992 Annual Technical Conference and Exhibition, Dallas, TX, October 6–9, 1991.
10. Emanuel, A. S. et al. "Reservoir Performance Prediction Methods Based on Fractal Geostatistics," *SPERE* (August 1989): 311–318.
11. Johnson, C. R. and T. A. Jones "Putting Geology Into Reservoir Simulations: A Three-Dimensional Modeling Approach." SPE Paper 18321 presented at the 1988 Annual Technical Conference and Exhibition, Houston, TX, October 2–5, 1988.

12. Harris, D. G. "The Role of Geology in Reservoir Simulation Studies," *JPT* (May 1975): 625–632.
13. Ghauri, W. K. et al "Changing Concepts in Carbonate Waterflooding—West Texas Denver Unit Project—An Illustrative Example," *JPT* (June 1974): 595–606.
14. Staggs, H. M. and E. F. Herbeck. "Reservoir Simulation Models—An Engineering Overview," *JPT* (December 1971): 1428–1436.
15. Robertson, J. D. "Reservoir Management Using 3-D Seismic Data," *Geophysics: The Leading Edge of Exploration* (February 1989): 25–31.
16. Nolen-Hoeksema, R. C. "The Future of Geophysics in Reservoir Engineering," *Geophysics: The Leading Edge of Exploration* (December 1990): 89–97.
17. Harris, D. G. and C. H. Hewitt. "Synergism in Reservoir Management—The Geologic Perspective," *JPT* (July 1977): 761–770.
18. Harris, D. G. "The Role of Geology in Reservoir Simulation Studies," *JPT* (May 1975): 625–632.
19. Haldorsen, H. H. and E. Damsleth. "Stochastic Modeling," *JPT* (April 1990): 404–412.
20. Hewitt, T. A. "Fractal Distributions of Reservoir Heterogeneity and Their Influence on Fluid Transport." SPE Paper 15386 presented at the SPE Annual Technical Conference and Exhibition, New Orleans, October 5–8, 1986.
21. Begg, S. H., R. R. Carter and P. Dranfield. "Assigning Effective Values to Simulator Gridblock Parameters for Heterogeneous Reservoirs," *SPERE* (November 1989): 455–463.
22. Journel, A. G. and F. G. Alabert. "New Method for Reservoir Mapping," *JPT* (February 1990): 212–218.
23. Hewett, T. A. and Behrens, R. A.: "Conditional Simulation of Reservoir Heterogeneity With Fractals," SPERE (September 1992): 217–225.
24. Saleri, N. G. et al. "Data and Data Hierarchy," *JPT* (December 1992): 1286–1293.
25. Craig, F. F. et al. "Optimized Recovery Through Continuing Interdisciplinary Cooperation," *JPT* (July 1977): 755–760.
26. Woofter, D. M. and J. MacGillivray. "Brassey Oil Field, British Columbia: Development of an Aeolian Sand—A Team Approach," *SPERE* (May 1992): 165–172.
27. "Texaco Sets Horizontal Well Marks," *Oil and Gas Journal* (July 6, 1992): 30–32.

CHAPTER 6

▼ ▼ ▼

Reservoir Performance Analysis and Forecast

A major reservoir management activity involves:

- Estimation of the original hydrocarbon in place in the reservoir.
- Analysis of the past and present performance of the reservoir.
- Prediction of the future performance under the prevailing reservoir conditions (i.e., estimation of the reserves and recovery rates).
- Additionally, estimation of the reserves and recovery rates under various other producing methods that are presently known, or that may become practical and economical in the future.
- Periodic updating of the previous items as the quality and quantity of the data improve during the life of the reservoir.

Even though reservoir engineers are mainly involved in this activity, geologists, petrophysicists, and production engineers are also responsible for providing data, reviewing, and verifying results of the reservoir performance analysis.

This chapter presents:

- A brief review of natural producing mechanisms which influence reservoir performance.
- Techniques used for reservoir performance analysis and reserves forecasts.

NATURAL PRODUCING MECHANISMS

Primary reservoir performance of oil and gas reservoirs is dictated by natural viscous, gravity, and capillary forces. It is characterized by variations in reservoir pressure, production rates, gas-oil and water-oil ratios, aquifer water influx, and gas cap expansion. Factors influencing the reservoir performance are geological characteristics, rock and fluid properties, fluid flow mechanisms, and production facilities. The quality of reservoir management is also very important because the same reservoir exploited by different engineering and operating personnel with different equipment and production practices would differ in its performance. The policies of the company's management and the government regulatory agencies would also be a contributing factor.

The natural producing mechanisms influencing the primary reservoir performance are listed as follows (see Figures 6–1 and 6–2):

Oil Reservoirs:

- Liquid and rock expansion (A–B).
- Solution gas drive (B–C).
- Gas cap drive.
- Aquifer water influx.
- Gravity segregation.
- Combination drives.

Gas Reservoirs:

- Gas depletion or expansion (D–E–F).
- Aquifer water influx.
- Combination drives.[1]

Figure 6–3 and Table 6–1 present important characteristics of natural producing mechanisms of oil reservoirs.

FIGURE 6–1. Hydrocarbon Phase Behavior

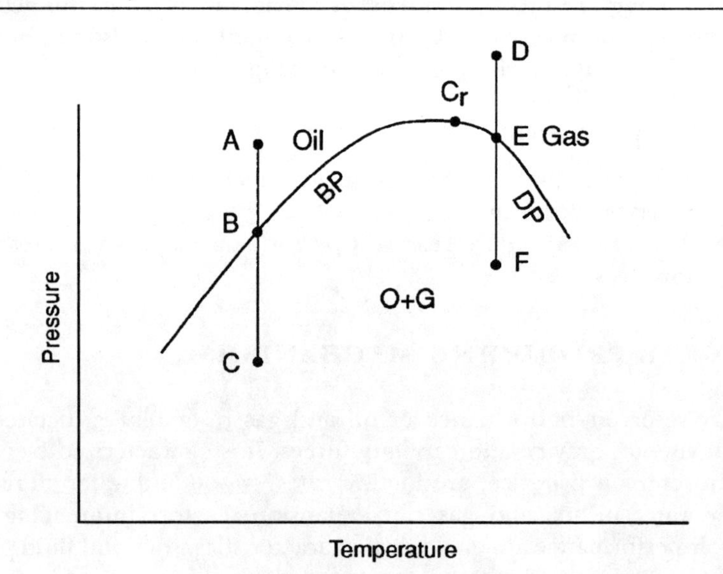

FIGURE 6–2. Hydrocarbon Reservoirs

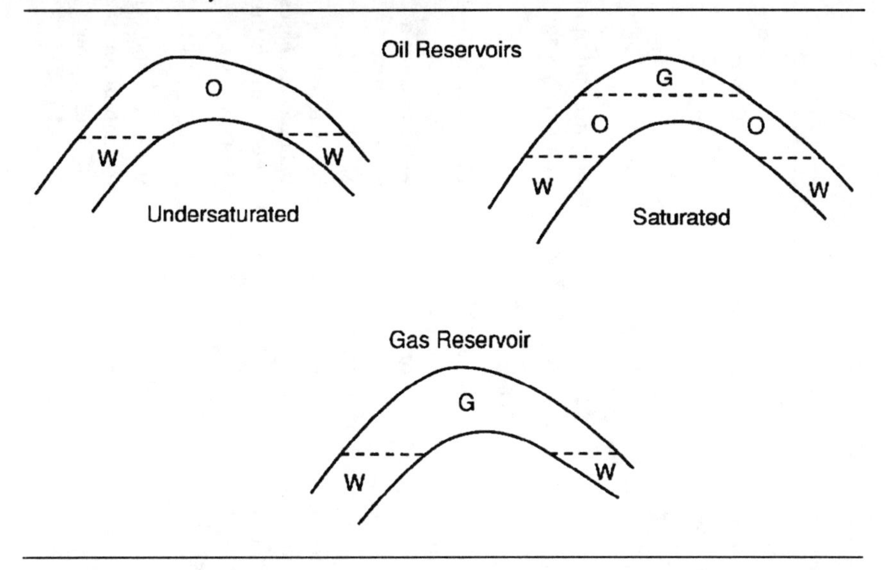

FIGURE 6–3. Influence of Primary Producing Mechanisms on Reservoir Pressure and Recovery Efficiency

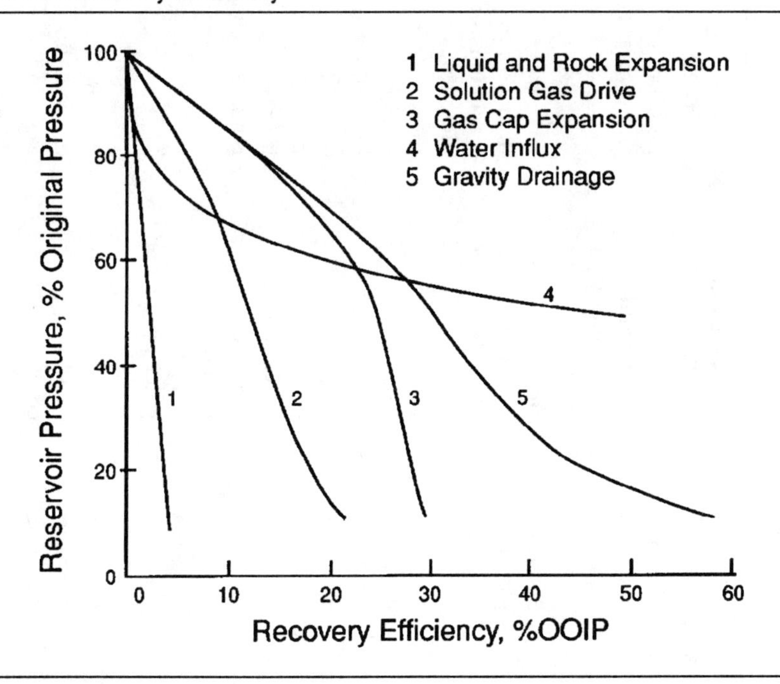

TABLE 6–1. Characteristics of Various Driving Mechanisms

Mechanisms	Characteristics				
	Reservoir Pressure	GOR	Water Production	Efficiency	Others
1. Liquid and rock expansion	Declines rapidly and continuously $p_i > p_b$	Remains low and constant	None (except in high S_w reservoirs)	1–10% Avg. 3%	
2. Solution gas drive	Declines rapidly and continuously	First low, then rises to maximum and then drops	None (except in high S_w reservoirs)	5–35% Avg. 20%	Requires pumping at an early stage
3. Gas cap drive	Falls slowly and continuously	Rises continuously in up-dip wells	Absent or negligible	20–40% Avg. 25% or more	Gas breakthrough at a down-dip well indicates a gas cap drive
4. Water drive	Remains high. Pressure is sensitive to the rate of oil, gas, and water production	Remains low if pressure remains high	Down-dip wells produce water early and water production increases to appreciable amount	35–80% Avg. 50%	N calculated by material balance increases when water influx is neglected
5. Gravity drainage	Declines rapidly and continuously	Remains low in down-dip wells and high in up-dip wells	Absent or negligible	40–80% Avg. 60%	When $k > 200$ mD, formation dip > 10° and μ_o low (< 5 cP)

RESERVES

Hydrocarbon reserve or "reserve" is defined as the future economically recoverable hydrocarbons from a reservoir. Ultimate recovery is given by

$$UR = OHCIP \times E_r \tag{6–1}$$

where:

UR = Ultimate recovery
$OHCIP$ = Original-hydrocarbon-in-place
E_r = Recovery efficiency

Then, considering prior production,

$$\text{Reserve} = UR - Q_p \tag{6–2}$$

where:

Q_p = cumulative production

Reserves are classified as proved, probable, and possible—depending upon the technological and economic certainty with which recovery can be made.

Commonly used reservoir performance analysis and reserve estimation techniques are:

- Volumetric.
- Decline curves.
- Material balance.
- Mathematical simulation.

These techniques with examples are described in this chapter; more discussions including equations are given in the Appendix C. Table 6–2 presents applicability, accuracy, data requirement, and results of these techniques.

VOLUMETRIC METHOD

Oil Reservoirs

The volumetric method for estimating recoverable reserves consists of determining the original oil in place (OOIP) and then multiplying OOIP by an estimated recovery efficiency factor. The original oil in place is given by the bulk volume of the reservoir, the porosity, the initial oil saturation, and the oil formation volume factor (see Equation C–1 in Appendix C). The bulk volume is determined from the isopach map of the reservoir, average porosity and oil saturation values from log and core analysis data, and oil-formation volume factor from laboratory tests or correlations.

TABLE 6–2. Comparison of Reservoir Performance Analysis and Reserves Estimation Techniques

	Volumetric	Decline Curve	Material Balance	Mathematical Models
1. *Applicability/Accuracy*				
Exploration	Yes/Questionable	No	Yes/Questionable	Yes/Questionable
Discovery	Yes/Questionable	No	Yes/Questionable	Yes/Questionable
Delineation	Yes/Questionable	No	Yes/Questionable	Yes/Fair
Development	Yes/Better	No	Yes/Better	Yes/Good
Production	Yes/Fair	Yes/Fair	Yes/Good	Yes/Very Good
2. *Data Requirements*				
Geometry	Area, thickness	No	Area, thickness homogeneous	Area, thickness heterogeneous
Rock	porosity, saturation	No	porosity, saturation, rel. permeability, compressibility homogeneous	porosity, saturation, rel. permeability, compressibility capillary pressure heterogeneous

Fluid	Form. Vol. Factors	No	PVT homogeneous	PVT heterogeneous
Well	No	No	PI for rate vs. time	Locations Perforations PI
Production & Injection	No	Production	Yes	Yes
Pressure	No	No	Yes	Yes
3. Results				
Original Hydrocarbon in Place	Yes	No	Yes	Yes
Ultimate Recovery	Yes with rec. eff.	Yes	Yes	Yes
Rate vs. Time	No	Yes	Yes with PI	Yes
Pressure vs. Time	No	No	Yes with PI	Yes

Once the original oil in place has been estimated, the reserve may be estimated using recovery efficiency factors. The oil recovery efficiency factor (STB/acre-ft or % OOIP) may be estimated from the performance data on similar and/or offset reservoirs. In the absence of these type of data, published API correlations for primary recovery efficiencies (see Equations C–4 and C–5 in Appendix C) can be used.[2,3]

Gas Reservoirs

The product of the original gas in place (OGIP) and recovery efficiency factor gives the recoverable gas reserves. The OGIP can be calculated by using Equation C–3 in Appendix C. The gas recovery efficiency factor may be estimated from performance data on comparable and/or offset reservoirs. For gas reservoirs without the aid of water influx, the recovery efficiency is usually high, 80–90% OGIP. Recovery from active water-drive gas reservoirs may be substantially less, 50–60% OGIP, attributable to higher abandonment pressure, gas entrapment, and water coning.

DECLINE CURVE METHOD[4,5]

Oil producers recognized a long ago that they were dealing with a depleting system, one whose producing rate is bound to decline. Since the graphical representation of production data eventually shows production curves decrease with time, the curves are known as "decline curves."

When sufficient production data are available and production is declining, the past production curves of individual wells, lease, or field can be extended to indicate future performance. The very important assumption in using decline curves is that all factors that influenced the curve in the past remain effective throughout the producing life. Many factors influence production rates and consequently decline curves. These are proration, changes in production methods, workovers, well treatments, pipeline disruptions, and weather and market conditions. Therefore, care must be taken in extrapolating the production curves in the future. When the shape of a decline curve changes, the cause should be determined, and its effect upon the reserves evaluated.

The commonly used decline curves for oil reservoirs are:

1. Log of production rate vs. time (see Figure 6–4).
2. Production rate vs. cumulative production (see Figure 6–5).
3. Log of watercut or oil cut vs. cumulative production (see Figures 6–6 and 6–7).
4. Oil-water contact (water level) or gas-oil contact (gas cap) vs. cumulative production (see Figure 6–8).

FIGURE 6–4. Log of Production Rate vs. Time

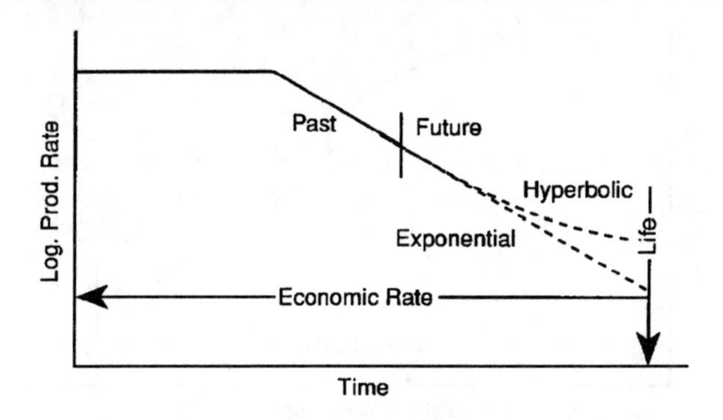

FIGURE 6–5. Production Rate vs. Cumulative Production

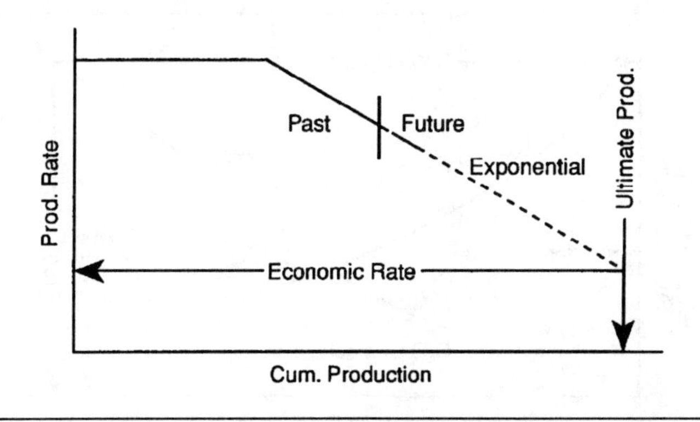

FIGURE 6–6. Log of Water Cut vs. Cumulative Oil Production

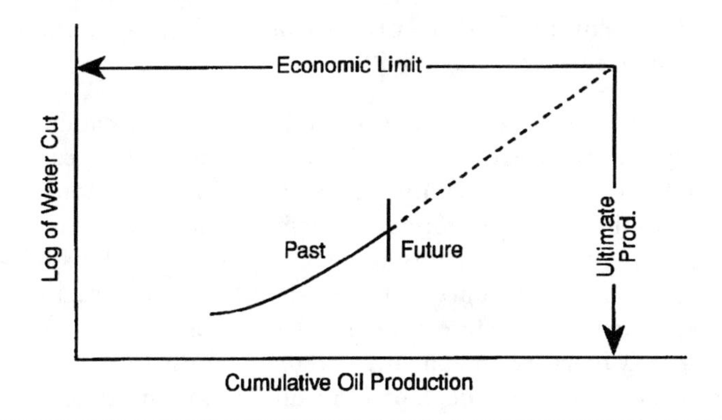

FIGURE 6–7. Log of Oil Cut vs. Cumulative Oil Production

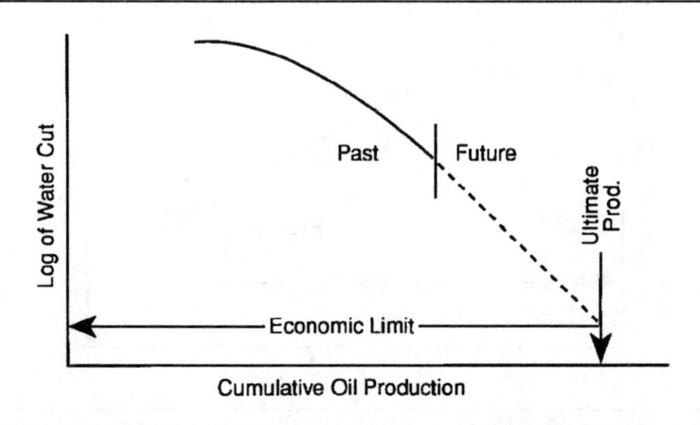

FIGURE 6–8. O/W Contact or G/O Contact vs. Cumulative Oil Production

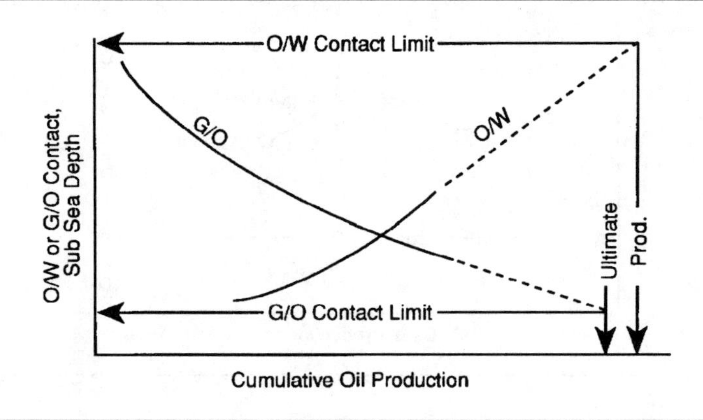

5. Log of cumulative gas production vs. log of cumulative oil production (see Figure 6–9).

When Type 1 and 2 plots are straight lines, they are called "constant rate" or "exponential decline curves." Since a straight line can be easily extrapolated, exponential decline curves are most commonly used. In case of "harmonic" or "hyperbolic rate decline," the plots show curvature. Both the exponential and harmonic decline curves are the special cases of the hyperbolic decline curves. Unrestricted early production from a well shows hyperbolic decline rate. However, constant or exponential decline rate may be reached at a latter stage of production. Mathematical expressions for the hyperbolic, harmonic and exponential decline curves are presented in Equations C–6 to C–17 in Appendix C.

FIGURE 6–9. Log Cumulative Gas vs. Log Cumulative Oil Production

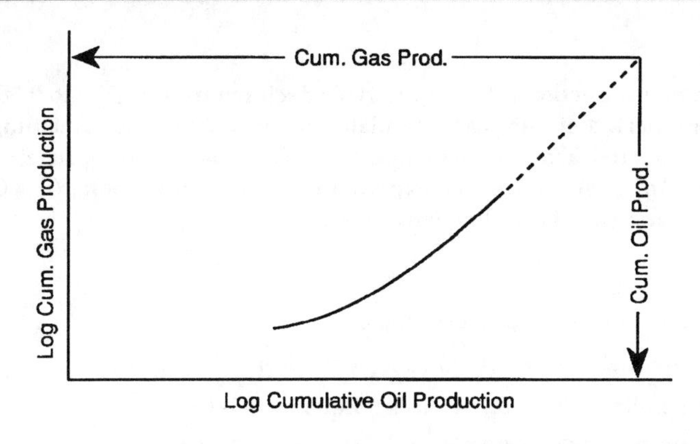

Type 3 curves are employed when economic production rate is dictated by the cost of water disposal. A straight line extrapolation of log of water cut versus cumulative oil production may not be reasonably done in the higher water cut levels. It may yield a conservative estimate of reserves. On the other hand, if oil cut data are used instead of water cut in the same levels, straight line extrapolation of log of oil cut versus cumulative oil production may deteriorate and lead to optimistic reserve estimates.

Type 4 curves are used for natural water or gas-cap drive reservoirs. Type 5 curves are used when the oil reserves are known, and the gas reserves are to be estimated or vice versa.

The following example problems illustrate the use of decline curve methods:

1. Prediction of reserves and remaining life of a reservoir using exponential, harmonic, and hyperbolic decline rate. (See Table 6–3.)

   This example shows the influence of the mode of decline on reserves and life of a reservoir.

2. Past history match and reserves forecasts. (See Figure 6–10.)

   This example shows three cycles of production and fluctuations in production rates in each cycle. The reasons should be investigated. Teamwork involving production and reservoir engineers would produce more realistic results.

3. Prediction of oil production rates and reserves based on volumetric original oil in place, API correlations for recovery efficiency and hyperbolic decline rate. (See Table 6–4, and Figure 6–11.)

   This example illustrates how volumetric and decline curve methods can be combined to give production data needed for economic analysis of a project.

TABLE 6–3. Decline Curve Prediction

Problem:

The oil production rate of a reservoir declined from 11,300 to 9,500 BOPD over a period of one year. Calculate the oil reserves and remaining life for an estimated abandonment rate 850 BOPD using appropriate equations from Appendix C. Use the exponential (n = 0), hyperbolic (n = 0.5), and harmonic (n = 1) decline curve methods.

Solution:

Calculate decline rates as follows:

Exponential	D	= 0.1735 using Equation C–8
Hyperbolic	Di	= 0.1813 using Equation C–13
Harmonic	Di	= 0.1895 using Equation C–16

Calculate reserves and remaining life as follows:

Mode of Decline	Reserves M bbl	Remaining Life Yr.-Mo.
Exponential	18,210 (Eqn. C–9)	13 – 11 (Eqn. C–8)
Hyperbolic	26,828 (Eqn. C–14)	25 – 10 (Eqn. C–13)
Harmonic	44,198 (Eqn. C–17)	53 – 8 (Eqn. C–16)

Note the influence of the mode of decline on the reserves and remaining life of the reservoir.

FIGURE 6–10. Decline Curve Analysis

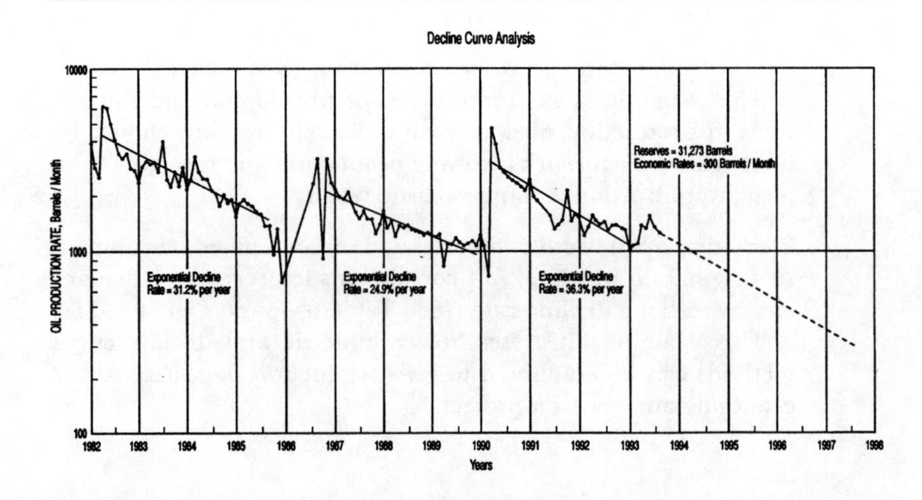

TABLE 6–4. Reservoir Performance

Problem:

Given the following data, calculate the primary oil reserves and production rates of a depletion drive reservoir. Use equations from Appendix C.

Data:

$A = 2{,}078$ acres; $h = 29$ ft; $\phi = 22.3\%$; $S_{wi} = 0.35$; $k = 83$ md;
$p_i = 3{,}450$ psia; $p_b = 2{,}805$ psia; $p_a = 250$ psia; $T_R = 140°F$;
$B_{oi} = 1.311$ RB/STB; $B_{ob} = 1.303$ RB/STB; $\mu_{oi} = 1.032$ cp; $\mu_{ob} = 1.011$ cp
$c_o = 12.4 \times 10^{-6}$ psi^{-1}; $c_r = 3.1 \times 10^{-6}$ psi^{-1}; $c_w = 3 \times 10^{-6}$ psi^{-1}
$q_i = 15{,}500$ BOPD; $q_e = 893$ BOPD

Solution:

Original oil in place (Eqn. C–1),

$N_o = 51.690$ MMBO

Effective compressibility (Eqn. C–29),

$C_e = 18.78 \times 10^{-6}$ psi^{-1}

Production above bubble point (Eqn. C–28),

$N_{p1} = 630$ MBO

Oil in place at bubble point,

$N_b = (51.690 - 0.630) \times 10^6 = 51.060$ MMBO

Recovery efficiency below bubble point (Eqn. C–4),

$E_R = 23.64\%$ OIP at bubble point

Oil recovery below bubble point,

$= 51.060 \times 0.2364 \times 10^6 = 12.071$ MMBO

Total Oil Recovery $= (0.630 + 12.071) \times 10^6$

$= 12.701$ MMBO

Using Eqn. C–14, $D_i = 67.7\%$ per yr (Assuming $n = 0.5$)

Then using Eqns. C–13 and C–14, oil production rates and cumulative oil productions are calculated as given below.

Time (yr)	Production Rate MBO/Day	Cum. Production MMBO
0.0	15.500	0.0
0.5	11.338	2.419
1.0	8.651	4.227
1.5	6.818	5.629
2.0	5.512	6.747
3.0	3.816	8.421
4.0	2.797	9.614
5.0	2.138	10.506
6.0	1.687	11.200
7.0	1.367	11.754
8.0	1.128	12.205
9.0	0.947	12.582

FIGURE 6–11. Performance Prediction

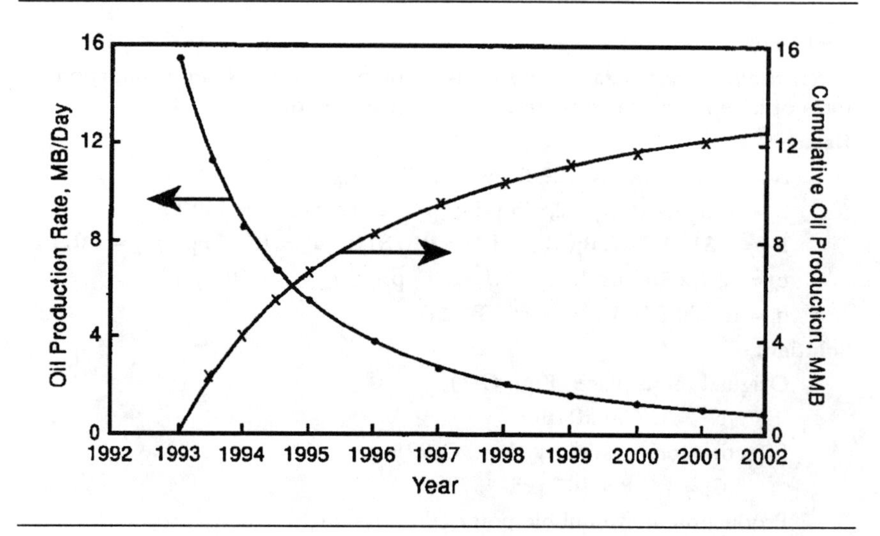

MATERIAL BALANCE METHOD

Classical material-balance method is more fundamental than the decline curve technique for analyzing reservoir production performance. The material balance method is used to estimate the original hydrocarbon in place and ultimate primary recovery from a reservoir. It is based upon the law of conservation of mass, which simply means that mass is conserved (i.e., neither created nor destroyed). The basic assumptions made in this technique are:

- Homogeneous tank model (i.e., rock and fluid properties are the same throughout the reservoir).
- Fluid production and injection occur at single production and single injection points.
- There is no direction to fluid flow.

However, the reservoirs are not homogeneous; production and injection wells are areally distributed and are activated at different times and fluid flows in definite directions. Nevertheless, the material balance method is widely used and it is found to be a very valuable tool for reservoir analysis with reasonably acceptable results.

The material balance equations are used for a history match of the past performance for estimating original hydrocarbon in place and also for predicting future performance. Applications of the material balance methods for oil and gas reservoirs are discussed.

Oil Reservoirs

History Match

General material balance equations for oil reservoirs (see Equations C–18 to C–29 in Appendix C) contain three unknowns: original oil in place, gas cap size and cumulative natural water influx. The equations include production and injection data, and rock and fluid properties that depend upon reservoir pressure.

Table 6–5 summarizes oil and gas material-balance methods and their plots.[6] There are five commonly used graphical methods to calculate original oil in place by history matching:[7–11]

1. FE method.
2. Gas cap method.
3. Havlena-Odeh method.
4. Campbell method.
5. Pressure Match method.

TABLE 6–5. Summary of Different Material Balance Methods *(after Litvack, B.L.[6])*

Oil Reservoirs	**Gas Reservoirs**
I. General Equations	
$F = N(E_o + mE_g + E_{fw}) + W_e$ $= N\,E_t + W_e$	$F = G(E_g + E_{fw}) + W_e$ $= GE_t + W_e$
II. Equations of Straight Line	
1. FE Method (solution gas drive) assume $W_e = 0$, $F = NE_t$ Plot: F vs E_t through origin N = Slope	1. FE Method (depletion) assume $W_e = 0$ $F = GE_t$ Plot: F vs E_t through origin G = Slope

A. With Gascap:
$m \neq 0$

F
m – TOO SMALL
SLOPE = N, CORRECT m
m – TOO LARGE
E_t

B. Without Gascap:
$m = 0$

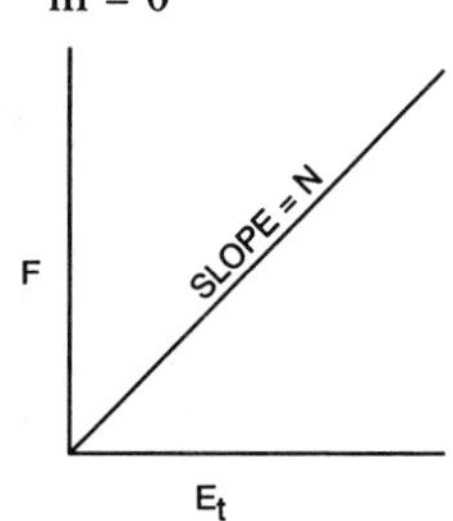

F
SLOPE = G
E_t

(Note: P/Z vs F plot is usually used instead of this)

TABLE 6–5. Summary of Different Material Balance Methods (continued)

2. GASCAP Method
(first gascap)
assume $W_e = E_{fw} = 0$
divided by E_o:

$$\frac{F}{E_O} = N + mN\left(\frac{E_g}{E_O}\right)$$

Plot: $\frac{F}{E_O}$ vs $\frac{E_g}{E_O}$
$N = Y$ intercept
$Nm = $ Slope

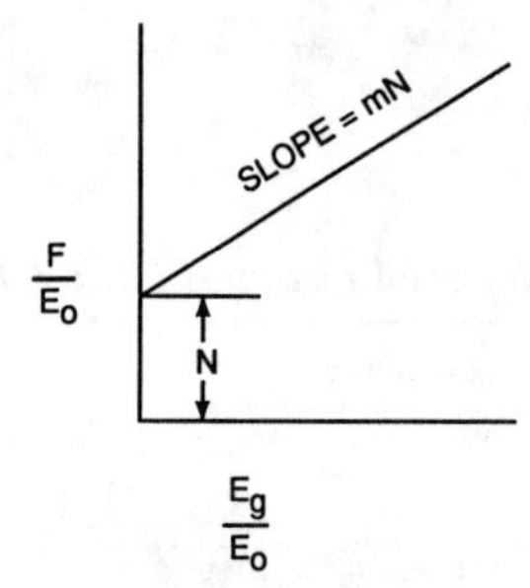

2. Divided by E_g

assume $E_{fw} = 0$

$$\frac{F}{E_g} = G + \frac{W_e}{E_g}$$

Plot: Same as in Method 3 if $E_{fw} = 0$

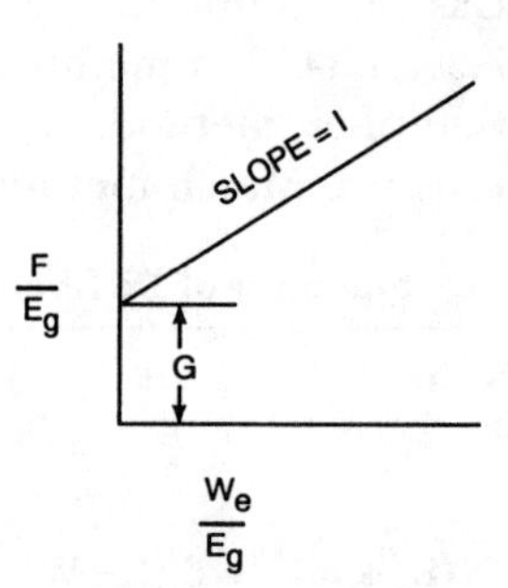

3. Havlena-Odeh Method[7,8]
(water drive)
divided by E_t:

$$\frac{F}{E_t} = N + \frac{W_e}{E_t}$$

$$= N + U\frac{S}{E_t}$$

Plot: $\frac{F}{E_t}$ vs $\frac{W_e}{E_t}\left(\text{or } \frac{S}{E_t}\right)$
$N = Y$ intercept
$(U = \text{Slope})$

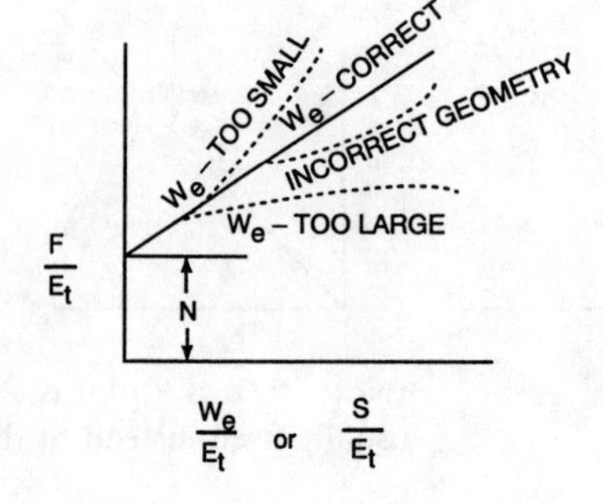

3. Havlena-Odeh Method[7,8]
(water drive)
divided by E_t:

$$\frac{F}{E_t} = G + \frac{W_e}{E_t}$$

$$= G + U\frac{S}{E_t}$$

Plot: $\frac{F}{E_t}$ vs $\frac{W_e}{E_t}\left(\text{or } \frac{S}{E_t}\right)$
$G = Y$ intercept
$(U = \text{Slope})$

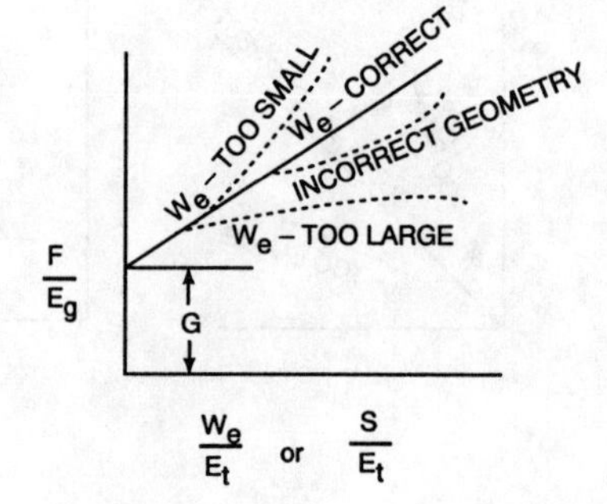

TABLE 6–5. Summary of Different Material Balance Methods (continued)

4. Campbell Method[10] (water drive)

Plot: $\frac{F}{E_t}$ vs F

N = Y intercept

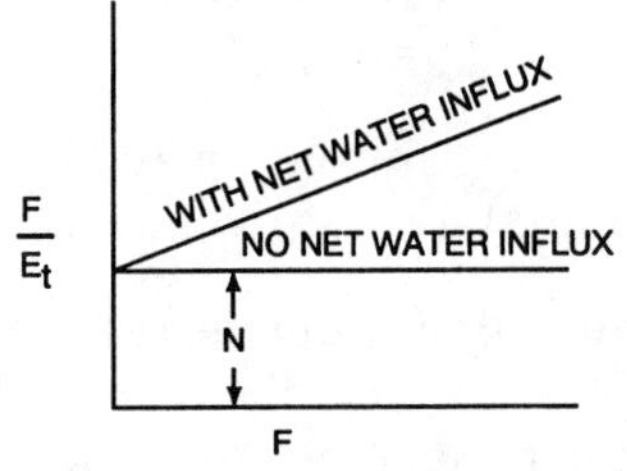

4. Cole Method[14] (water drive)

Plot: $\frac{F}{E_t}$ vs F

G = Y intercept

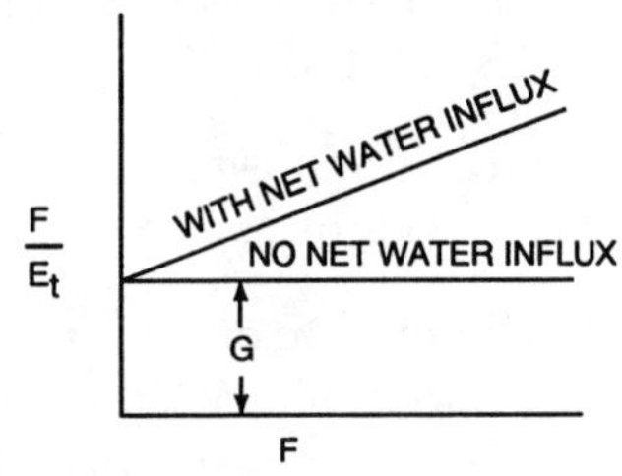

III. Pressure Analysis Methods

5. Pressure Match Method (any reservoirs)

Plot: P vs N_p

Correct N, m, and W_e for best match

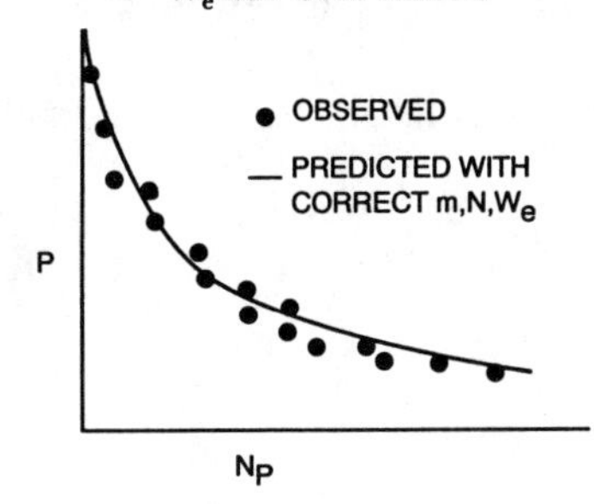

5. Pressure Match Method (any reservoirs)

Plot: P vs G_p

Correct G and W_e for best match

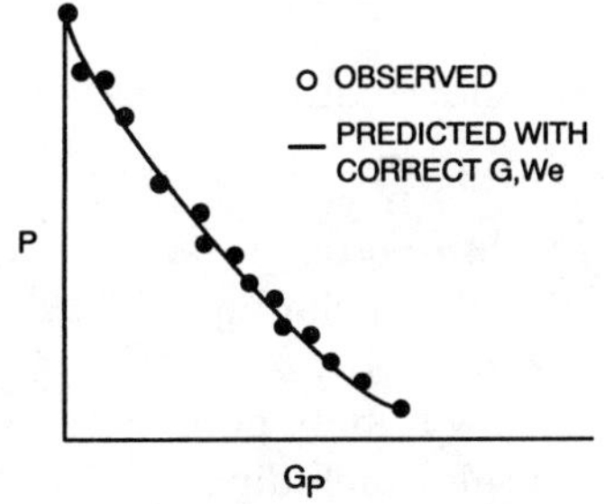

1. $\frac{P}{Z}$ Method (any reservoirs)

$$\frac{P}{Z} = \left(1 - \frac{G_p}{G}\right)\frac{P_i}{Z_i}$$

Plot: $\frac{P}{Z}$ vs G_p; G = X intercept

Plot is linear if $W_e = E_{fw} = 0$

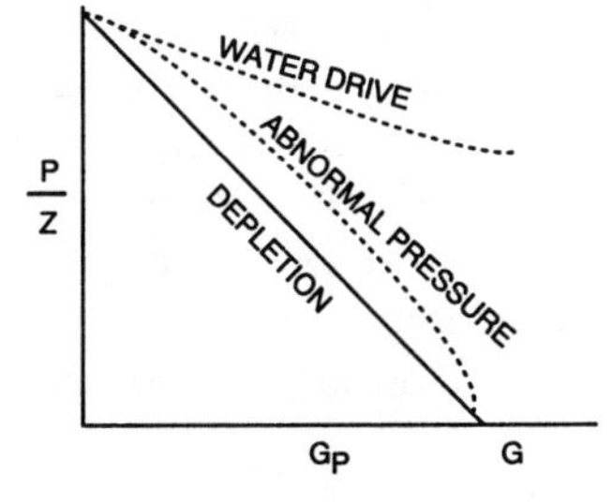

Each method applies to a specific type of reservoir. The following straight line methods should be used for different reservoirs:

- Solution Gas-Drive oil reservoirs (unknown is N)
  Use method (1).
- Gas cap Drive oil reservoirs (unknowns are N, m)
  Use method (2).
- Water Drive oil reservoirs (unknowns are N, W_e)
  Use methods (3), (4).

All three unknowns can be adjusted during pressure matching. Three types of reservoir data are required for history match:

1. Cumulative oil, water, and gas productions from the reservoir for a series of time "points."
2. Average reservoir pressures at the corresponding time "points" of (1).
3. PVT data for the reservoir fluids over the pressure ranges expected in the reservoir.

To illustrate the various oil reservoir material-balance methods, results of computer analyses of the following cases are given:[9,12,13]

1. An undersaturated Texas oil reservoir. Results are summarized below:

	Method	OOIP MMSTB	Std. Deviation %	Figure
A.	FE plot including points above and below bubble point	105.33	3.83	6–12
B.	FE plot including points above the bubble point	74.63	4.58	6–13
C.	Campbell plot	76.28	19.17	6–14

Figure 6–15 shows the pressure match. Even though the Campbell method gives very high standard deviation, this analysis indicates small water influx. Material balance results showed that primary producing mechanisms in this reservoir were rock and fluid expansions above the bubble point and solution gas drive below the bubble point. The original oil in place in this reservoir was calculated by volumetric method at 61 MMSTBO, which is much lower than given by the material balance methods. It will be presented in the later section that history match by reservoir simulation gave original oil in place comparable to the FE method including only the data above the bubble point.

FIGURE 6–12. Depletion Drive Oil Reservoir (F vs. Etot. OIP = 105.33 MMSTB, S () = 3.83%)

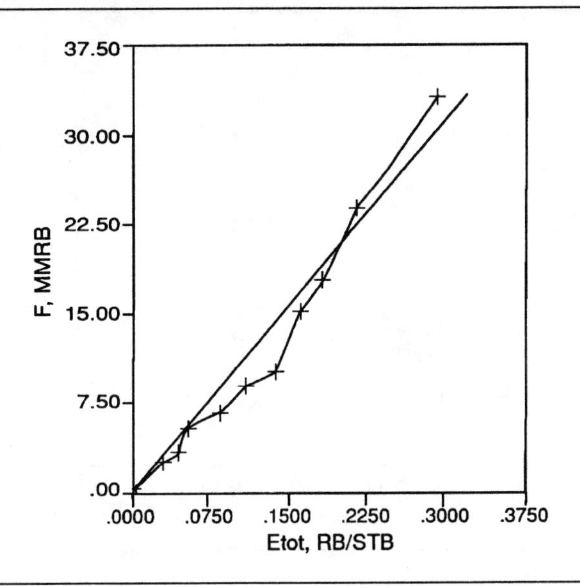

FIGURE 6–13. Depletion Drive Oil Reservoir (F vs. Etot. OIP = 74.63 MMSTB, S () = 4.58%)

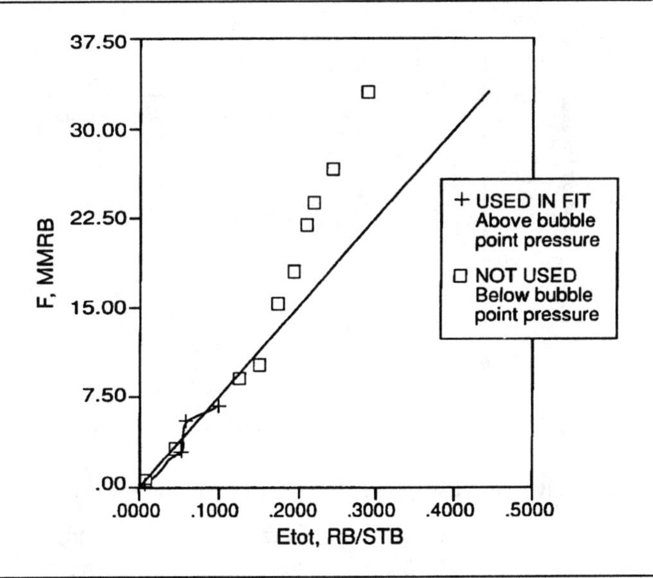

FIGURE 6–14. Depletion Drive Oil Reservoir (Campbell Plot. OIP = 76.28 MMSTB, S () = 19.17%)

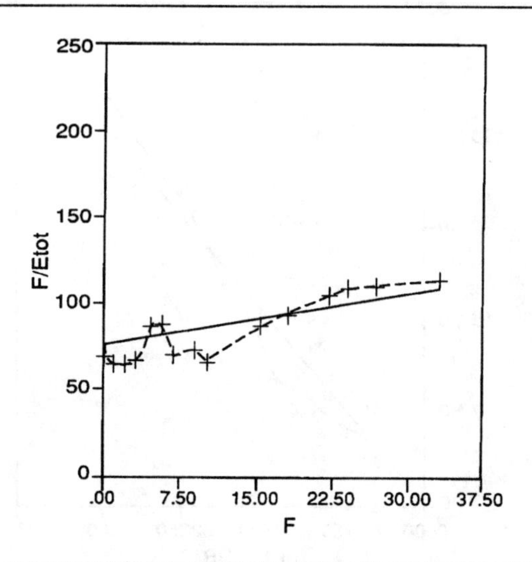

FIGURE 6–15. Depletion Drive Oil Reservoir (Comparison of Pressure Matches)

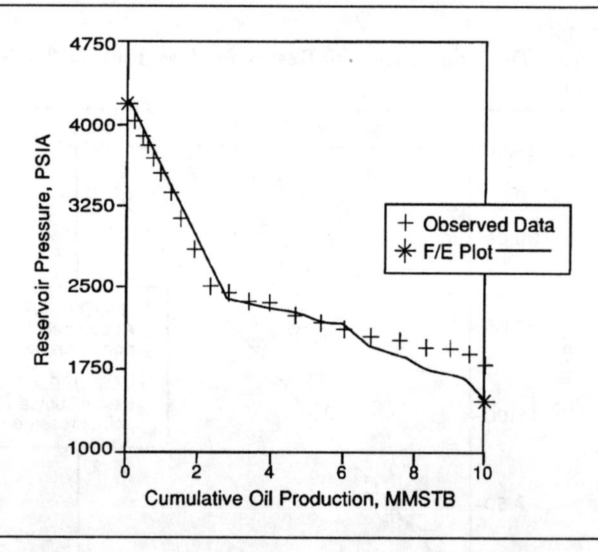

2. A gas-cap drive oil reservoir. Results are given below:

Method	OOIP MMSTB	Gas Cap Size, Frac.	Std. Dev. %	Figure
F/E_o vs. E_g/E_o plot	108.71	0.5412	5.32	6–16
F vs. Etot	111.63	0.4	0.49	6–17

FIGURE 6–16. Gas Cap Drive Oil Reservoir (F/Eo vs. Eg/Eo. OIP = 108.71 MMSTB, Gascap Fraction = .5412, S () = 5.32%)

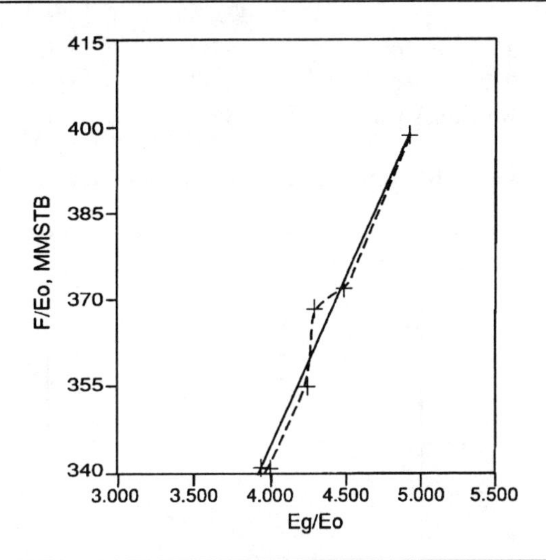

FIGURE 6–17. Gas Cap Drive Oil Reservoir (F vs. Etot. OIP = 111.63 MMSTB, S () = .49%)

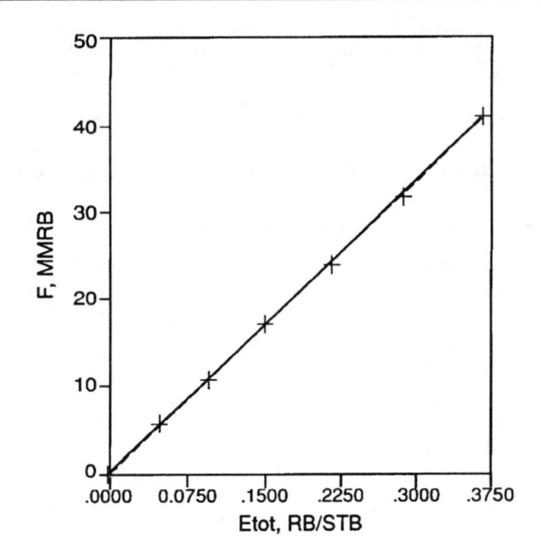

The original oil in place by volumetric method was 115 MMSTBO with a 0.4 gas cap size, which agrees with F vs. Etot method.

3. A natural water-drive oil reservoir with a small gas cap (m = 0.07). Results are summarized in the table on the following page:

Method	OOIP MMSTB	Std. Deviation %	Figure
Campbell Plot	47.75	9.34	6–18
Hurst Steady State Aquifer	46.89	4.33	6–19
Schilthuis Steady State Aquifer	52.53	4.93	6–20
Unsteady State Radial Aquifer	38.81	3.75	6–21

FIGURE 6–18. Water Drive Oil Reservoir (Campbell Plot. OIP = 47.75 MMSTB, S () = 9.34%)

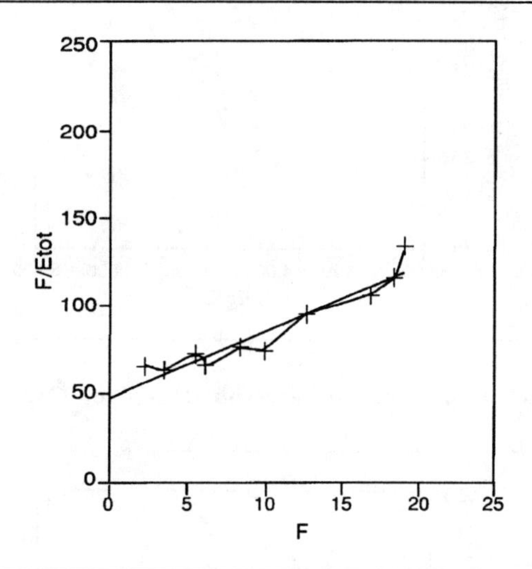

FIGURE 6–19. Water Drive Oil Reservoir (Hurst Steady-State Aquifer. OIP = 46.89 MMSTB, S () = 4.33%)

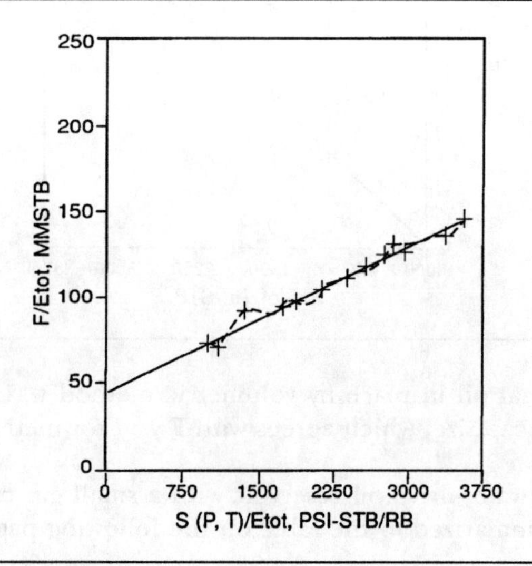

FIGURE 6–20. Water Drive Oil Reservoir (Schilthuis Steady-State Aquifer. OIP = 52.53 MMSTB, S () = 4.93%)

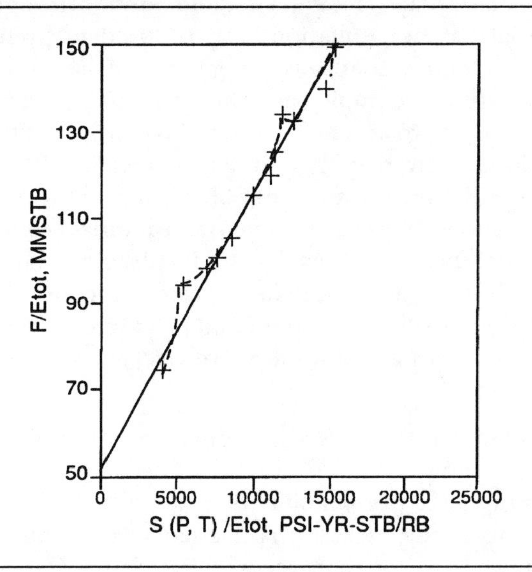

FIGURE 6–21. Water Drive Oil Reservoir (Unsteady-State Radial Aquifer – Min S (). OIP = 38.81, ra/ro = 15.0, tdf = .20, S () = 3.75%)

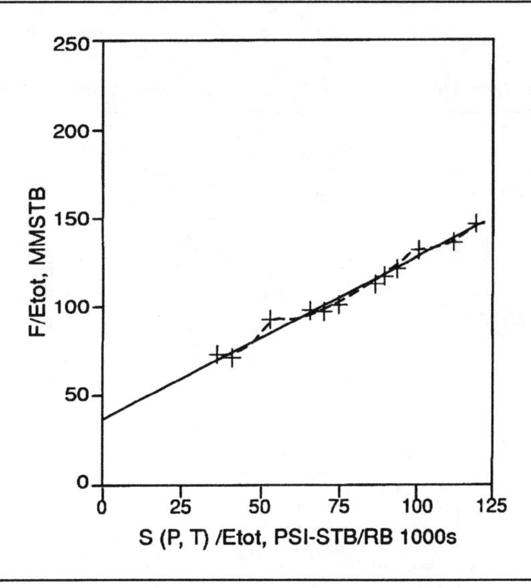

The solutions illustrate how difficult it can be to differentiate between several possible solutions. The small pressure drop over a period of 10 years indicated strong water drive. Campbell and steady-state solutions gave OOIP close to the volumetric OOIP.

Performance Prediction

If the original oil in place, gas cap size, and aquifer strength and size are known, material balance equations can be used to predict the future performance. However, solutions for gas cap drive and natural water drive reservoirs are very complex and these will not be discussed. Material balance equations above and below the bubble point for the solution gas drive reservoirs are given in Equations C–30 to C–33 in Appendix C. Performance prediction above the bubble point is rather straightforward. However, prediction below the bubble point requires simultaneous solutions of material balance equation and subsidiary liquid saturation, produced gas-oil ratio, and cumulative gas production equations. Trial-and-error solutions assuming incremental production give reservoir pressure and instantaneous gas-oil ratio vs. recovery. Types of reservoir data needed for the solution are:

- PVT data for the reservoir fluids over the expected reservoir pressure range.
- Gas-oil relative permeability data.
- Original oil in place if it is needed to calculate cumulative oil production rather than recovery efficiency (i.e., cumulative oil production/original oil in place).

Figure 6–22 presents results of an example solution drive reservoir performance.

FIGURE 6–22. Example Solution Gas Drive (Reservoir Performance)

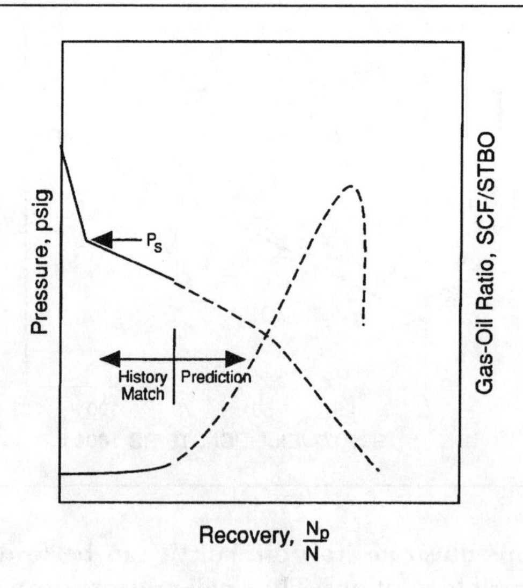

After material balance prediction has been made, reservoir performance can be related to time by the use of productivity index data (i.e., oil production rate/pressure drop [see Equations D–6 and D–7 in Appendix D]). The procedure is based upon time step by considering incremental oil production and the production rate in the time interval estimated from the productivity index data.

Gas Reservoir

History Match

General material-balance equations for gas reservoirs (see Equations C–34 to C–44 in Appendix C) contain two unknowns: original gas in place and cumulative water influx. The equations include production data and fluid properties that depend upon reservoir pressure.

There are four commonly used graphical methods to calculate original gas in place:[7,8,9,12,14]

1. p/z method.
2. Cole method.
3. Havlena and Odeh method.
4. Pressure method.

The straight line method (1) or (4) can be used to estimate original gas in place for depletion drive reservoirs. Methods (2)–(4) can be employed for estimating original gas in place and water influx in case of water drive reservoirs. Types of data required for history match are:

1. Cumulative gas, condensate and water productions from the reservoir for a series of time "points."
2. Average reservoir pressures at the corresponding time "points" of (1).
3. PVT data for the reservoir fluids over the expected reservoir pressure ranges.

To illustrate the various gas reservoir material-balance methods, the results of computer analyses of the following cases are given:[9,12,13]

1. Depletion drive Gulf Coast gas reservoir.

P/Z plot (see Figure 6–23) shows OGIP = 615.188 BCF with a standard deviation = 4.98% as compared to volumetric OGIP = 430 BCF. Other plots did not indicate natural water drive of any significance. It will be presented in the later section that production history match by a reservoir simulator gave original gas in place comparable to the material balance method.

FIGURE 6–23. Depletion Drive Gas Reservoir (p/z method2: fit line not pass (pzi,0) OGIP = 615.188 BCF, S () = 4.98%)

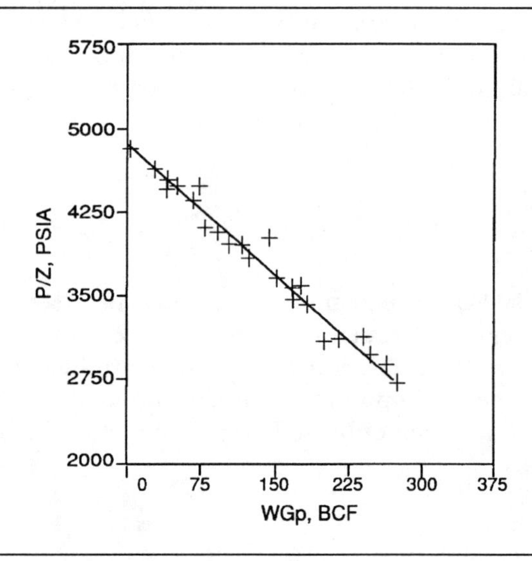

FIGURE 6–24. Abnormally Pressured Gas Reservoir (Ramagost p/z plot: abnormal p reservoir. OGIP = 449.270 BCF, S () = 2.99%)

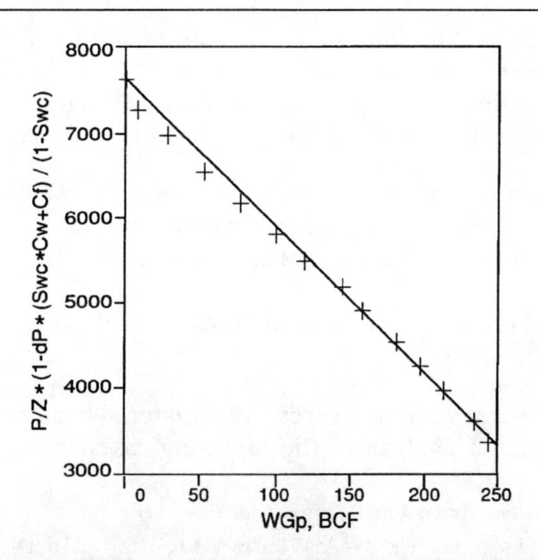

2. Abnormally pressured gas reservoir.

This reservoir was recognized as an abnormally high pressure reservoir because of its higher pressure gradient (0.86 psi/ft) as opposed to normal 0.5 psi/ft in this area. Ramagost p/z plot (see Figure 6–24) with rock compressibility = 19×10^{-6} psi^{-1} gave OGIP = 449.27 BCF, which compared favorably with the volumetric OGIP.

3. A gas reservoir with an infinite linear aquifer.

The Cole plot (see Figure 6–25) with OGIP = 347.837 BCF and standard deviation = 9.76% is a good indication of strong water drive. The unsteady-state infinite linear aquifer gave OGIP = 309.050 BCF with standard deviation = 3.10% (see Figure 6–26).

FIGURE 6–25. Water Drive Gas Reservoir (Cole Plot. OGIP = 347.837 BCF, S () = 9.76%)

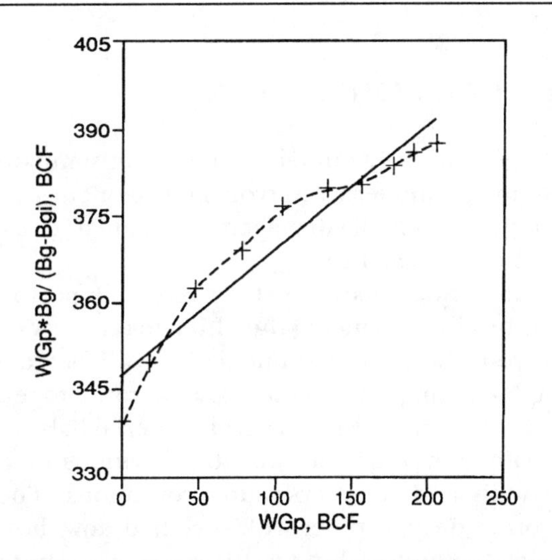

FIGURE 6–26 Water Drive Gas Reservoir (Unsteady-state infinite linear aquifer. OGIP = 309.050 BCF, S () = 3.10%)

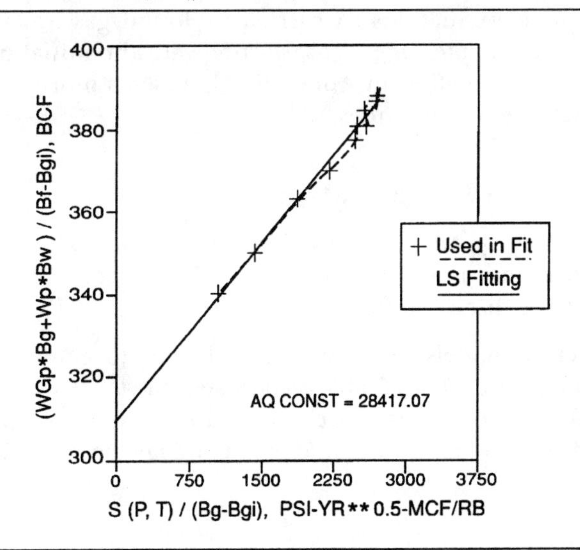

Performance Prediction

The future performance of gas reservoirs without water influx and water production can be calculated directly from a materal balance equation giving a straight line relationship between p/z and cumulative gas production (see Equation C–44 in Appendix C). Data needed for calculating gas production are gas compressibility (z) vs. pressure (p) and original gas in place. Performance prediction for a water-drive gas reservoir is complex and will not be discussed.

MATHEMATICAL SIMULATION

As discussed in Chapter 5, numerical reservoir simulators play a very important role in the modern reservoir management process. They are used to develop a reservoir management plan and to monitor and evaluate reservoir performance.

Simulators are widely used to study reservoir performance and to determine methods for enhancing the ultimate recovery of hydrocarbons from the reservoir. Numerical simulation is still based upon material balance principles, taking into account reservoir heterogeneity and direction of fluid flow. Unlike the classical material-balance approach, a reservoir simulator takes into account the locations of the production and injection wells and their operating conditions. The wells can be turned on or off at desired times with specified downhole completions. The well rates or the bottom hole pressures, or even both the rates and pressures, can be set as desired.

The reservoir is divided into many small tanks, cells, or blocks to take into account reservoir heterogeneity. Computations using material balance and fluid flow equations are carried out for oil, gas, and water phases for each cell at discrete time steps, starting with the initial time. Mathematical simulation section in Appendix C presents more on numerical reservoir simulation techniques.[6,15–19]

Types of Reservoir Simulators

Reservoir simulators are generally classified as black oil, compositional, thermal and chemical, depending upon fluid flow, mass and heat transport behavior as discussed below:

- Black-oil models are most frequently used to simulate isothermal, simultaneous flow of oil, gas and water due to viscous, gravitational and capillary forces. "Black oil" is a term used to signify that the hydrocarbon phase is considered as a single liquid and gas and not by chemical composition. The phase composition is constant even though the gas solubility in oil and water is taken into account.

- Compositional simulators account for variation of phase composition with pressure in addition to flow of the phases. They are used for performance studies of volatile-oil and gas-condensate reservoirs.
- Thermal simulators account for both fluid flow and heat transport and chemical reactions. They are used for simulating steamflood and in-situ combustion processes.
- Chemical simulators account for fluid flow and mass transport due to dispersion, absorption, partitioning, and complex phase behavior. They are used for surfactant, polymer, and alkaline flooding.

Historically, reservoir simulators have been used for studying large scale or segmental fields and reservoirs undergoing complex recovery processes. These studies, requiring highly trained professionals, are costly, time consuming, and have not been justified for modeling small reservoirs. A minisimulation technique offers a middle ground between the very sophisticated fine-tuning reservoir simulation method and the classical reservoir-engineering approach.[20] Assisted by an interactive simulator on mainframe computer, workstation or personal computer, minisimulation can play an important role in studying small reservoirs that were previously thought to be uneconomic to model. Computer-assisted studies can be made quickly and inexpensively to provide answers to the many operational problems facing practicing reservoir engineers.

Model Characteristics

Black oil simulators are characterized by the number of (fluid) phases flowing in the system, the number of directions of flow, and the type of solution used for the finite difference equations.

The simulator is called:

- "Single phase" when oil or gas only is flowing.
- "Two phase" when oil and gas or oil and water or gas and water are flowing.
- "Three phase" when oil, gas, and water are flowing.

It is called (see Figure 6–27):

- "1-dimensional linear" or "radial" when the flow is only in one direction.
- "2-dimensional areal" or "cross-sectional" or "radial cross-sectional" when the flow is in x-y or x-z or r-z directions.
- "3-dimensional" when the flow occurs in x-y-z directions.

FIGURE 6–27. Typical models used in reservoir simulation: (a) tank, (b) 1D, (c) 1D radial, (d) cross-sectional, (e) areal, (f) radial cross-sectional, and (g) 3D. *(After SPE Monograph Series 13, Richardson, TX, 1990[15])*

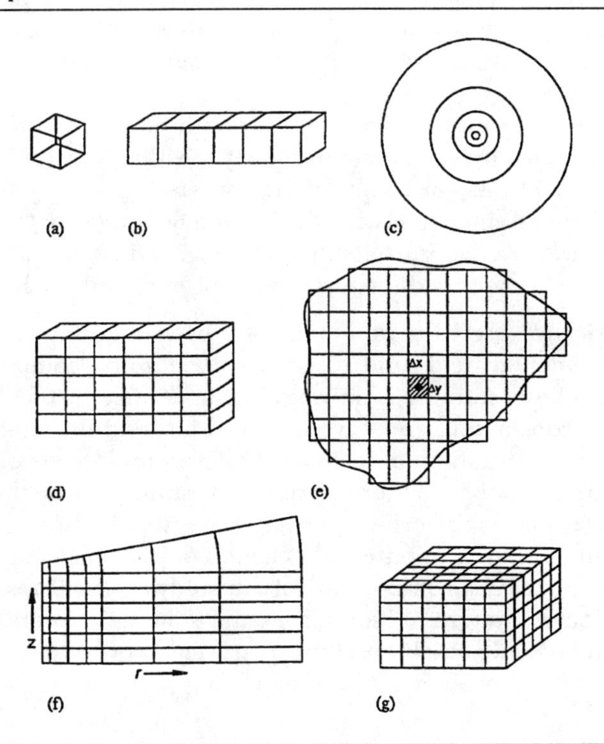

The simulator is called "impes," "implicit" or "fully implicit" depending upon the solution used to solve the finite difference equations.

Simulation Process

An outline of reservoir simulation process is shown in Figure 5–1.[19] The prerequisite to reservoir simulation is basic reservoir engineering analysis, which will greatly aid selection of a proper simulation model. Because of the complexity and the volume of data needed for simulation, it is advisable to begin a simulation study using the simplest model possible. The accuracy of the simulation results depends upon the quality of the input to the "black box" and not necessarily upon the sophistication of the model.

One of the acronyms most quoted in the computer industry and applicable to reservoir simulation is GIGO or "Garbage In; Garbage Out." Unfortunately, expensive decisions are made on the garbage out of a simulation study. It is obvious to everyone concerned that a large part of a reservoir study should be devoted to gathering and verifying the reservoir data. However, it seems that once the actual simulation work begins, changes are made to reservoir properties in pursuit of a perfect history match without regard to the physical realities of the reservoir. The successful history match carries more credence than core measurements, log analysis, or common sense even though engineers know that their match is a nonunique solution. The moral is to obtain the best, most sensible reservoir data possible. If the simulation work indicates that something must be changed, then tread cautiously.

In general, reservoir simulation process can be divided into three main phases:[6]

1. Input data gathering.
2. History matching.
3. Performance prediction.

Input Data

Input data for a black oil simulation generally consists of:

- General data for the entire reservoir—dimensions, grid definition, number of layers, original reservoir pressure, initial water-oil and gas-oil contacts. These data are obtained from base maps, log and core analyses, and well pressure tests.
- Rock and fluid data—relative permeabilities, capillary pressures, rock compressibilities, and PVT data that are obtained from laboratory tests or correlations.
- Grid data—geological data including elevations, gross and net thicknesses, permeabilities, porosities, and initial fluid saturations. These data are obtained from well log and core analyses and from well pressure and well productivity tests.
- Production/injection and well data—oil, water and gas production or injection history, and future production and injection schedule for each well, well location, productivity index, skin factor, and perforation intervals for each well.

Gathering the needed data, which can be very time consuming and expensive, requires integrated team efforts involving geoscientists and engineers. Ascertaining the reliability of the available data and information is vital for successful reservoir modeling.

History Matching

History matching of past production and pressure performance consists of adjusting the reservoir parameters of a model until the simulated performance matches the observed or historical behavior. This is a necessary step before the prediction phase because the accuracy of a prediction can be no better than the accuracy of the history match. However, it must be recognized that history matches are not unique. The reservoir parameters that may be adjusted must be identified, and the degree of adjustment determined. Some reservoir data are known with a greater degree of accuracy than others. For example, it is usually assumed that the fluid properties are valid, provided careful laboratory measurements were made. On the other hand, reservoir formation properties (i.e., porosity, permeability and capillary pressure, etc.) are known only at the locations where the wells have penetrated the formation, and even these may be subject to error. In the interwell regions, the formation properties must be interpolated from geological and petrophysical correlations. Thus, if the well values are not precise, then the results of the simulation may also be inaccurate.

A "step-wise" history matching procedure is given below:

- Initialization verifying input of initial data.
- Pressure matching by specifying production/injection for the wells and adjusting parameters affecting original hydrocarbon in place.
- Saturation matching by adjusting relative permeability curves, vertical permeabilities, water-oil and gas-oil contacts, etc.
- Well pressure matching by modifying productivity indices.

Performance Prediction

Predicting future performance of a reservoir under existing operating conditions and/or some alternative development plan such as infill drilling, waterflood after primary, and so forth is the final phase of a reservoir simulation study. The main objective is to determine the optimum operating condition in order to maximize economic recovery of hydrocarbon from the reservoir.

It should be noted that simulation studies can be useful even before production starts. This can be the case in new reservoirs. In this situation, reservoir simulation can be used to perform sensitivity studies to identify reservoir parameters that most influence hydrocarbon recovery. This can help in planning development strategies and identifying additional data that should be gathered.

Reservoir Simulation Examples

A number of examples are presented to illustrate the applications of black oil simulators as follows:

A Combination Drive Oil Reservoir[21]

In this example engineering studies of G-1/G-2 and G-2/G-3 sands of Meren field, offshore Nigeria, were made to investigate various operating schemes for optimizing oil recovery from these reservoirs. Geologic evaluation, material balance calculations, and three-phase, two-dimensional (areal and cross-sectional) reservoir simulation models were used. Available reservoir data for this study are presented in Chapter 4.

Material balance calculations, which were made to obtain the reservoir characteristics and aquifer properties, served to check the validity of the basic geologic and engineering data. It was established that G-1 and G-2 sands in Fault Blocks A and C are in communication, and they can be combined as a single-producing interval. It was also found that G-2 and G-3 sands in Fault Block B can be combined as a single-producing formation.

A cross-sectional simulation study with 3 layers (Sand G-1 at the top, Sand G-2 in the middle, and the aquifer at the bottom layer) was made with a water injection well in the middle layer and a producing well at the top layer. The objective was to verify the applicability of an areal representation of the reservoirs by varying vertical to horizontal permeability ratios from 1/1 to 1/1000. The cumulative oil productions for the three cases run were essentially the same.

Finally, areal simulation models with 31 × 16 grids (see Figure 6–28) were developed to match past performance history and predict future performance of G-1/G-2 and G-2/G-3 reservoirs under a variety of

FIGURE 6–28. Total Grid System *(Copyright © 1990, SPE, from* JPT*, April 1982*[21]*)*

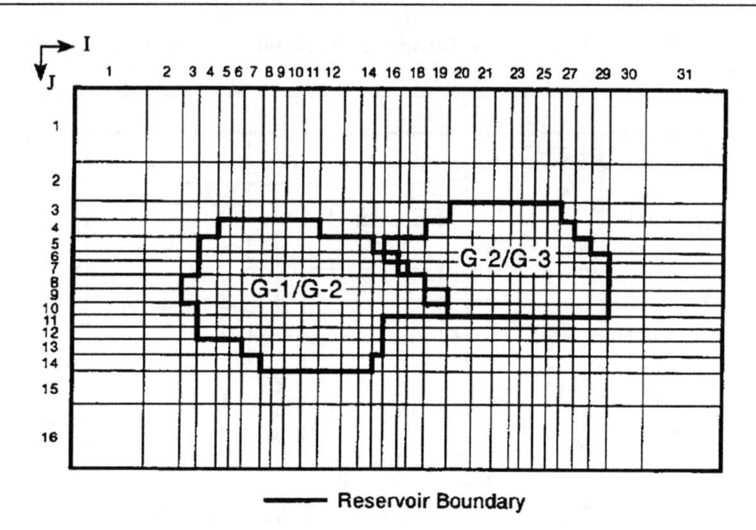

operating schemes. A good match of historical reservoir performance was achieved for both reservoirs. The results of G-1/G-2 reservoir is illustrated in Figures 4–7 and 4–8.

Prediction runs on each reservoir were made with the history-matched models as follows:

1. Natural depletion with existing wells and the current operating conditions (see Figure 6–29).
2. Water injection with existing wells.

FIGURE 6–29. Natural Depletion Prediction, G-1/G-2 Reservoirs *(Copyright © 1990, SPE, from* JPT, *April 1982*[21]*)*

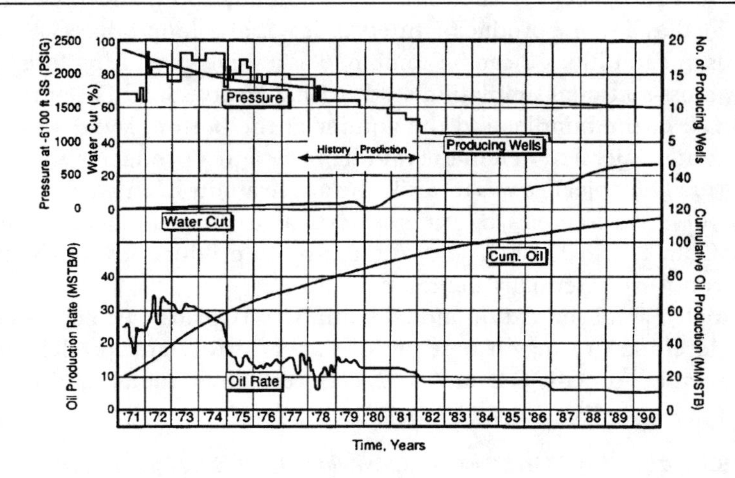

FIGURE 6–30. Waterflood Performance Prediction, G-1/G-2 Reservoirs *(Copyright © 1990, SPE, from* JPT, *April 1982*[21]*)*

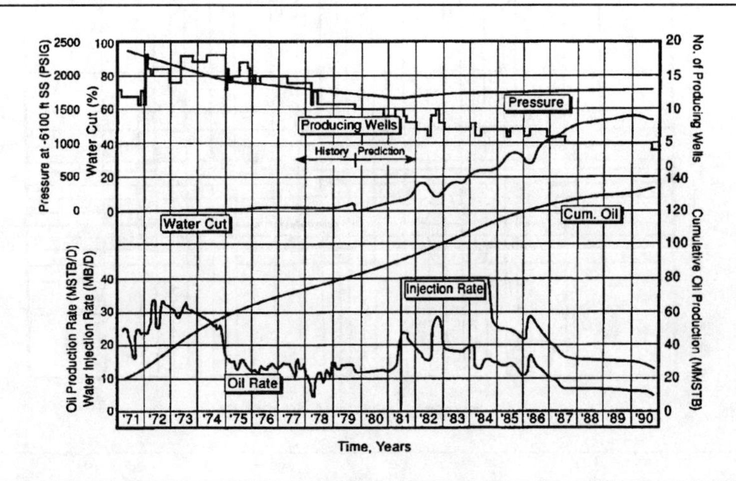

3. Water injection with existing wells, workovers and a new production well (see Figure 6–30).
4. Gas injection with existing wells.
5. Gas injection with existing wells, workovers, and a new production well.

The prediction runs showed that water injection in G-1/G-2 reservoir increases recovery by about 5% over natural depletion. The increased recovery in G-2/G-3 reservoir was nearly 8%. Gas injection did not increase recovery over natural depletion as much as water injection.

An Undersaturated Oil Reservoir

The previous example undersaturated Texas oil reservoir was used for simulation. The material balance analysis which showed depletion drive production mechanism and higher OOIP than the volumetric OOIP was very useful to build a simulation model for this reservoir.

This reservoir—with more than 100 wells—has been producing for several years. Considering the very high permeability (571 md) and thin pay of the reservoir, it was justified to build a single block, single layer tank model (1 × 1 × 1) with a single well accounting for the composite production from all the wells.[20]

Initial PC minisimulation runs showed a reasonable pressure history match using the composite past oil production rates and the original oil-in-place of 73 MMSTBO. However, the laboratory gas-oil relative permeability data had to be adjusted in order to obtain a reasonable match of the produced gas-oil ratio. The relative permeability adjustment was made by calculating average oil saturation and gas-oil relative permeability ratio from the produced oil and gas. Given the laboratory oil relative permeability data, gas relative permeability data were derived. Then the calculated gas relative permeability data were used as a guide to adjust the laboratory gas relative permeability curve. The future oil production was simulated by setting the bottom hole well pressure at 500 psia.

Figures 6–31, 6–32 and 6–33 show the pressure, oil production rate, and produced gas-oil ratio performances of the reservoir.

Depletion Drive Gas Reservoir

The previous example Gulf Coast depletion drive gas reservoir was also used for simulation. This field was discovered and developed more than 20 years ago. Five of the six originally completed wells are currently producing. The objective of this study was to history match the past production performance of the reservoir followed by prediction of its future performance using a PC simulator. Material balance analysis using the available data showed that the producing mechanism of this reservoir was due to pressure depletion with little or no natural water influx.

FIGURE 6–31. Final History Match & Prediction, Pressure vs. Cumulative Produced Oil

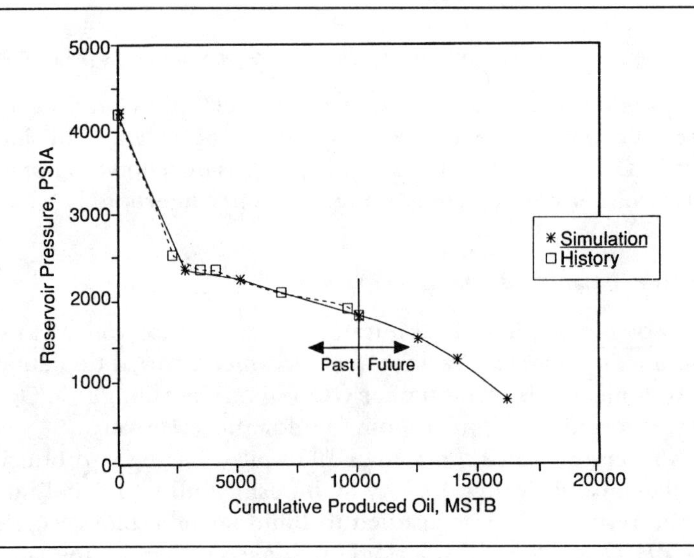

FIGURE 6–32. Final History Match & Prediction, Oil Rate vs. Time

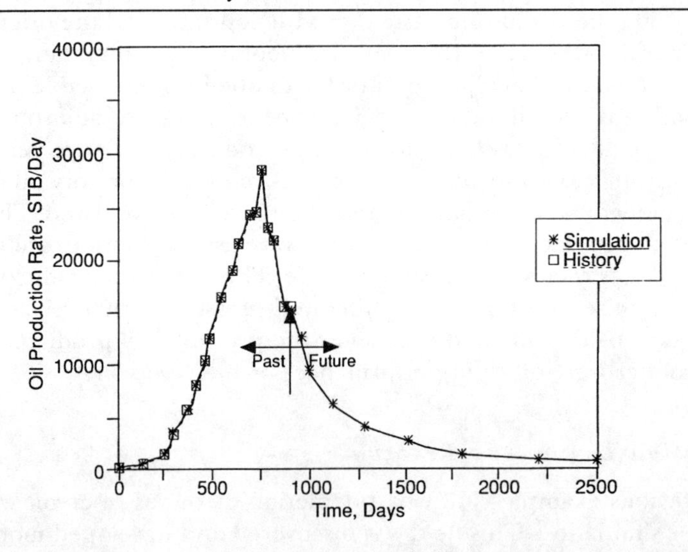

In consideration of high reservoir permeability (562 md) showing good pressure equilibrium throughout the reservoir, a single layer, single-block tank model (1 × 1 × 1) was built to simulate the performance of this reservoir.[20] The history match model was based upon composite gas production through a single well while neglecting condensate pro-

FIGURE 6–33. Final History Match & Prediction, Gas/Oil Ratio vs. Cumulative Produced Oil

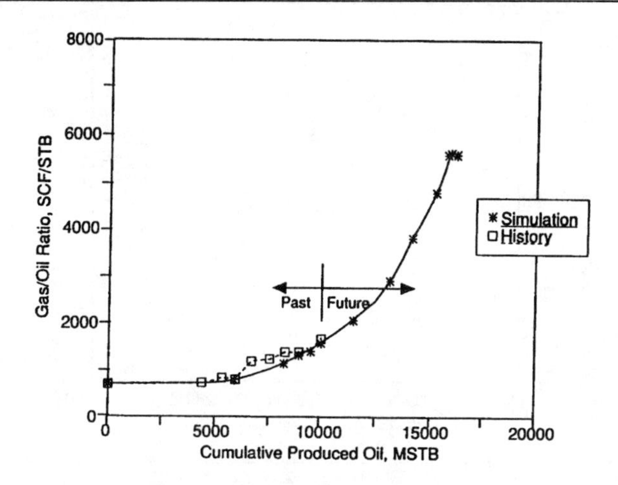

duction (less than 5% of the cumulative gas production) and water production (less than 2% of the initial oil in place).

The results of the initial history match run showed a lower computed reservoir pressure than the actual case as shown in Figure 6–34. Since natural water drive was discounted by the material balance analysis, it was evident that the initially estimated volumetric original gas-in-place (430 BSCF) was lower than the actual. Therefore, the model was rerun increasing the reservoir thickness and consequently the original gas-in-place from 430 to 615 as given by the material balance calculation. A reasonable pressure history match was obtained as shown in Figure 6–35. The future gas production was predicted by setting the bottom hole pressure at 1,000 psia. Figure 6–36 shows the past and simulated future production rate.

Abuse of Reservoir Simulation

The use of reservoir simulation has grown steadily over the past 20 years because of the constant improvement in simulator software and computer hardware. The rapid growth and acceptance of simulation has led to some confusion and occasional misuse of this reservoir engineering tool as follows:

- Unrealistic expectations—Many times there are unrealistic expectations of what reservoir simulation can do. There may be a tendency to believe that "just because the answer came out of the computer" it is infallible.
- Insufficient justification for simulation—Sometimes simulation studies are commissioned that are much more detailed and costly

FIGURE 6–34. Trial History Match, Pressure vs. Cumulative Produced Gas

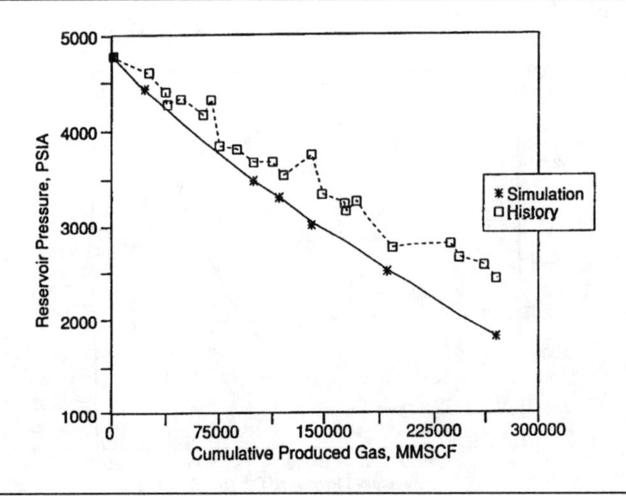

FIGURE 6–35. Final History Match & Prediction, Pressure vs. Cumulative Produced Gas

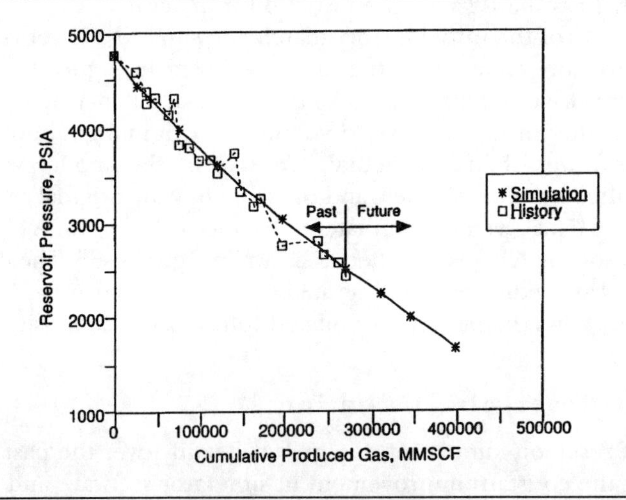

than can be justified with the available data. Sometimes the objectives of the study are not clear. Also, it may be that conventional reservoir engineering tools such as material balance would provide as much information as reservoir simulation.

- Unrealistic reservoir description for simulation—In an effort to achieve a history match for a reservoir, sometimes unrealistic

FIGURE 6–36. Final History Match & Prediction, Gas Production Rate vs. Time

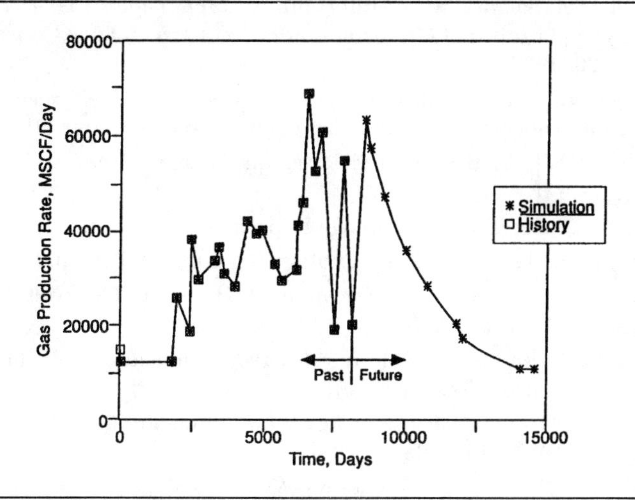

reservoir descriptions are used because of drastic adaptations of parameters such as relative permeabilities or hydrocarbon fluid properties.

It is useful to first go through the logic of building a data input for the reservoir simulator model when a study is being considered. This can be helpful in clarifying the purpose of the study, identifying additional reservoir data required, judging the reliability of reservoir data, and estimating the cost of a full study.

REFERENCES

1. Clark, N. J. *Elements Of Petroleum Reservoirs,* Henry L. Doherty Series, *SPE of AIME*: Dallas, Texas (1969): 66–83.
2. Arps, J. J., et al. "A Statistical Study of Recovery Efficiency," *API Bulletin D14* (1967).
3. Doscher, T. M., et al. "Statistical Analysis of Crude Oil Recovery and Recovery Efficiency," *API Bulletin D14* (1984).
4. Arps, J. J. "Estimation of Decline Curves," *Trans. AIME* (1945): Vol 160: 228–247
5. Arps, J. J. "Estimation of Primary Oil Reserves," *Trans. AIME* 207 (1956): 182–191.
6. Litvak, B. L., Texaco E&P Technology Department: Personal Contact.
7. Havlena, D. and Odeh, A. S. "The Material Balance as an Equation of Straight Line," *J. Pet. Tech.* (August 1963): 896–900.
8. Havlena, D. and Odeh., A. S. "The Material Balance as an Equation of Straight Line—part II, Field Cases," *J. Pet. Tech.* (July 1964): 815–822.

9. Wang, B. and T. S. Teasdale. "GASWAT-PC: A Microcomputer Program for Gas Material Balance with Water Influx." SPE Paper 16484, Petroleum Industry Applications of Microcomputers, Del Lago on Lake Conroe, Texas, June 23–26, 1987.
10. Campbell, R. A. and J. M. Campbell, Sr. *Mineral Property Economics,* vol. 3: Petroleum Property Evaluation, Campbell Petroleum Series (1978).
11. Schilthuis, R. J. "Active Oil and Reservoir Energy," *Trans. AIME* vol. 118, (1936): 33–52.
12. Wang, B., Texaco E&P Technology Department: Personal Contact.
13. Frizzell, D. F., Texaco E&P Technology Department: Personal Contact.
14. Cole, F. W. *Reservoir Engineering Manual.* Houston, TX: Gulf Publishing Company, (1969): 284–288.
15. Mattax, C. C. and R. L. Dalton. "Reservoir Simulation, SPE Monograph Series," Richardson, Texas (1990): 13.
16. Coats, K. H. "Use and Misuse of Reservoir Simulation Models," *J. Pet. Tech.* (November 1969): 1391–98.
17. Odeh, A. S. "Reservoir Simulation—What Is It?" *J. Pet. Tech.* (November 1969): 1383–88.
18. Coats, K. H. "Reservoir Simulation: State of the Art," *J. Pet. Tech.* (August 1982): 1633–42.
19. Saleri, N. G. and R. M. Toronyi. "Engineering Control in Reservoir Simulation," SPE Paper 18305, SPE 63rd Annual Tech. Conf. & Exhib., Houston, TX, October 2–5, 1988.
20. Satter, A., D. F. Frizzell, and J. E. Varnon. "The Role of Mini-Simulation in Reservoir Management," Indonesian Petroleum Association, Annual Convention, Jakarta, Indonesia, 1991.
21. Thakur, G. C. "Engineering Studies of G-1, G-2, and G-3 Reservoirs, Meren Field, Nigeria," *J. Pet. Tech.* (April 1982): 721–732.

CHAPTER 7

▼ ▼ ▼

Reservoir Management Economics

Reservoir management requires economic evaluation and analysis of the property and associated projects throughout the life of the reservoir. Making a sound business decision requires that a project will be economically viable (i.e., it will produce profits satisfying the economic yardsticks of the company).

A list of references is provided for general knowledge on economic evaluation and investment decision process.[1–7] This chapter provides a working knowledge of project economics and not of property evaluation (i.e., fair market value (FMV) estimates). Since companies operating in various countries have their own particular situations for calculating income taxes, this discussion will be limited to economic evaluation before taxes.

The tasks in project economic analysis require team efforts consisting of (see Figure 3–6):

1. Setting an economic objective based on the company's economic criteria. Production and/or reservoir engineers are responsible for developing economic justification with the input from management.
2. Formulating scenarios for project development. Engineers and geologists are the primary contributors with management guidance.
3. Collecting production, operation and economic data (see Table 7–1).
4. Making economic calculations. Engineers and geologists are primarily responsible.
5. Making risk analysis and choosing optimum project. Both engineers and geologists are primarily responsible for analysis. Engineers, geologists, operations staff, and management work together to decide on the optimum project.

TABLE 7–1. Economic Data

Data	Source/Comment
Oil and gas production rates vs. time	Reservoir and production engineers Unique to each project
Oil and gas prices	Finance and economic professionals Strategic planning interpretation
Capital investment (tangible, intangible) and operating costs	Facilities, operations and engineering professionals Unique to each project
Royalty/production sharing	Unique to each project
Discount and inflation rates	Finance and economics professionals Strategic planning interpretation
State and local taxes (production, severance, ad valorem, etc.)	Accountants
Federal income taxes, depletion and amortization schedules	Accountants

ECONOMIC CRITERIA

Making a sound business decision requires yardsticks for measuring the value of proposed investments and financial opportunities. Each company has its own economic criteria with required minimum values to fit its strategy for doing business profitably. Acceptance or rejection of individual proposals are largely governed by the company's economic criteria. Commonly used criteria are reviewed as follows:

Payout Time

The time needed to recover the investment is defined as the *payout time.* It is the time when the cumulative undiscounted or discounted cash flow (CF = revenue – capital investment – operating expenses) is equal to zero.

The shorter the payout time (2 to 5 years), the more attractive the project. Although it is an easy and simple criterion, it does not give the ultimate lifetime profitability of a project, and it should not be used solely for assessing the economic viability of a project.

Discounted cash flow means that a deferment or discount factor is used to account for the time value of money by converting the future value or worth of money to the present worth (PW) in accordance to the specified discount rate. The time value of money is not recognized in case of undiscounted cash flow.

Considering that revenues are received once a year at the midpoint of the year, the discount factor is given by

$$DF = \frac{1}{(1+i)^{t-0.5}} \tag{7.1}$$

where t is the time, and i is the discount rate in fraction.

Profit-to-Investment Ratio

Profit-to-investment ratio is the total undiscounted cash flow without capital investment divided by the total investment. Unlike the payout time, it reflects total profitability; however, it does not recognize the time value of money.

Present Worth Net Profit (PWNP)

Present worth net profit is the present value of the entire cash flow discounted at a specified discount rate.

Investment Efficiency or Present Worth Index or Profitability Index

Investment efficiency or *present worth index* or *profitability index* is the total discounted cash flow divided by the total discounted investment. The value of this parameter in the range of 0.5 to 0.75 is considered favorable.

Discounted Cash Flow Return on Investment or Internal Rate of Return

Discounted cash flow return on investment or *internal rate of return* is the maximum discount rate that needs to be charged for the investment capital to produce a break-even venture (i.e., the discount rate at which the present worth net profit is equal to zero). This can be also expressed as the discount rate at which the total discounted cash flow excluding investments is equal to the discounted investments over the life of the project.

SCENARIOS

Economic optimization is the ultimate goal of sound reservoir management. It involves more than one scenario or alternative approaches to picking the best solution. For example, possible choices and questions concerning the recovery scheme and development plan for a newly discovered offshore reservoir are:

1. Recovery scheme—natural depletion, fluid injection or natural depletion augmented by fluid (water or gas injection).
2. Well spacing—number of wells, and platforms.

The economic analyses and comparisons of the results of the various choices can provide the answer to make the best business decision to maximize profits.

DATA

The data required for economic analysis can be generally classified as production, investment and operating costs, financial, and economic data. Table 7–1 provides a list of the pertinent data.

ECONOMIC EVALUATION

Procedure for economic calculation before income tax (BIT) is outlined below:

1. Calculate annual revenues using oil and gas sales from productions and unit sales prices.
2. Calculate year-by-year total costs including capital, drilling/completion, operating, and production taxes.
3. Calculate annual undiscounted cash flow by subtracting total costs from the total revenues.
4. Calculate annual discounted cash flow by multiplying the undiscounted cash flow by the discount factor at a specified discount rate.

In order to illustrate the computational procedure, example spread sheet economic calculations are shown in Table 7–2. The values of a number of economic parameters for this example at 12% discount rate are given below:

1. Payout time = 3.43 years.
2. Profit to investment ratio = 578.09/224.941 = 2.57
   where: $ 578.09 MM (353.149 + 224.941) is the total undiscounted cash flow without the total undiscounted investment of $ 224.941.
3. Present worth net profit = $ 164.599 MM.
4. Present worth index = 164.599/169.807 = 0.97
   where $ 169.807 MM is the total discounted investment.
5. Discounted cash flow return on investment (DCFROI), found by trial and error (see Table 7–2) = 64.23 %.

TABLE 7–2. Economic Evaluation Example

Year	(1) Oil Prod. (MSTB)	(2) Oil Price ($/STB)	(3) Oil Revenue ($MM) (1) × (2)	(4) Gas Prod. (MMSCF)	(5) Gas Price ($/MSCF)	(6) Gas Revenue ($MM) $\frac{(4) \times (5)}{1000}$	(7) Total Revenue ($MM) (3) + (6)
1992	0	20.00	0.00	0	1.50	0.00	0.00
1993	0	20.00	0.00	0	1.50	0.00	0.00
1994	5505.3	20.00	110.11	3276	1.50	4.91	115.02
1995	10079.1	20.00	201.58	11934	1.50	17.90	219.48
1996	5524.2	20.00	110.48	13208.4	1.50	19.81	130.30
1997	2098.8	20.00	41.98	5848.2	1.50	8.77	50.75
1998	1020.6	20.00	20.41	2968.2	1.50	4.45	24.86
1999	1184.4	20.00	23.69	2031.3	1.50	3.05	26.73
2000	2211.3	20.00	44.23	2178.9	1.50	3.27	47.79
2001	2653.2	20.00	53.06	3468.6	1.50	5.20	58.27
2002	1976.4	20.00	39.53	4762.8	1.50	7.14	46.67
2003	972	20.00	19.44	3364.2	1.50	5.05	24.49
2004	619.2	20.00	12.38	220.3	1.50	3.33	15.71
2005	257.4	20.00	5.15	1087.2	1.50	1.63	6.78
Total	34101.9		682.04	56348.1		84.52	766.56

(continued)

TABLE 7–2. Economic Evaluation Example (continued)

Year	(8) Capital Cost ($MM)	(9) Operating Cost ($MM)	(10) Prod. Tax ($MM)	(11) Total Cost ($MM) (8) + (9) + (10)	(12) Undiscounted Cash Flow ($MM) (7) – (11)
1992	5.715	0.000	0.000	5.715	–5.715
1993	64.680	0.000	0.000	64.680	–64.680
1994	143.977	3.825	11.502	159.304	–44.284
1995	4.205	10.359	21.948	36.512	182.971
1996	0.000	9.922	13.030	22.952	107.345
1997	0.000	9.922	5.075	14.997	35.751
1998	0.000	9.922	2.486	12.408	12.456
1999	0.000	9.922	2.673	12.595	14.139
2000	0.000	9.922	4.749	14.671	32.823
2001	0.000	9.922	5.827	15.749	42.518
2002	0.000	9.922	4.667	14.589	32.083
2003	0.000	9.922	2.449	12.371	12.116
2004	0.000	9.922	1.571	11.493	4.221
2005	6.364	8.332	0.678	15.374	–8.595
Total	224.941	111.814	76.656	413.41	353.149

Year	(13) Discount Factor @ 12%	(14) Discounted Cash Flow @ 12%, $MM	(15) Discount Factor @ 20%	(16) Discounted Cash Flow @ 20%, $MM	(17) Discount Factor @ 30%	(18) Discounted Cash Flow @ 30%, $MM	(19) Discount Factor @ 64.23%	(20) Discounted Cash Flow @ 64.23%, $MM
1992	0.9449	–5.400	0.9129	–5.217	0.8771	–5.012	0.7803	–4.460
1993	0.8437	–54.569	0.7607	–49.204	0.6747	–43.637	0.4751	–30.732
1994	0.7533	–33.358	0.6339	–28.073	0.5190	–22.932	0.2893	–12.812
1995	0.6726	123.060	0.5283	96.660	0.3992	73.043	0.1762	32.233
1996	0.6005	64.462	0.4402	47.257	0.3071	32.964	0.1073	11.515
1997	0.5362	19.169	0.3669	13.116	0.2362	8.445	0.0653	2.335
1998	0.4787	5.963	0.3057	3.808	0.1817	2.263	0.0398	0.495
1999	0.4274	6.044	0.2548	3.602	0.1398	1.976	0.0242	0.342
2000	0.3816	12.526	0.2123	6.968	0.1075	3.529	0.0147	0.484
2001	0.3407	14.488	0.1769	7.522	0.0827	3.517	0.0090	0.382
2002	0.3042	9.761	0.1474	4.730	0.0636	2.041	0.0055	0.175
2003	0.2716	3.291	0.1229	1.489	0.0489	0.593	0.0033	0.040
2004	0.2425	1.024	0.1024	0.432	0.0376	0.159	0.0020	0.009
2005	0.2165	–1.861	0.0853	–0.733	0.0290	–0.249	0.0012	–0.011
Total		164.599		102.357		56.650		–0.004

RISK AND UNCERTAINTIES

The nature of economic evaluation entails risk taking and uncertainties involving technical, economic, and political conditions. The results of the analysis are subjected to many restrictive assumptions in forecasting recoveries, oil and gas prices, investment and operating costs, and inflation rate. Unforeseen national and world economic and political climates can also severely affect the outcome of the projects. Figures 7–1 and 7–2 show the sensitivities of DCFROI and PWNP to the oil price, oil production, and operating costs. The analysis shows that DCFROI and PWNP are affected more drastically by both oil price and oil production than the operating costs.

The Monte Carlo risk-analysis technique—based upon probability distribution of uncertainties in the variables used for economic evaluation—can provide a more reliable estimate of the range of expected results. However, the discussion of this technique is beyond the scope of this book.

FIGURE 7–1. DCFROI Sensitivity Analysis

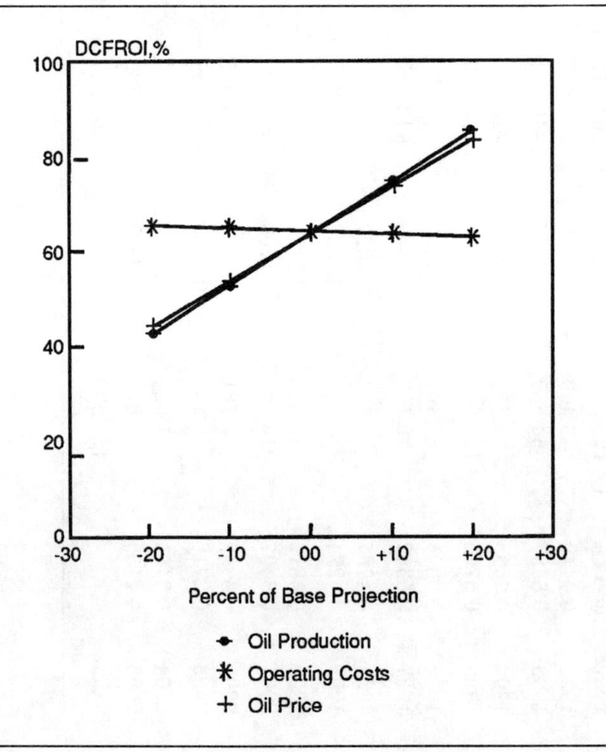

FIGURE 7–2. NPV Sensitivity Analysis

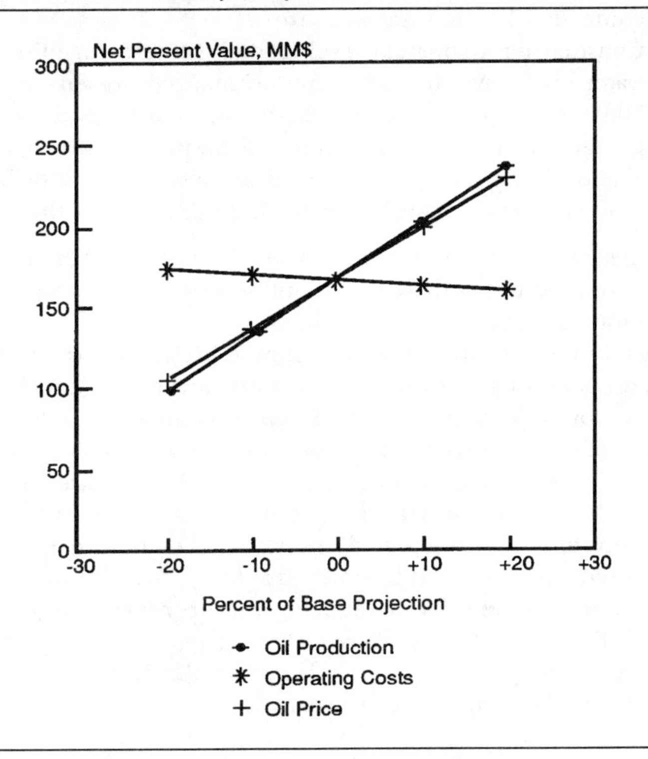

ECONOMIC OPTIMIZATION EXAMPLE

This section presents an example of an approach to optimize economics in a West Texas CO_2 flood.[8] The study showed how optimizing the operations can significantly improve the economics of existing CO_2 floods.

The purpose of the study was to develop operating schemes to maximize economic performance of Amoco's four existing CO_2 floods in an environment of relatively low oil prices. The primary and secondary economic criteria were selected to be present worth and near term cash flow, respectively. The approach used in the study was to develop production/injection profiles for different operating schemes by using a numerical simulator and running economics on each of the profiles.

The four major economic relationships in designing a CO_2 flood are: oil price, gas-processing costs, CO_2 purchase costs, and operating costs. The three major technical parameters are CO_2 and water half-cycle slug size, gas-water ratio and ultimate CO_2 slug size. An analysis showed:

- The ultimate incremental tertiary recovery was predominately controlled by the total slug size.
- Considering a constant level loaded gas rate, an initial 1:1 gas-water ratio gave the maximum normalized present worth.
- 700 MCFD per pattern was the optimum level loaded gas-production rate based on normalized present worth.
- The optimum ultimate CO_2 slug size was at a 45% hydrocarbon pore volume maximizing normalized present worth.

The near-term cash flow, present worth, and production/injection profiles were used to compare a new optimization scheme with a continued operation scheme as discussed below.

Figure 7–3 shows normalized discounted present worth optimized vs. continued operations for two different discount rates A and B, where A represents a lower discount rate than B. The present worth of A was used to normalize the plots. The optimized case had a present worth approximately 7% higher than the continued operations case for both discount rates.

Figure 7–4 shows normalized annual cash flows and cumulative net cash flow optimized vs. continued operations. The first-year cash flow of the optimized case was used to normalize the plots. The plots show the importance of the benefits of reducing the near-term gas-water ratios. Yearly cash flows for the first two years are greater for the optimized case than the continued operations case. However, the differences are diminished significantly in the next five years.

FIGURE 7–3. Four West Texas CO_2 Floods (Normalized Discounted Present Worth Optimized vs. Continued Operations) *(Copyright © 1991, SPE, from paper 22022[8])*

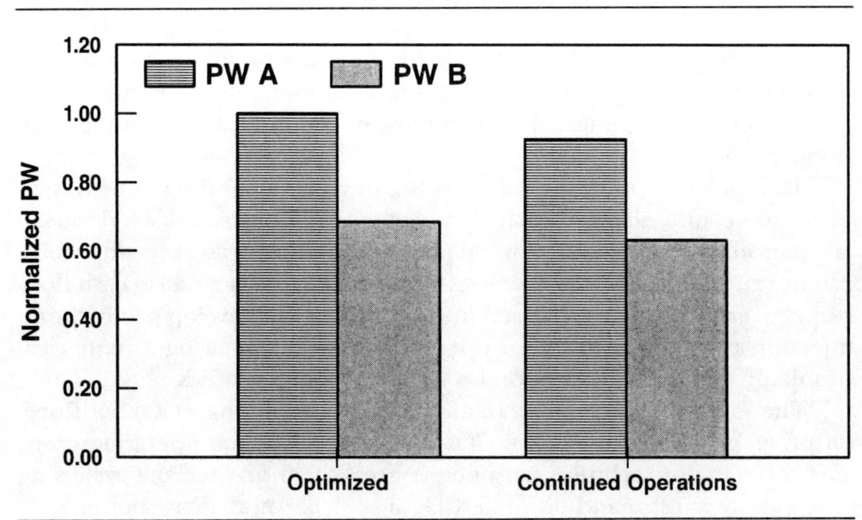

FIGURE 7–4. Four West Texas CO_2 Floods (Normalized Annual Cash Flows and Cumulative Net Cash Flow Optimized vs. Continued Operations) *(Copyright © 1991, SPE, from paper 22022[8])*

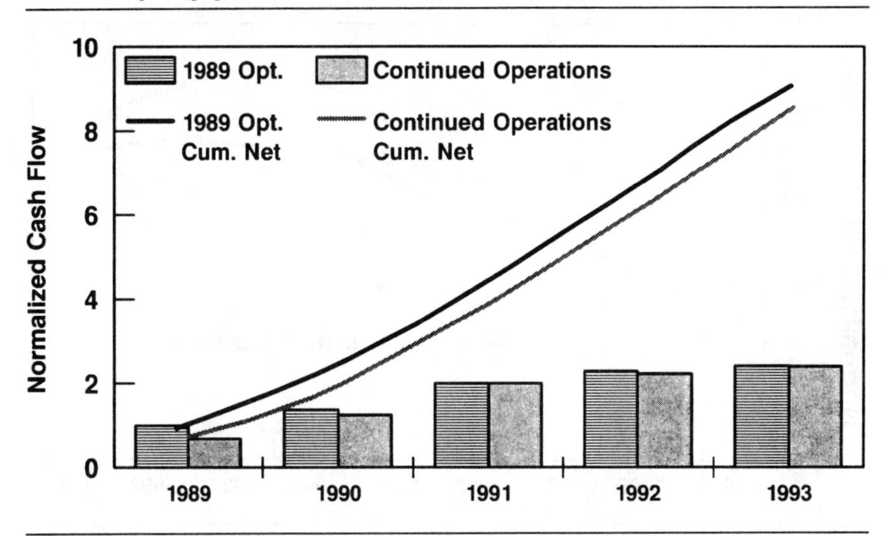

Figures 7–5, 7–6, and 7–7 show yearly average normalized oil production, gas production and CO_2 purchases, respectively for optimized and continued operations cases. The first-year optimized case was used to normalize the plots. The results show that a significant percentage of oil production is deferred during the early years of the new optimized operating scheme. However, larger reductions are being experienced in gas production and CO_2 purchases.

The economic sensitivity analyses under varying oil prices, gas processing costs, and CO_2 purchase prices confirmed that the new optimization scheme was the proper course of action to take for the four CO_2 floods. The controllable operating expenses and capital investments were reduced by implementing the new operating scheme. The optimization process improved project profitability, deferring or eliminating additional facility investments, and thus opened up other CO_2 flood investment opportunities.

An additional benefit of the study was a great deal of insight into the design of new CO_2 flood projects. The parameters that need to be included for optimizing the design of new CO_2 flood projects are plant and surface facilities, pattern size and configuration, reservoir quality, and CO_2 purchase contracts.

Plant design is a major critical factor in designing a new CO_2 flood project. A design process to handle multiple-design considerations such as plant type, size, location, construction and operating costs, liquid hydrocarbon recoveries, and CO_2 product purity are illustrated in Figure 7–8.

FIGURE 7–5. Four West Texas CO_2 Floods (Yearly Average Normalized Oil Production) *(Copyright © 1991, SPE, from paper 22022[8])*

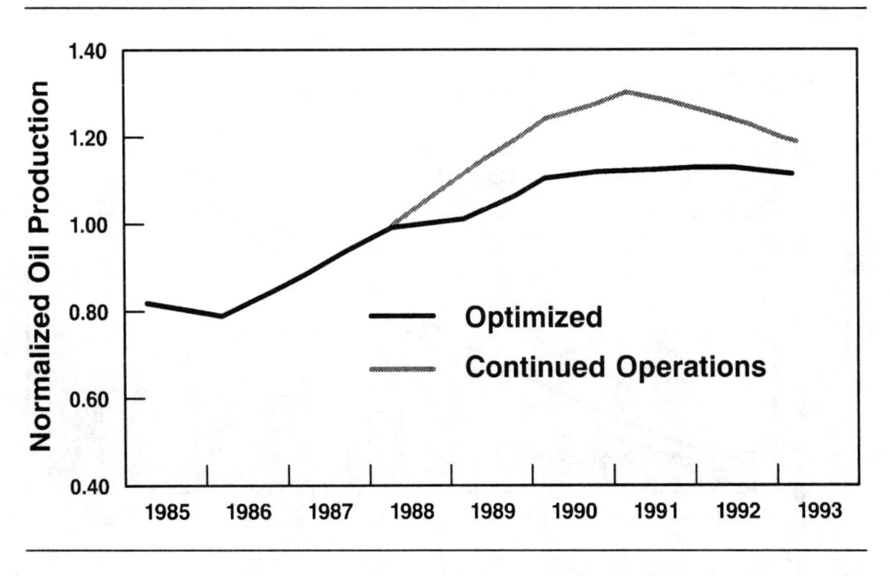

FIGURE 7–6. Four West Texas CO_2 Floods (Yearly Average Normalized Gas Production) *(Copyright © 1991, SPE, from paper 22022[8])*

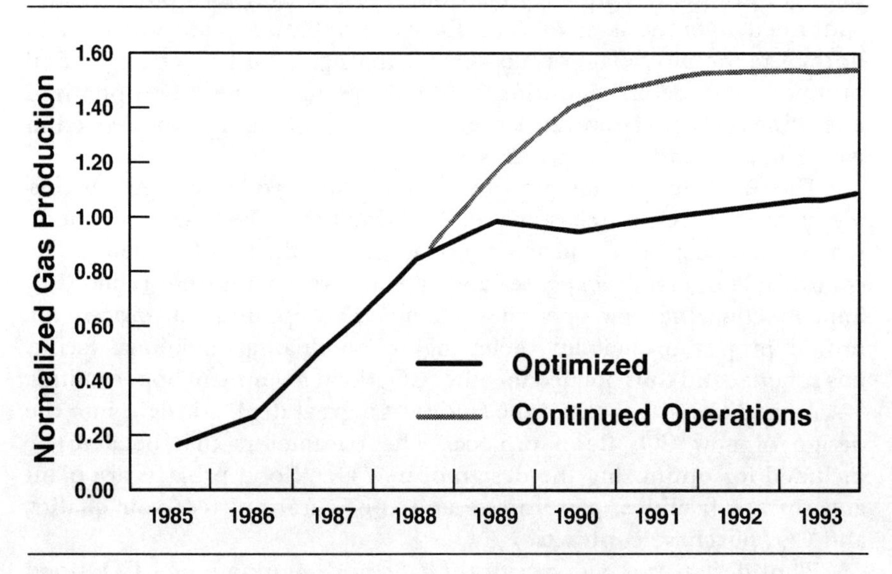

FIGURE 7–7. Four West Texas CO_2 Floods (Yearly Average Normalized CO_2 Purchases) *(Copyright © 1991, SPE, from paper 22022[8])*

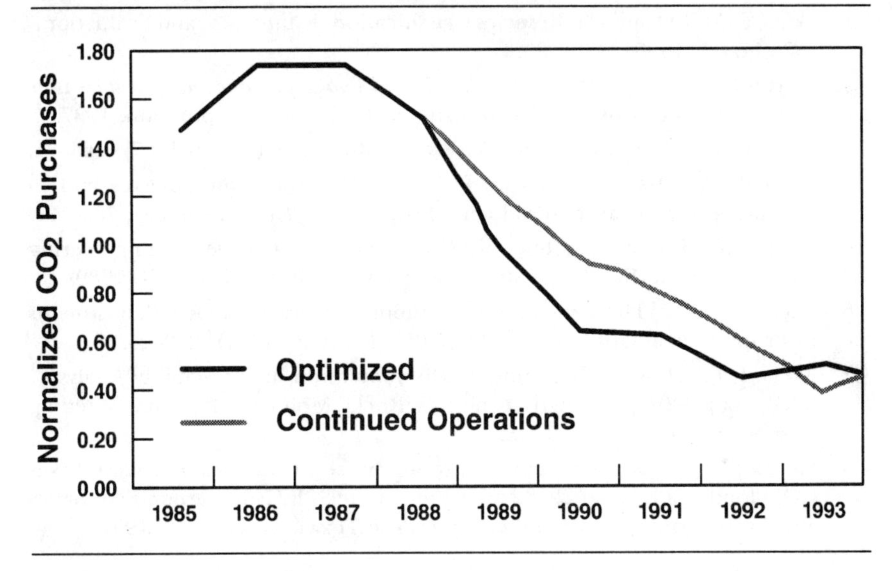

FIGURE 7–8. CO_2 Flood Plant Size of Opitimization Process *(Copyright © 1991, SPE, from paper 22022[8])*

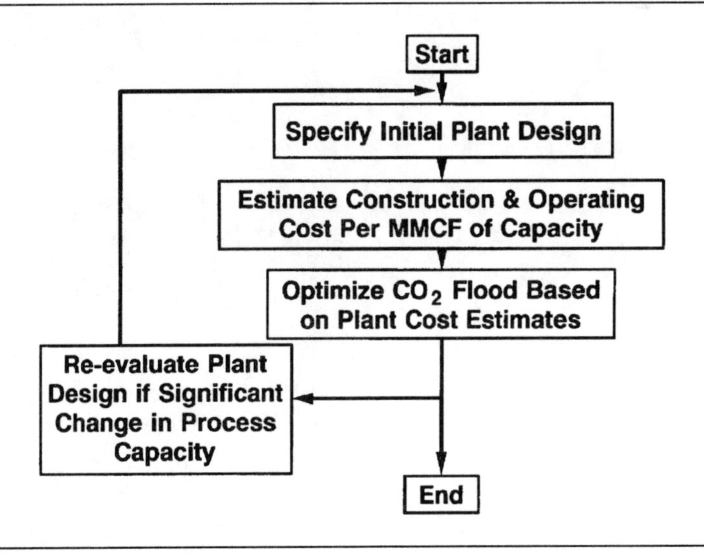

REFERENCES

1. Garb, F. A. "Oil and Gas Reserves Classification, Estimation, and Evaluation," *JPT* (March 1985): 373–390.
2. Stermole, F. J. and J. M. Stermole. *Economic Evaluation and Investment Decision Methods,* 6th ed. Golden, CO: Investment Evaluation Corporation, 1987.
3. Seba, R. D. "Determining Project Profitability," *JPT* (March 1987): 263–71.
4. Garb, F. A. "Assessing Risk In Estimating Hydrocarbon Reserves and in Evaluating Hydrocarbon-Producing Properties," *JPT* (June 1988): 765–68.
5. Rose, S. C., J. F. Buckwalter, and R. J. Woodhall. "The Design Engineering Aspects of Waterflooding," SPE Monograph 11, Richardson, TX, 1989.
6. Hickman, T. S. "The Evaluation of Economic Forecasts and Risk Adjustments in Property Evaluation in the U.S.," *JPT* (February 1991): 220–25.
7. Evers, J. F. "How to Use Monte Carlo Simulation in Profitability Analysis." SPE paper 4401 presented at the SPE Rocky Mountain Regional Meeting, Casper, WY, May 15–16, 1973.
8. Pariani, G. J. et al. "An Approach to Optimize Economics in a West Texas CO_2 Flood." SPE Paper 22022 presented at the SPE Hydrocarbon Economics and Evaluation Symposium held in Dallas, Texas, April 11–12, 1991.

CHAPTER 8

▼ ▼ ▼

Improved Recovery Processes

Primary methods that use natural reservoir energy (i.e., liquid and rock expansion drive, solution gas drive, gas cap drive, natural water influx, and combination drive processes) and secondary methods that augment natural energy by fluid injection (i.e., gas, water and gas-water combination floods) leave behind one-third to one-half or more of the original oil in place. This means that more oil will be left unrecovered than has been or will be produced by primary and secondary methods. Moreover, with certain heavy oil reservoirs, tar sands, and oil shales, there is negligible recovery by primary or secondary methods. Thus, it is apparent that enhanced oil recovery (EOR) techniques ultimately must be employed to acquire these enormous energy resources.

EOR processes include all methods that use external sources of energy and/or materials to recover oil that cannot be produced economically by conventional means. EOR processes can be classified broadly as:

1. Thermal—steamflooding, hot waterflooding, and in situ combustion.
2. Nonthermal—chemical floods, miscible floods, and gas drives.

This chapter provides reservoir engineering aspects of waterflooding and an overview of enhanced oil recovery processes. Basic knowledge of these recovery methods and their applications are needed for sound reservoir management. For background information, this chapter also presents a review of the theories and concepts concerning fluid flow in reservoirs (viz., steady state flow, unsteady state flow, natural water influx, and immiscible displacement [see Appendix D]).

WATERFLOODING

Waterflooding is the most widely used post-primary recovery method in the United States and it contributes substantially to current production and reserves. The waterflood technology broadly encompasses both reservoir and production engineering.[1,2] Reservoir engineers are responsible for waterflood design, performance prediction, and reserves determination. They share responsibilities with the production engineers for the implementation, operation, and evaluation of the waterflood project.

This section presents a review of reservoir engineering aspects of waterflooding.

Flood Pattern

The commonly used flood patterns (i.e., injection-production well arrangements) are shown in Figure 8–1, and their characteristics are given in Table 8–1. Injectors positioned around the periphery of a reservoir, peripheral injection, and along the crest of small reservoirs with sharp structural features, crestal injection, are also used.

Reservoir Heterogeneity

Reservoirs are not uniform in their properties such as permeability, porosity, pore size distribution, wettability, connate water saturation, and fluid properties. The variations can be areal and vertical. The heterogeneity of the reservoirs is attributed to the depositional environments and subsequent events, as well as to the nature of the particles constituting the sediments. The performances of the reservoirs, whether primary or waterflood, are greatly influenced by their heterogeneities.

Commonly used methods to characterize vertical permeability stratification are:

1. Flow capacity distribution (permeability × thickness), which is evaluated from a plot of the cumulative capacity versus cumula-

TABLE 8–1. Characteristics of Waterflood Patterns

Pattern	P/I Regular	P/I Inverted	d/a	E_A, %
Direct Line Drive	1	–	1	56
Staggered Line Drive	1	–	1	78
4-Spot	2	1/2	0.866	–
5-Spot	1	1	1/2	72
7-Spot	1/2	2	0.866	–
9-Spot	1/3	3	1/2	~80

P = number of production wells
I = number of injection wells
d = distance from an injector to the line connecting two producing wells
a = distance between wells in line in regular pattern
E_A = areal sweep efficiency at water breakthrough at a producing well for a water-oil mobility ratio = 1

FIGURE 8–1. Flood Patterns

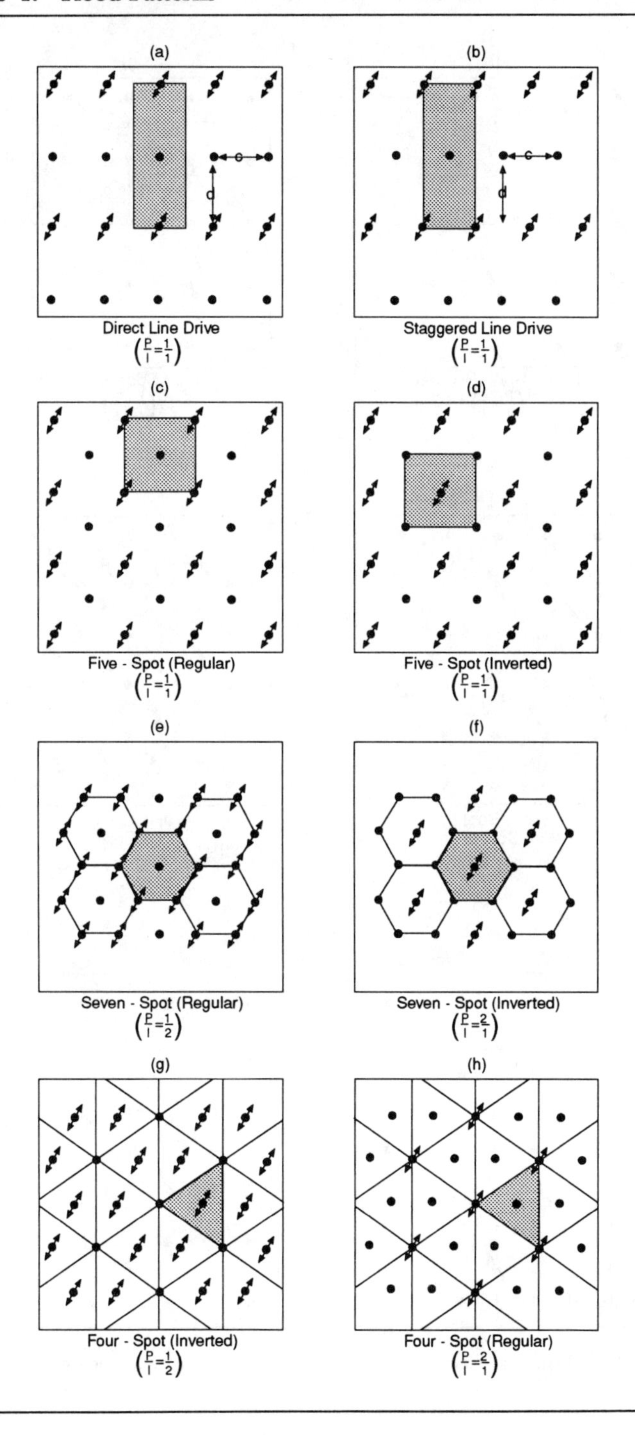

FIGURE 8–1. Flood Patterns (continued)

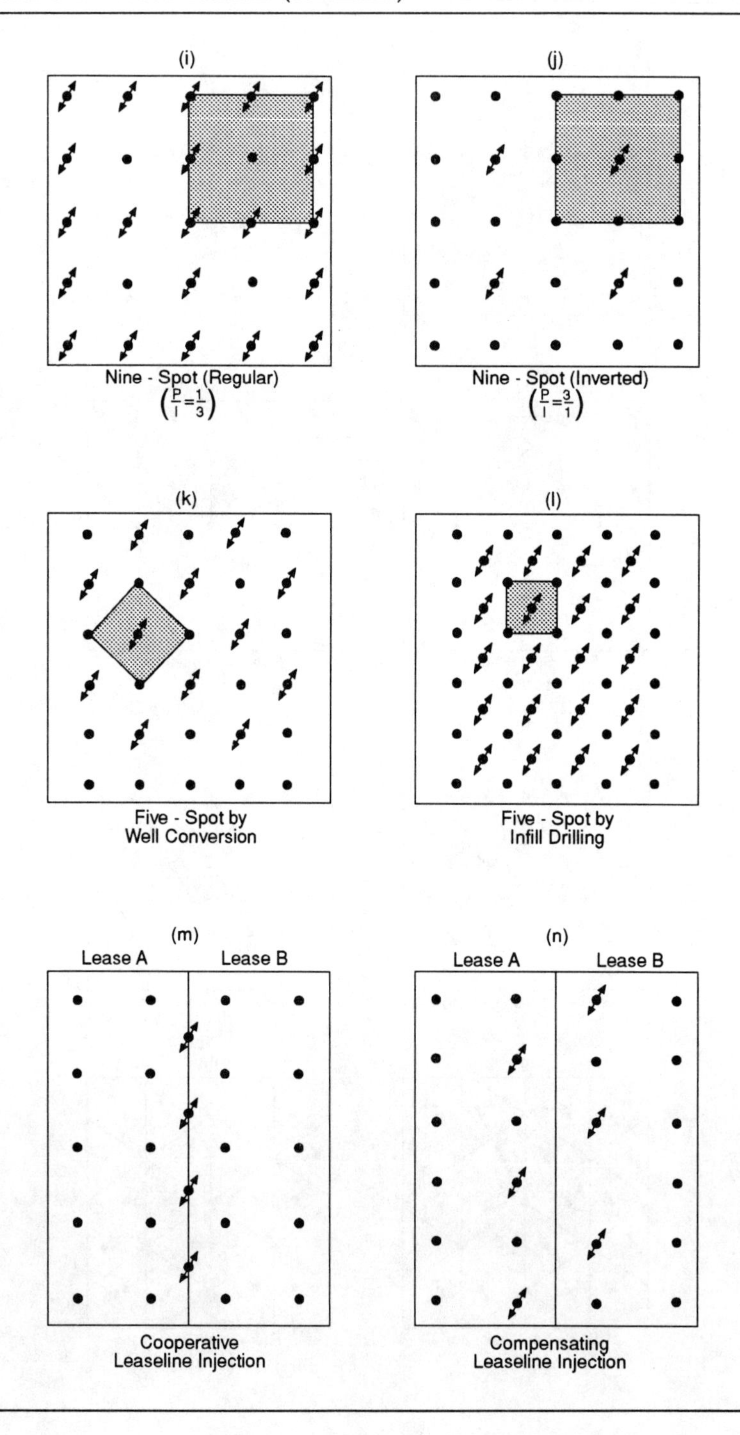

tive thickness, of a reservoir having layered permeability (see Figure 8–2). For a uniform permeability, the capacity distribution would plot as the straight line. Deviation from this straight line is a measure of the heterogeneity due to permeability variation.

2. Lorenz coefficient, which is based upon the flow capacity distribution, is a measure of the contrast in permeability from the homogeneous case. It is defined by the ratio of the area ABCA to the area ADCA (see Figure 8–2), and it ranges from 0 (uniform) to 1 (extremely heterogeneous). It is not a unique measure of reservoir heterogeneity, since several different permeability distributions can yield the same value of Lorenz coefficient.
3. Dykstra-Parsons permeability variation factor[3] is based upon the log normal permeability distribution (see Figure 8–3). Statistically, it is defined as:

$$V = \frac{\bar{k} - k_{\sigma}}{\bar{k}} \tag{8–1}$$

where:

$\bar{k}$ = mean permeability (i.e., permeability at 50% probability)
k_{σ} = permeability at 84.1% of the cumulative sample

The permeability variation ranges from 0 (uniform) to 1 (extremely heterogeneous), and it is widely used to characterize reservoir heterogeneity.

FIGURE 8–2. Flow Capacity Distribution

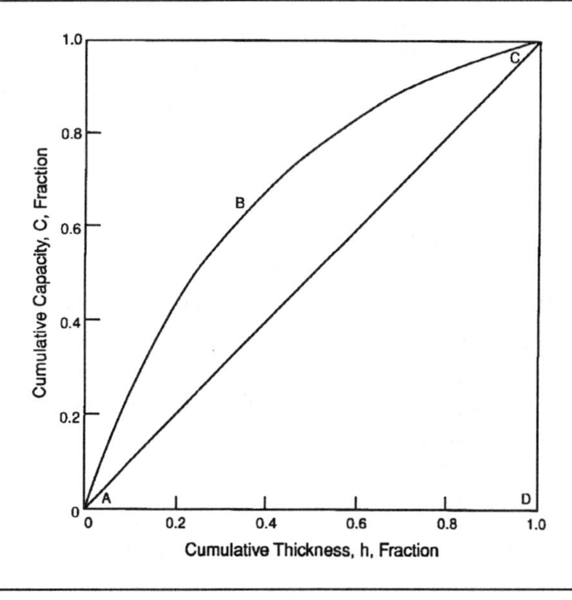

FIGURE 8–3. Permeability Variation

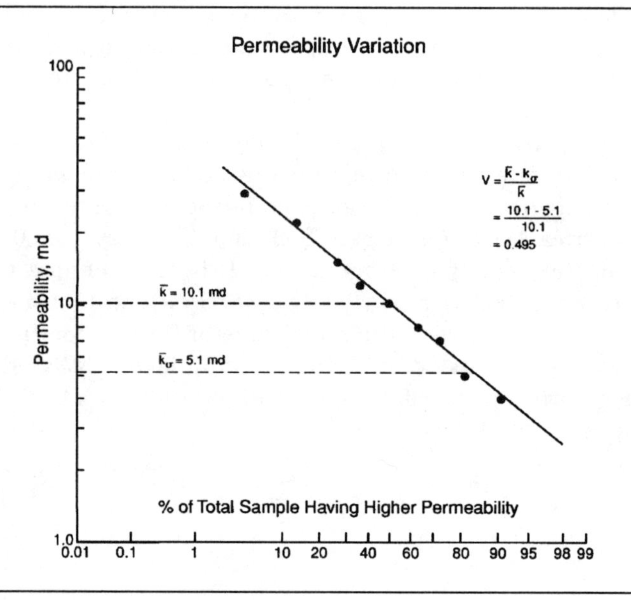

Mobility Ratio

Mobility ratio is defined as

$$M = \frac{\lambda_w \textit{ in the water contacted portion}}{\lambda_o \textit{ in the oil bank}} \tag{8–2}$$

$$= \frac{\dfrac{k_{rw}}{\mu_w}}{\dfrac{k_{ro}}{\mu_o}}$$

where:

λ = mobility = k/μ
k_r = relative permeability
μ = viscosity, cp
w,o = subscripts denoting oil and water, respectively

The relative permeabilities are for two different and separate regions in the reservoir. Craig[1] suggested calculating mobility ratio prior to water breakthrough (i.e., k_{rw} at the average water saturation in the swept region) and k_{ro} in the unswept zone.

Recovery Efficiency

The overall waterflood recovery efficiency is given by (see Figure 8–4):

$$E_R = E_D \times E_V \tag{8–3}$$

FIGURE 8–4. Oil Displacement by Waterflood

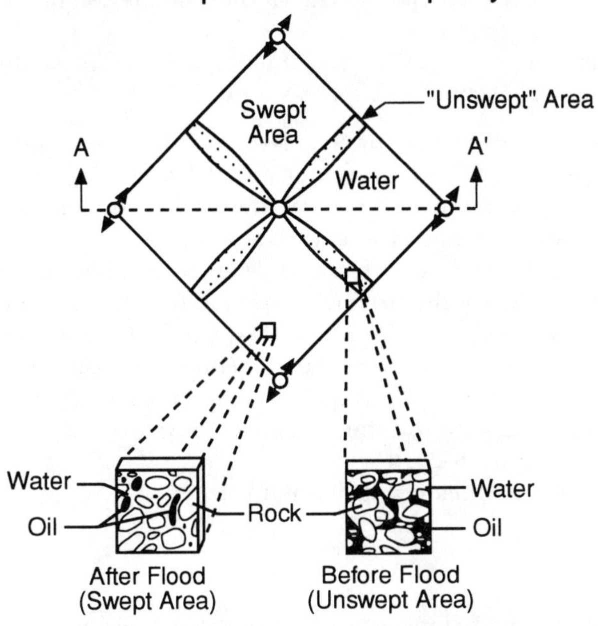

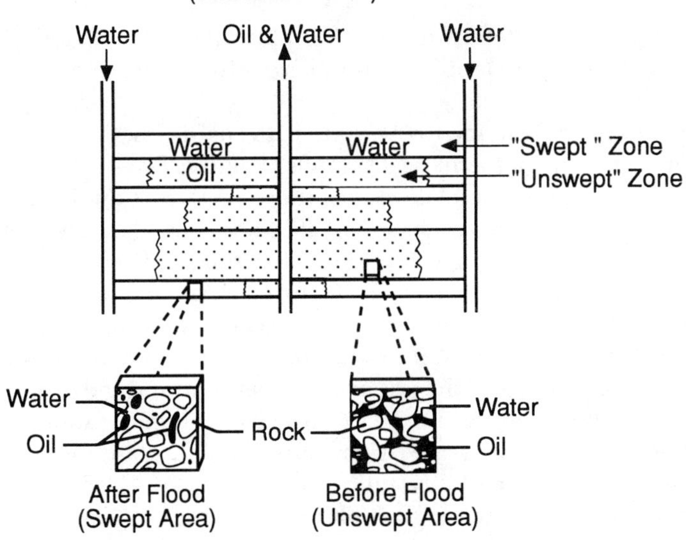

where:

E_R = overall recovery efficiency, fraction or percent
E_D = displacement efficiency within the volume swept by water, fraction, or percent
E_V = volumetric sweep efficiency, fraction or percent of the reservoir volume actually swept by water,

Displacement efficiency (Equations D–17 and D–18 in Appendix D) is influenced by rock and fluid properties, and throughput (pore volumes injected). It can be determined by (1) laboratory core floods, (2) frontal advance theory, and (3) empirical correlations.

Laboratory core floods, ideally using representative formation cores and actual reservoir fluids, are the preferred method for obtaining of S_{or} and E_D. Fractional flow theory[4] (see Equations D–12 to D–18 in Appendix D) can be used to estimate S_{or} and E_D, but it requires measured water-oil relative permeability curves. Alternately, empirical correlations such as Croes and Schwarz[5] based upon the results of laboratory waterfloods can be also used (see Figure 8–5).

Volumetric sweep efficiency is defined by:

$$E_V = E_A \times E_I \tag{8–4}$$

where:

E_A = areal sweep efficiency—fraction of the pattern area or intended flood area that is swept by the displacing fluid (i.e., water)
E_I = vertical or invasion sweep efficiency—fraction of the pattern thickness or intended thickness that is swept by the displacing fluid (i.e., water)

The factors that determine areal or pattern sweep efficiency are the flooding pattern type, mobility ratio, throughput and reservoir heterogeneity.

Areal sweep efficiency for various patterns has been studied using both physical and mathematical models. For the five-spot pattern, the most frequently used correlations are those by Dyes, Caudle, and Erickson.[6] Figure 8–6 presents areal sweep efficiency correlated with mobility ratio, M, and water cut as a fraction of the total flow coming from the swept portion of the pattern, f_D. In Figure 8–7, areal sweep efficiency is related to mobility ratio and the displaceable volumes injected. The displaceable volume, V_D, is defined as the cumulative injected water as a fraction of the product of the pattern pore volume and the displacement efficiency of the flood. It should be noted in the correlations that the areal sweep efficiency increases after water breakthrough for all mobility ratios. For favorable mobility ratios (i.e., $M \le 1$), 100% areal sweep efficiency can be obtained with prolonged injection of water.

FIGURE 8–5. Experimental Waterflood Performance *(Copyright © 1955, SPE, from Trans. AIME, 1955.[5])*

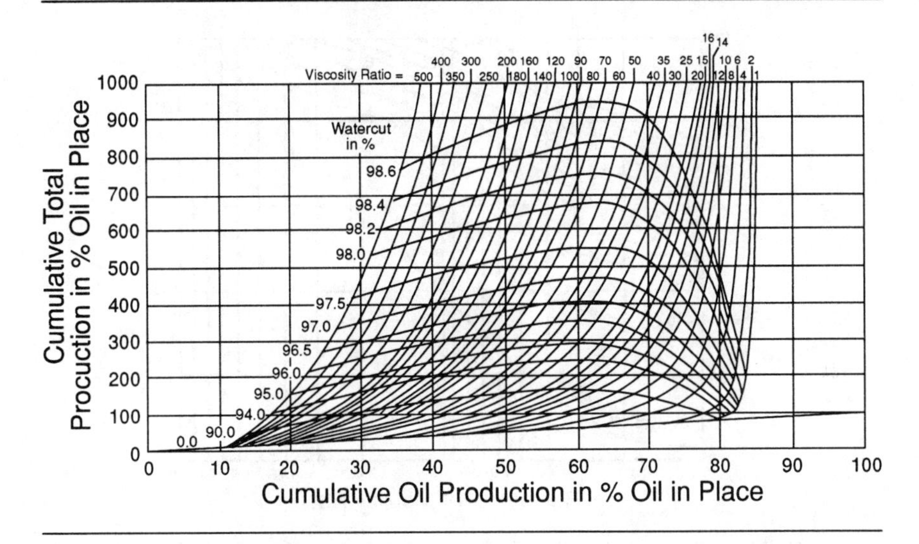

FIGURE 8–6. Effect of Mobility Ratio on Water Cut for the Five-Spot Pattern *(Copyright © 1954, SPE, from Trans. AIME, 1954.[6])*

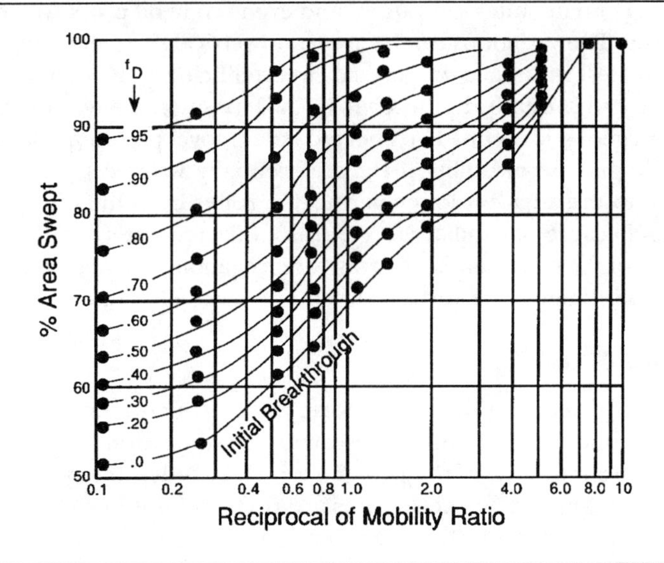

FIGURE 8–7. Effect of Mobility Ratio on the Displaceable Volumes Injected for the Five-Spot Pattern *(Copyright © 1954, SPE, from Trans. AIME, 1954[6])*

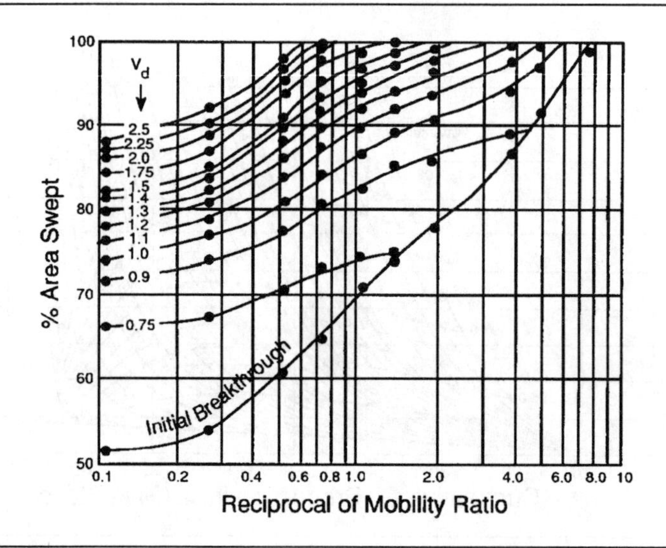

Vertical (or invasion) sweep efficiency is influenced by reservoir heterogeneity, mobility ratio, cross-flow, gravity, and capillary forces. Reservoir properties such as permeability, porosity, pore size distribution, wettability, connate water saturation, and even crude oil properties can vary significantly. The variations can be areal and vertical. Permeability variation has the greatest influence on vertical sweep efficiency. Horizontal permeabilities vary with depth due to change in depositional environments and subsequent geologic events. The injected water moves preferentially through zones of higher permeability. In a preferentially water-wet rock, water is imbibed into the adjacent lower permeable zones from the higher permeable zones because of capillary forces. Also, injected water tends to flow to the bottom of the reservoir due to gravity segregation. The net effect of these factors is to influence the vertical sweep efficiency of a waterflood project.

Injection Rates

The rate of oil recovery and, therefore, the life of a waterflood depends upon water injection rate into a reservoir. The injection rate, which can vary throughout the life of the project, is influenced by many factors. The variables affecting the injection rates are rock and fluid properties, areas and fluid mobilities of the swept and unswept regions, and the oil geometry (i.e., pattern, spacing and wellbore radii).

Muskat[7] and Deppe[8] provided injection rate equations for regular patterns with unit mobility ratio and no free gas saturation. Craig[1] presented a table of these equations in his monograph.

The water injectivity is defined as the injection rate per unit pressure difference between the injection and producing wells. A drastic decline in water injectivity occurs during the early period of injection into a reservoir depleted by solution gas drive (Figure 8–8). After fill-up, the injectivity variation depends upon the mobility ratio. It remains constant in the case of unit mobility ratio, increases if M > 1 (unfavorable), and decreases if M < 1 (favorable).

Performance Prediction Methods

There are many published classical methods for predicting waterflood performance.[1] Commonly used prediction methods, primarily concerned with reservoir heterogeneity but considering piston-like displacement, are:

- Dykstra-Parsons Method[3] based upon a correlation between waterflood recovery and both mobility ratio and permeability variation factor (Figure 8–9).
- Stiles Method[9] accounting for the different flood-front positions in liquid-filled, linear-insulated layers having different permeabilities. Permeability variation of the layers and the layer flow capacities are used to derive oil recovery and watercut equations.
- Prats-Matthews-Jewett-Baker Method[10] based upon a correlation of oil recovery, including the combined effects of mobility ratio and areal sweep efficiency, and considering the presence of free

FIGURE 8–8. Water Injectivity Variation in a Radial System *(Copyright © 1971, SPE, from SPE Monograph 3, Richardson, TX, 1971[1])*

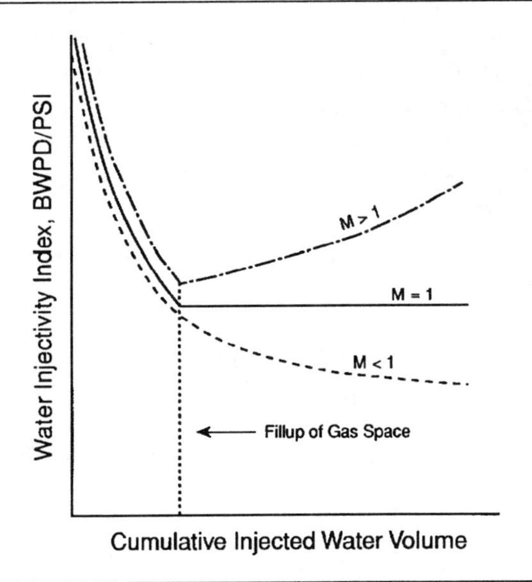

FIGURE 8–9. Dykstra-Parsons Waterflood Recovery Correlations *(After Secondary Recovery of Oil in the United States, second edition, API, 1950[3])*

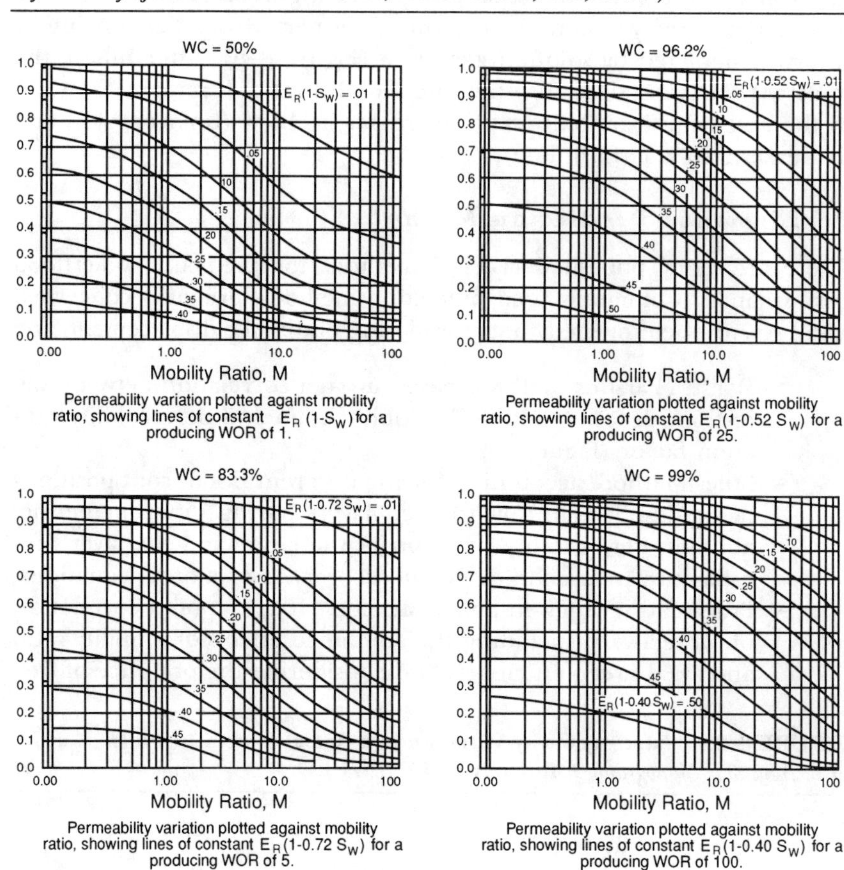

Permeability variation plotted against mobility ratio, showing lines of constant $E_R(1-S_W)$ for a producing WOR of 1.

Permeability variation plotted against mobility ratio, showing lines of constant $E_R(1-0.52\ S_W)$ for a producing WOR of 25.

Permeability variation plotted against mobility ratio, showing lines of constant $E_R(1-0.72\ S_W)$ for a producing WOR of 5.

Permeability variation plotted against mobility ratio, showing lines of constant $E_R(1-0.40\ S_W)$ for a producing WOR of 100.

gas prior to waterflooding and variation in injectivity through the life a flood.

Commonly used prediction methods primarily concerned with displacement mechanisms are:

- Buckley-Leverett Method[4] considering the immiscible displacement of oil by water in a linear or a radial system. Welge[11] modification to the frontal advance equation greatly simplified its use.
- Craig-Geffen-Morse Method[1] based upon a modified Welge equation and correlations of areal sweep efficiency at and after breakthrough.

The Craig, et al. method is one of the most thorough and practical prediction methods available for 5-spot pattern.

Table 8–2 lists the features of Dykstra-Parsons, Stiles, Prats, et al., Buckley-Leverett and Craig, et al. prediction methods. It can be seen that these methods are based upon many restrictive assumptions.

The most comprehensive waterflood performance prediction tool is a reservoir simulator. Today, black-oil model simulators are used extensively for waterflood performance prediction. These models permit inclusion of detailed reservoir description and laboratory-measured rock and fluid properties for more accurate predictions. One of the advantages is the ease of studying the effects of alternate operating strategies. Pattern type and size, infill drilling, effect of irregular patterns, injection rate scheduling, lifting capacities, zonal completions, can be varied simply, and the effects observed. Many years of project life can be repeated under different operating strategies in a few seconds of high-speed computer time.

Figure 8–10 compares simulated oil recovery curves with laboratory 5-spot flood results reported by Douglas, et al.[12]

A 40-acre, 5-spot pattern performance was simulated for partial primary depletion followed by waterflood. The rock and fluid properties were comparable to the example waterflood problem given in Craig's monograph.[1] The simulated results for homogeneous and layered cases with and without vertical permeability are shown in Figures 8–11 – 8–15. Figure 8–16 compares simulation and classical pattern waterflood solutions. Even though the ultimate recoveries calculated by the various methods compared favorably, the simulated results show considerably

TABLE 8–2. Features of Classical Waterflood Prediction Methods

	Dykstra-Parsons	Stiles	Prats et al.	Buckley-Leverett	Craig et al.
Linear Flow					
Piston-Like	x	x	x		
Frontal				x	x
Cross-Flow					
Areal	no	no	no	no	no
Vertical	no	no	no	no	no
Initial Gas Saturation	no	no	yes	no	yes
Sweep					
Areal	no	no	yes	no	yes
Vertical	yes	yes	yes	no	yes
Mobility Ratio	yes	1.0	yes	yes	yes
Stratification	yes	yes	yes	no	yes
Pattern					
5-Spot	no	no	yes	no	yes
Other	no	no	no	no	no

FIGURE 8–10. Comparison of Mathematical Model Results with Laboratory Five-Spot Waterflood Data

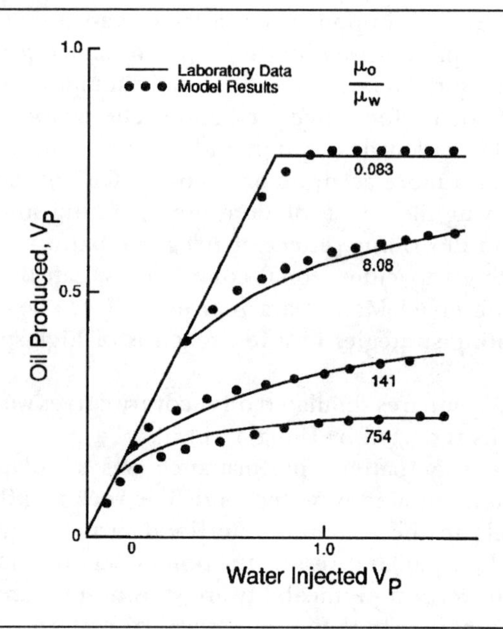

FIGURE 8–11. Primary Followed by Waterflood, Oil Production Rate vs. Time

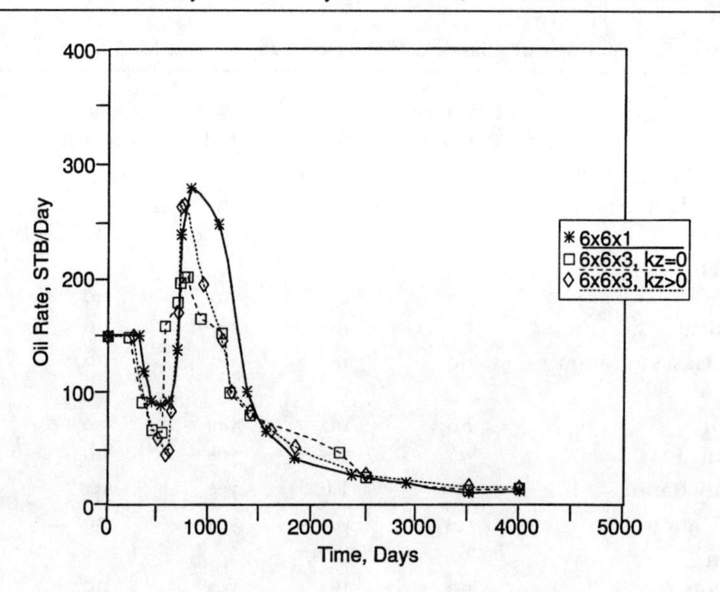

FIGURE 8–12. Primary Followed by Waterflood, Cumulative Oil Production vs. Time

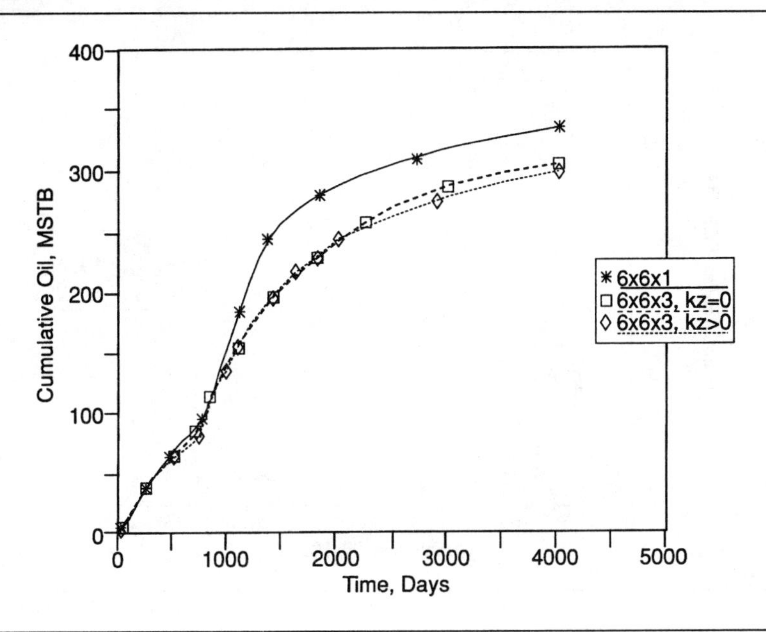

FIGURE 8–13. Primary Followed by Waterflood, Pressure vs. Cumulative Oil Production

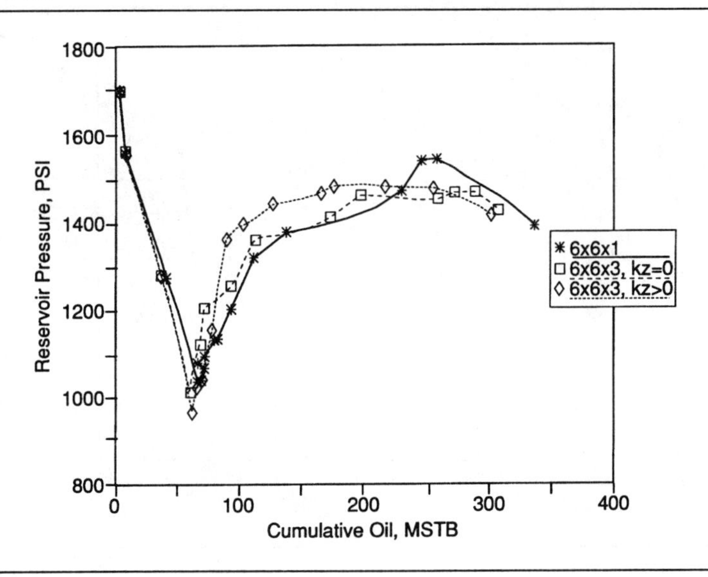

FIGURE 8–14. Primary Followed by Waterflood, Gas-Oil Ratio vs. Cumulative Oil

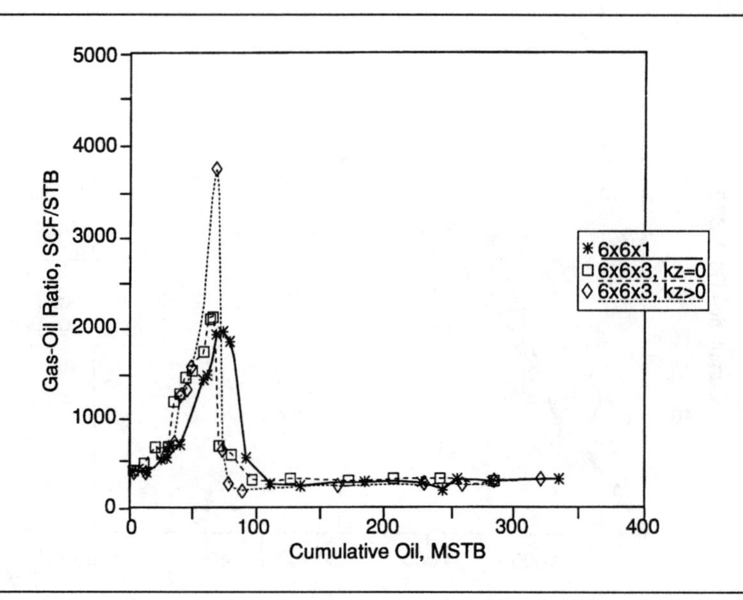

FIGURE 8–15. Primary Followed by Waterflood, Water-Oil Ratio vs. Cumulative Oil

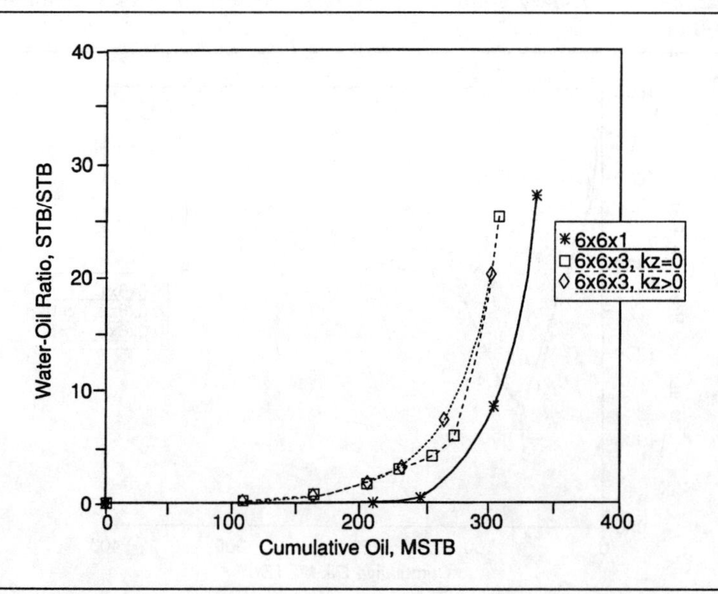

FIGURE 8–16. Comparison of Simulation and Classical Pattern Waterflood Solutions

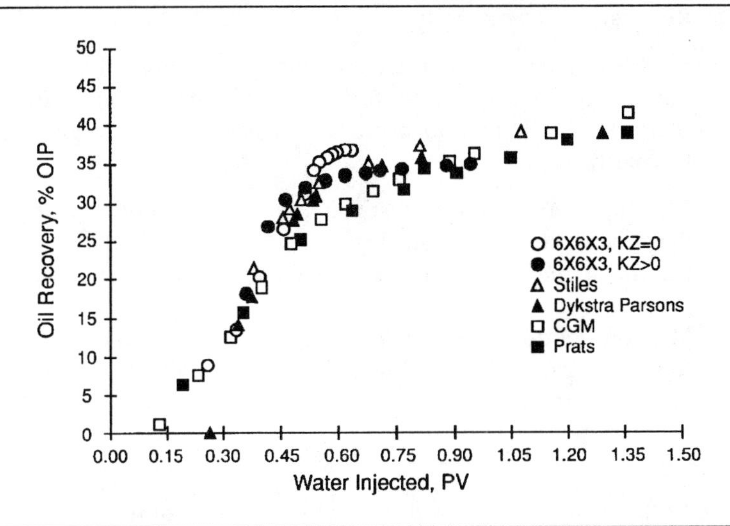

less total water injection than the other methods. Less water injection translates into shorter project life and favorable economics.

ENHANCED OIL RECOVERY PROCESSES

Over the past four decades, the petroleum industry has been engaged in the research and development of various enhanced oil recovery (EOR) processes needed to produce oil left behind by conventional methods. In general, conventional processes leave behind from one-third to one-half of the original oil-in-place. Moreover, most of the 300 billion barrels or more of hydrocarbon resources in the United States is not recoverable by conventional methods. Exploitation of this enormous untapped energy source is the greatest challenge ever faced by the oil industry.

Conventional methods of recovering crude oil include:

- Primary methods that use natural reservoir energy (i.e., liquid and rock expansion drive, solution gas drive, gas cap drive, natural water influx, and combination drive processes).
- Secondary methods that augment natural energy by fluid injection (i.e., gas, water and gas-water combination floods).

There are basically three physical factors that lead to high remaining oil saturation after primary and secondary recovery:

1. High oil viscosity.
2. Interfacial forces.
3. Reservoir heterogeneity.

Enhanced oil recovery processes include all methods that use external sources of energy and/or materials to recover oil that cannot be produced economically by conventional means. EOR processes can be classified broadly as:

- Thermal methods: steam stimulation, steamflooding, hot water drive, and in-situ combustion.
- Chemical methods: polymer, surfactant, caustic, and micellar/polymer.
- Miscible methods: hydrocarbon gas, CO_2 and nitrogen. In addition, flue gas and partial miscible/immiscible gas flood may be also considered.

More elaborate recovery mechanisms and methods are presented in Table 8–3.[13]

The EOR processes could substantially increase domestic production during the next 20 years or more. Tables 8–4 and 8–5 present a list of

TABLE 8–3. EOR Recovery Mechanisms *(from* OGJ *(OGJ Special), April 1992*[13]*)*

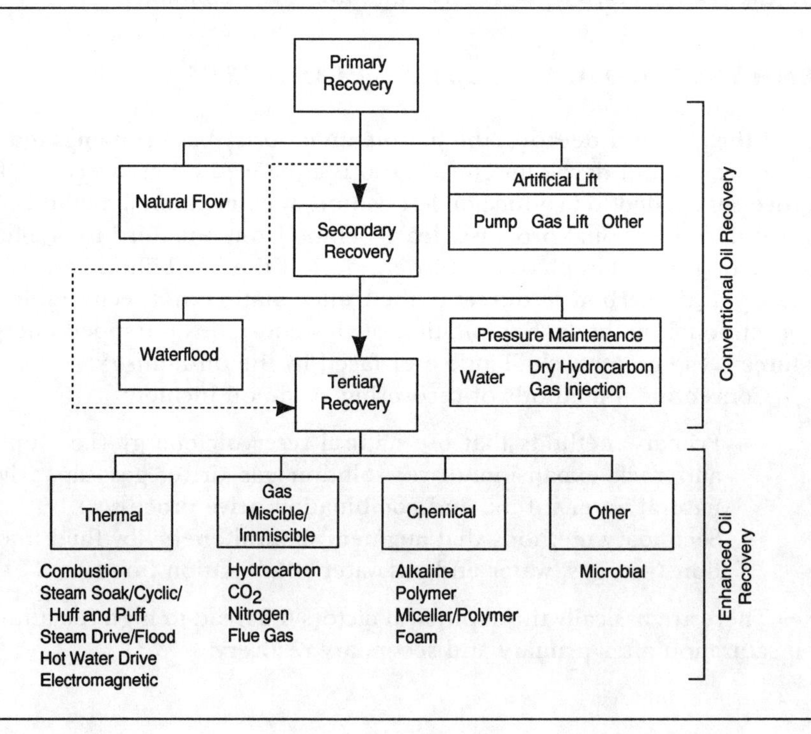

TABLE 8–4. Active U.S. EOR Projects *(from* OGJ *(OGJ Special), April 1992*[13]*)*

	Number of Projects											Change from 1990, %
	1971	1974	1976	1978	1980	1982	1984	1986	1988	1990	1992	
Thermal												
Steam	53	64	85	99	133	118	133	181	133	137	117	–17.1
Combustion in-Situ	38	19	21	16	17	21	18	17	9	8	8	0.0
Hot water								3	10	9	6	–50.0
Total thermal	91	83	106	115	150	139	151	201	152	154	131	–17.6
Chemical												
Micellar-polymer	5	7	13	22	14	20	21	20	9	5	3	–66.7
Polymer	14	9	14	21	22	55	106	178	111	42	23	–82.6
Caustic/alkaline		2	1	3	6	10	11	8	4	2		
Surfactant										1		
Total chemical	19	18	28	46	42	85	138	206	124	50	26	–92.3
Gas												
Hydrocarbon miscible/immiscible	21	12	15	15	9	12	16	26	22	23	25	8.0
CO_2 miscible	1	6	9	14	17	28	40	38	49	52	52	0.0
CO_2 immiscible						1	18	28	8	4	2	–100.0
Nitrogen miscible/immiscible					1	4	7	9	9	9	7	–28.6
Flue gas (miscible and immiscible)					3	3	3	3	2	3	2	–50.0
Other					4	2					1	100.0
Total gas	22	18	24	29	34	50	84	104	90	91	89	–2.2
Other												
Carbonated waterflood												
Microbial								1			2	
Total other	0	0	0	0	0	0	0	1	0	0	2	
GRAND TOTAL	132	119	158	190	226	274	373	512	366	295	248	–19.0

TABLE 8–5. U.S. EOR Production *(from* OGJ *(OGJ Special), April 1992*[13]*)*

	b/d							Change from
	1980	**1982**	**1984**	**1986**	**1988**	**1990**	**1992**	**1990, %**
Thermal								
Steam	243,477	288,396	358,115	488,692	455,484	444,137	454,009	2.2
Combustion in-Situ	12,133	10,228	6,445	10,272	6,525	6,090	4,702	–29.5
Hot water				705	2,896	3,985	1,980	–101.3
Total thermal	255,610	298,624	364,560	479,669	464,905	454,212	460,691	1.4
Chemical								
Micellar-polymer	930	902	2,832	1,403	1,509	617	254	–142.9
Polymer	924	2,927	10,232	15,313	20,992	11,219	1,940	–478.3
Caustic/alkaline	550	580	334	185				
Surfactant						20		
Total chemical	2,404	4,409	13,398	16,901	22,501	11,856	2,194	–440.4
Gas								
Hydrocarbon miscible/ immiscible	15,448	12,515	14,439	33,767	25,935	55,386	113,072	51.0
CO_2 miscible	21,532	21,953	31,300	28,440	64,192	95,591	144,973	34.1
CO_2 immiscible		490	702	1,349	420	95	95	0.0
Nitrogen miscible/ immiscible	2,027	1,400	7,170	18,510	19,050	22,260	22,580	1.4
Flue gas (miscible and immiscible)	35,200	35,200	29,400	26,150	21,400	17,300	11,000	–57.3
Other	600	370					6,300	
Total gas	74,807	72,028	83,011	108,216	130,997	190,632	298,020	36.0
Other								
Carbonated waterflood								
Microbial							2	
Total other							2	
GRAND TOTAL	332,821	375,061	460,969	604,786	618,403	656,700	760,907	13.7

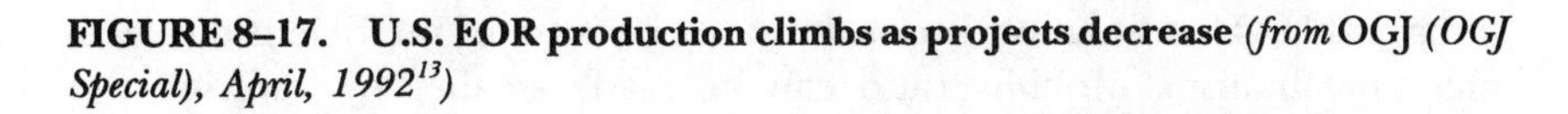

FIGURE 8–17. U.S. EOR production climbs as projects decrease *(from* OGJ *(OGJ Special), April, 1992*[13]*)*

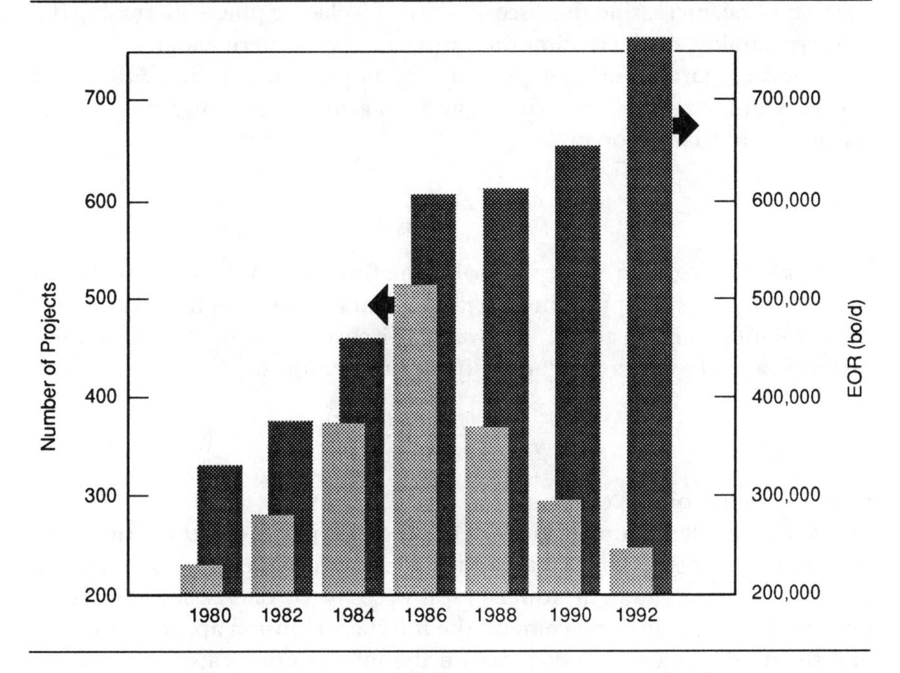

active U.S. EOR projects and production, respectively. While the projects are decreasing, EOR production is increasing (see Figure 8–17). Estimated 1992 primary, waterflood, and EOR oil recoveries are 3.9, 4.0 and, 0.761 millions of barrels per day, respectively.

This chapter presents an overview of the EOR processes in order to provide a basic understanding of these processes.[14–24]

EOR PROCESS CONCEPTS

The goal of EOR processes is to mobilize the "residual" oil throughout the entire reservoir. This is achieved by enhancing microscopic oil displacement and volumetric sweep efficiencies. Oil displacement efficiency is increased by decreasing oil viscosity (e.g., thermal floods) or by reducing capillary forces or interfacial tension (e.g., chemical floods). Volumetric sweep efficiency is improved by decreasing the drive water mobility (e.g., polymer floods).

If mobility (permeability/viscosity) of the displacing phase is greater than the phase being displaced, the mobility ratio (see Equation 8–2) is unfavorable. This situation is not desirable because of inefficient displacement mechanism. Mobility ratio can be made smaller by lowering the viscosity of oil, increasing the viscosity of the displacing phase, increasing the oil permeability, or decreasing the displacing phase permeability.

Another parameter that plays a very important role in EOR is the capillary number which is a dimensionless group expressing the ratio of viscous to interfacial forces:[25]

$$N_{cap} = \frac{\mu_w v}{\sigma \cos\theta} \tag{8–5}$$

where v is the frontal velocity $(q/A\phi)$, q the flow rate, A the cross-sectional area, ϕ the porosity, σ the interfacial tension between oil and water with θ their contact angle, and μ_w the water's viscosity. Abrams[26] included the effect of a water-wet core by modifying this group to:

$$N_{cap} = \frac{v\mu_w}{\sigma \cos\theta}\left(\frac{\mu_w}{\mu_o}\right)^{0.4} \tag{8–6}$$

where μ_o is the oil viscosity.

As the capillary number in an EOR process is increased, the residual oil saturation decreases. An increase in the capillary number is obtained by increasing pressure gradient, lowering oil viscosity, or decreasing interfacial tension. For miscible displacement, the interfacial tension approaches zero, and the oil displacement efficiency on the microscopic scale is very good.

There is not a single process that can be considered a "cure all" for recovering additional oil from every reservoir. Each process has its specific application. Before initiating an EOR process, reservoir rock and fluid properties and past production history should be analyzed. It is also important to review the preceding secondary recovery process in order to determine the principal reasons why the residual oil was left in that reservoir. Factors that strongly affect the success of a waterflood will usually also affect the success of a subsequent tertiary project.

The following methods are discussed including process description, production mechanisms, limitations, problems, and technical screening guides:

- Thermal Methods
  - Steam Stimulation
  - Steamflooding (Table 8–6)
  - In-Situ Combustion (Table 8–7)
- Chemical Methods
  - Polymer Flooding (Table 8–8)
  - Surfactant/Polymer Flooding (Table 8–9)
  - Caustic Flooding or Alkaline Flooding (Table 8–10)

TABLE 8–6. Steamflooding *(Copyright © 1983, SPE, from paper 12069[14])*

Description

The steam drive process or steamflooding involves the continuous injection of about 80% quality steam to displace crude oil towards producing wells. Normal practice is to precede and accompany the steam drive by a cyclic steam stimulation of the producing wells (called huff and puff).

Mechanisms

Steam recovers crude oil by:

- Heating the crude oil and reducing its viscosity.
- Supplying pressure to drive oil to the producing well.

TECHNICAL SCREENING GUIDES

Crude Oil	
Gravity	< 25° API (normal range is 10–25° API)
Viscosity	> 20 cp (normal range is 100–5,000 cp)
Composition	Not critical but some light ends for steam distillation will help
Reservoir	
Oil Saturation	> 500 bbl/acre-ft (or > 40–50% PV)
Type of Formation	Sand or sandstone with high porosity and permeability preferred
Net Thickness	> 20 feet
Average Permeability	> 200 md (see Transmissibility)
Transmissibility	> 100 md ft/cp
Depth	300–5,000 ft
Temperature	Not critical

Limitations

- Oil saturations must be quite high, and the pay zones should be more than 20 feet thick to minimize heat losses to adjacent formations.
- Lighter, less viscous crude oils can be steamflooded, but normally they will not be if the reservoir will respond to an ordinary waterflood.
- Steamflooding is primarily applicable to viscous oils in massive, high permeability sandstones or unconsolidated sands.
- Because of excessive heat losses in the wellbore, steamflooded reservoirs should be as shallow as possible as long as pressure for sufficient injection rates can be maintained.
- Steamflooding is not normally used in carbonate reservoirs.
- Since about one-third of the additional oil recovered is consumed to generate the required steam, the cost per incremental barrel of oil is high.
- A low percentage of water-sensitive clays is desired for good injectivity.

Problems

- Adverse mobility ratio and channeling of steam.

TABLE 8–7. In-situ Combustion *(Copyright © 1983, SPE, from paper 12069[14])*

Description

In-Situ combustion or fireflooding involves starting a fire in the reservoir and injecting air to sustain the burning of some of the crude oil. The most common technique is forward combustion in which the reservoir is ignited in an injection well, and air is injected to propagate the combustion front away from the well. One of the variations of this technique is combination of forward combustion and waterflooding (COFCAW). A second technique is reverse combustion in which a fire is started in a well that will eventually become a producing well, and air injection is then switched to adjacent wells. However, no successful field trials have been completed for reverse combustion.

Mechanisms

In-situ combustion recovers crude oil by:

- The application of heat that is transferred downstream by conduction and convection, thus lowering the viscosity of the crude.
- The products of steam distillation and thermal cracking that are carried forward to mix with and upgrade the crude.
- Burning coke that is produced from the heavy ends of the crude oil.
- The pressure supplied to the reservoir by the injected air.

TECHNICAL SCREENING GUIDES

Crude Oil

Gravity	< 40° API (Normally 10–25°)
Viscosity	< 1000 cp
Composition	Some asphaltic components to aid coke deposition

Reservoir

Oil Saturation	> 500 bbl/acre-ft (or < 40–50% PV)
Type of Formation	Sand or sandstone with high porosity
Net Thickness	> 10 ft
Average Permeability	> 100 md
Transmissibility	> 20 md ft/cp
Depth	> 500 ft
Temperature	> 150°F preferred

Limitations

- If sufficient coke is not deposited from the oil being burned, the combustion process will not be sustained.
- If excessive coke is deposited, the rate of advance of the combustion zone will be slow, and the quantity of air required to sustain combustion will be high.
- Oil saturation and porosity must be high to minimize heat loss to rock.
- Process tends to sweep through upper part of reservoir so that sweep efficiency is poor in thick formations.

Problems

- Adverse mobility ratio.
- Complex process, requiring large capital investment, is difficult to control.
- Produced flue gases can present environmental problems.
- Operational problems such as severe corrosion caused by low pH hot water, serious oil-water emulsions, increased sand production, deposition of carbon or wax, and pipe failures in the producing wells as a result of the very high temperatures.

TABLE 8–8. Polymer Flooding *(Copyright © 1983, SPE, from paper 12069*[14]*)*

Description

The objective of polymer flooding is to provide better displacement and volumetric sweep efficiencies during a waterflood. Polymer augmented waterflooding consists of adding water soluble polymers to the water before it is injected into the reservoir. Low concentrations (often 250–2,000 mg/L) of certain synthetic or biopolymers are used; properly sized treatments may require 15–25% reservoir PV.

Mechanisms

Polymer improve recovery by:

- Increasing the viscosity of water.
- Decreasing the mobility of water.
- Contacting a larger volume of the reservoir.

TECHNICAL SCREENING GUIDES

Crude Oil

Gravity	> 25° API
Viscosity	< 150 cp (preferably < 100)
Composition	Not critical

Reservoir

Oil Saturation	> 10% PV mobile oil
Type of Formation	Sandstone preferred but can be used in carbonate
Net Thickness	Not critical
Average Permeability	> 10 md (as low as 3 md in some cases)
Depth	< about 9000 ft (see Temperature)
Temperature	< 200°F to minimize degradation

Limitations

- If oil viscosities are high, a higher polymer concentration is needed to achieve the desired mobility control.
- Results are normally better if the polymer flood is started before the water-oil ratio becomes excessively high.
- Clays increase polymer adsorption.
- Some heterogeneities are acceptable but for conventional polymer flooding, reservoirs with extensive fractures should be avoided. If fractures are present, the crosslinked or gelled polymer techniques may be applicable.

Problems

- Lower injectivity than with water can adversely affect oil production rate in the early stages of the polymer flood.
- Acrylamide-type polymers lose viscosity due to sheer degradation, or it increases in salinity and divalent ions.
- Xanthan gum polymers cost more, are subject to microbial degradation, and have a greater potential for wellbore plugging.

TABLE 8–9. Surfactant/Polymer Flooding *(Copyright © 1983, SPE, from paper 12069[14])*

Description

Surfactant/polymer flooding, also called micellar/polymer or microemulsion flooding, consists of injecting a slug that contains water, surfactant, electrolyte (salt), usually a cosolvent (alcohol), and possibly a hydrocarbon (oil). The size of the slug is often 5–15% PV for a high surfactant concentration system and 15–50% PV for low concentrations. The surfactant slug is followed by polymer-thickened water. Concentrations of the polymer often ranges from 500–2,000 sg/L; the volume of polymer solution injected may be 50% PV, more or less, depending on the process design.

Mechanisms

Surfactant/polymer flooding recovers oil by:

- Lowering the interfacial tension between oil and water.
- Solubilization of oil.
- Emulsification of oil and water.
- Mobility enhancement.

TECHNICAL SCREENING GUIDES

Crude Oil

Gravity	> 25° API
Viscosity	< 30 cp
Composition	Light intermediates are desirable

Reservoir

Oil Saturation	> 30% PV
Type of Formation	Sandstone preferred
Net Thickness	> 10 ft
Average Permeability	> 20 md
Depth	< about 8,000 ft (see Temperature)
Temperature	< 175°F

Limitations

- An areal sweep of more than 50% on waterflood is desired.
- Relatively homogeneous formation is preferred.
- High amounts of anhydrite, gypsum, or clays are undesirable.
- Available systems provide optimum behavior over a very narrow set of conditions.
- With commercially available surfactants, formation water chlorides should be < 20,000 ppm and divalent ions (Ca^{++} and Mg^{++}) < 500 ppm.

Problems

- Complex and expensive system.
- Possibility of chromatographic separation of chemicals.
- High adsorption of surfactant.
- Interactions between surfactant and polymer.
- Degradation of chemicals at high temperatures.

TABLE 8–10. Alkaline Flooding *(Copyright © 1983, SPE, from paper 12069[14])*

Description

Alkaline or caustic flooding involves the injection of chemicals such as sodium hydroxide, sodium silicate, or sodium carbonate. These chemicals react with organic petroleum acids in certain crudes to create surfactants in situ. They also react with reservoir rocks to change wettability. The concentration of the alkaline agent is normally 0.2 to 5%; slug size is often 10 to 50% PV, although one successful flood only used 2% PV, (but this project also included polymers for mobility control). Polymers may be added to the alkaline mixture, and polymer-thickened water can be used following the caustic slug.

Mechanisms

Alkaline flooding recovers crude oil by:

- A reduction of interfacial tension resulting from the produced surfactants.
- Changing wettability from oil-wet to water-wet.
- Changing wettability from water-wet to oil-wet.
- Emulsification and entrainment of oil.
- Emulsification and entrapment of oil to aid in mobility control.
- Solubilization of rigid oil films at oil-water interfaces.

Not all mechanisms are operative in each reservoir.

TECHNICAL SCREENING GUIDES

Crude Oil

Gravity	13° to 35° API
Viscosity	< 200 cp
Composition	Some organic acids required

Reservoir

Oil Saturation	Above waterflood residual
Type of Formation	Sandstones preferred
Net Thickness	Not critical
Average Permeability	> 20 md
Depth	< about 9,000 ft (see Temperature)
Temperature	< 200°F preferred

Limitations

- Best results are obtained if the alkaline material reacts with the crude oil; the oil should have an acid number of more than 0.2 mg KOH/g of oil.
- The interfacial tension between the alkaline solution and the crude oil should be less than 0.01 dyne/cm.
- At high temperatures and in some chemical environments, excessive amounts of alkaline chemicals may be consumed by reaction with clays, minerals, or silica in the sandstone reservoir.
- Carbonates are usually avoided because they often contain anhydrite or gypsum, which interact adversely with the caustic chemical.

Problems

- Scaling and plugging in the producing wells.
- High caustic consumption.

- Miscible Methods
    - Hydrocarbon Miscible Flooding (Table 8–11)
    - Carbon Dioxide Flooding (Table 8–12)
    - Nitrogen and Flue Gas Flooding (Table 8–13)

THERMAL METHODS[14,17-20]

Many reservoirs contain quite viscous crude oil. Figure 8–18 shows a variation of oil viscosity with oil gravity. Attempts to produce such oils with waterflooding will yield very poor recoveries. Often the oil is too viscous to flow, or it requires pressures high enough to fracture the reservoir. Even if the oil is movable by waterflooding, the resultant fingering caused by the unfavorable viscosity ratio provides poor recovery. Application of heat is often the only feasible solution to such reservoirs. Crude oil viscosity is very sensitive to temperature, as can be seen in Figure 8–19. Thermal methods are primarily used for heavy viscous oil (10–20°API) and tar sands. About 60% of all EOR oil production was due to thermal recovery (see Table 8–5).

Steam Stimulation

Of all the EOR methods, steam stimulation, also known as "steam soak," "cyclic steam injection," or "huff-and-puff" is the most successful process. It involves a single well, until communication between wells develops. Steam is injected into the well at a high rate for a short period of time (a few weeks); next the steam is allowed to soak in for a few days, then the well is allowed to flow back and pumped. The oil rate increases initially, then drops off. When the rate becomes low, the entire process is repeated. This process is repeated many times until the well becomes uneconomic; or in some cases, it is converted from steam stimulation to steamflooding.

In the stimulation process, the steam fingers through the oil around the wellbore and heats the oil. The soak period permits the oil to be heated even further. During the production cycle, the mobilized oil flows into the wellbore, as a result of pressure drop, gravity, and other mechanisms.

This process is most effective in highly viscous oils with a good reservoir permeability. The performance of this method drops as more and more cycles are carried out. Oil recovery is generally very small in this process because only a fraction of the formation is affected.

TABLE 8–11. Hydrocarbon Miscible Flooding *(Copyright © 1983, SPE, from paper 12069[14])*

Description

Hydrocarbon miscible flooding consists of injecting light hydrocarbons through the reservoir to form a miscible flood. Three different methods are used. One method uses about 5% PV slug of liquified petroleum gas (LPG) such as propane, followed by natural gas or gas and water. A second method, called Enriched (Condensing) Gas Drive, consists of injecting a 10–20% PV slug of natural gas that is enriched with ethane through hexane (C_2 to C_6), followed by lean gas (dry, mostly methane) and possibly water. The enriching components are transferred from the gas to the oil. The third method, called High Pressure (Vaporizing) Gas Drive, consists of injecting lean gas at high pressure to vaporize $C_2 - C_6$ components from the crude oil being displaced.

Mechanisms

Hydrocarbon miscible flooding recovers crude oil by:

- Generating miscibility (in the condensing and vaporizing gas drive).
- Increasing the oil volume (swelling).
- Decreasing the viscosity of the oil.

TECHNICAL SCREENING GUIDES

Crude Oil

Gravity	> 35° API
Viscosity	< 10 cp
Composition	High percentage of light hydrocarbons ($C_2 - C_7$)

Reservoir

Oil Saturation	> 30% PV
Type of Formation	Sandstone or carbonate with a minimum of fractures and high permeability streaks
Net Thickness	Relatively thin unless formation is steeply dipping
Average Permeability	Not critical if uniform
Depth	> 2,000 ft (LPG) to > 5,000 ft (High Pressure Gas)
Temperature	Not critical

Limitations

- The minimum depth is set by the pressure needed to maintain the generated miscibility. The required pressure ranges from about 1,200 psi for the LPG process to 3,000–5,000 psi for the High Pressure Gas Drive, depending on the oil.
- A steeply dipping formation is very desirable to permit some gravity stabilization of the displacement that normally has an unfavorable mobility ratio.

Problems

- Viscous fingering results in poor vertical and horizontal sweep efficiency.
- Large quantities of expensive products are required.
- Solvent may be trapped and not recovered.

TABLE 8–12. Carbon Dioxide Flooding *(Copyright © 1983, SPE, from paper 12069[14])*

Description

Carbon dioxide flooding is carried out by injecting large quantities of CO_2 (15% or more of the hydrocarbon PV) into the reservoir. Although CO_2 is not truly miscible with the crude oil, the CO_2 extracts the light-to-intermediate components from the oil, and, if the pressure is high enough, develops miscibility to displace the crude oil from the reservoir.

Mechanisms

CO_2 recovers crude oil by:

- Generation of miscibility.
- Swelling the crude oil.
- Lowering the viscosity of the oil.
- Lowering the interfacial tension between oil and the CO_2 – oil phase in the near-miscible regions.

TECHNICAL SCREENING GUIDES

Crude Oil	
Gravity	> 26° API (preferably > 30°)
Viscosity	< 15 cp (preferably < 10 cp)
Composition	High percentage of intermediate hydrocarbons ($C_5 - C_{20}$), especially $C_5 - C_{12}$
Reservoir	
Oil Saturation	> 30% PV
Type of Formation	Sandstone or carbonate with a minimum of fractures and high permeability streaks
Net Thickness	Relatively thin unless formation is steeply dipping
Average Permeability	Not critical if sufficient injection rates can be maintained
Depth	Deep enough to allow high enough pressure (> about 2,000 ft.), pressure required for optimum production (sometimes called minimum miscibility pressure) ranges from about 1,200 psi for a high gravity (> 30° API) crude at low temperatures to over 4,500 psi for heavy crudes at higher temperatures.
Temperature	Not critical but pressure required increases with temperature

Limitations

- Very low viscosity of CO_2 results in poor mobility control.
- Availability of CO_2.

Problems

- Early breakthrough of CO_2 causes several problems.
- Corrosion in the producing wells.
- The necessity of separating CO_2 from saleable hydrocarbons.
- Repressuring of CO_2 for recycling.
- A high requirement of CO_2 per incremental barrel produced.

TABLE 8–13. Nitrogen and Flue Gas Flooding *(Copyright © 1983, SPE, from paper 12069[14])*

Description

Nitrogen and flue gas flooding are oil recovery methods which use these inexpensive non-hydrocarbon gases to displace oil in systems which may be either miscible or immiscible depending on the pressure and oil composition. Because of their low cost, large volumes of these gases may be injected. Nitrogen or flue gas are also considered for use as chase gases in hydrocarbon-miscible and CO_2 floods.

Mechanisms

Nitrogen and flue gas flooding recover oil by:

- Vaporizing the lighter components of the crude oil and generating miscibility if the pressure is high enough.
- Providing a gas drive where a significant portion of the reservoir volume is filled with low-cost gases.

TECHNICAL SCREENING GUIDES

Crude Oil	
Gravity	> 24° API (> 35 for nitrogen)
Viscosity	< 10 cp
Composition	High percentage of light hydrocarbons (C_1 – C_7)
Reservoir	
Oil Saturation	> 30% PV
Type of Formation	Sandstone or carbonate with few fractures and high permeability streaks
Net Thickness	Relatively thin unless formation is dipping
Average Permeability	Not critical
Depth	> 4,500 ft
Temperature	Not critical

Limitations

- Developed miscibility can only be achieved with light oils and at high pressures; therefore, deep reservoirs are needed.
- A steeply dipping reservoir is desired to permit gravity stabilization of the displacement, which has a very unfavorable mobility ratio.

Problems

- Viscous fingering results in poor vertical and horizontal sweep efficiency.
- Corrosion can cause problems in the flue gas method.
- The nonhydrocarbon gases must be separated from the saleable produced gas.

FIGURE 8–18. General Trend for Viscosity of Gas-Free Crude Oil at 100°F and Atmospheric Pressure

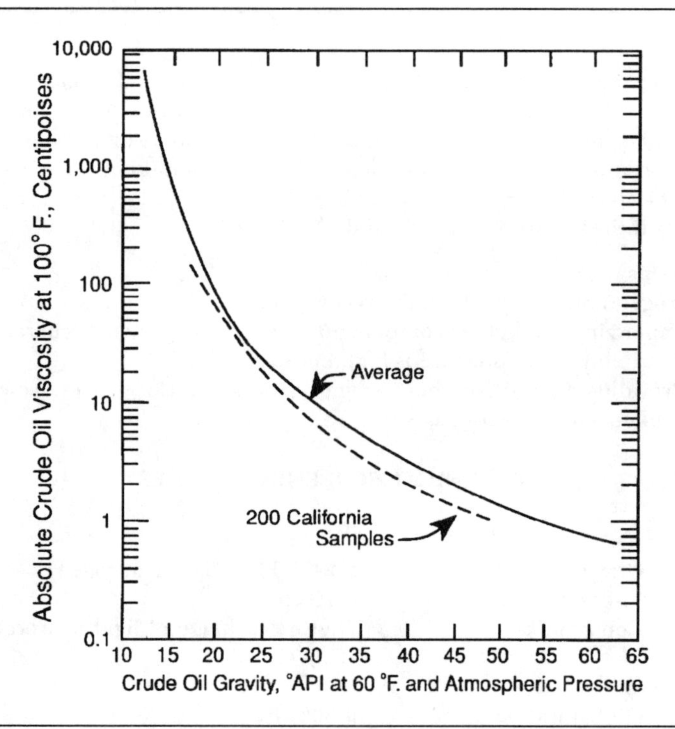

Steamflooding

In the steamflood (see Figure 8–20 and Table 8–6), steam is continuously introduced into injection wells to reduce the oil viscosity and to mobilize oil towards the producing wells. The injected steam forms a steam zone that advances slowly. The injected steam at the surface may contain about 80% steam and 20% water (i.e., 80% steam quality). When steam is injected into the reservoir, heat is transferred to the oil-bearing formation, the reservoir fluids, and some of the adjacent cap and base rock. Due to this heat loss, some of the steam condenses to yield a mixture of steam and hot water.

Ahead of the steam zone, an oil bank forms and moves towards the producing well. In many cases, the injected steam overrides the oil due to gravity. This behavior can create some problems. When steam breakthrough occurs, the steam injection rate is reduced by recompletion of wells or shutting off steam producing intervals. Steam reduces the oil saturation in the steam zone to very low values (about 10±%). Some oil is transported by steam distillation.

FIGURE 8–19. Viscosity Reduction of Oils and Water

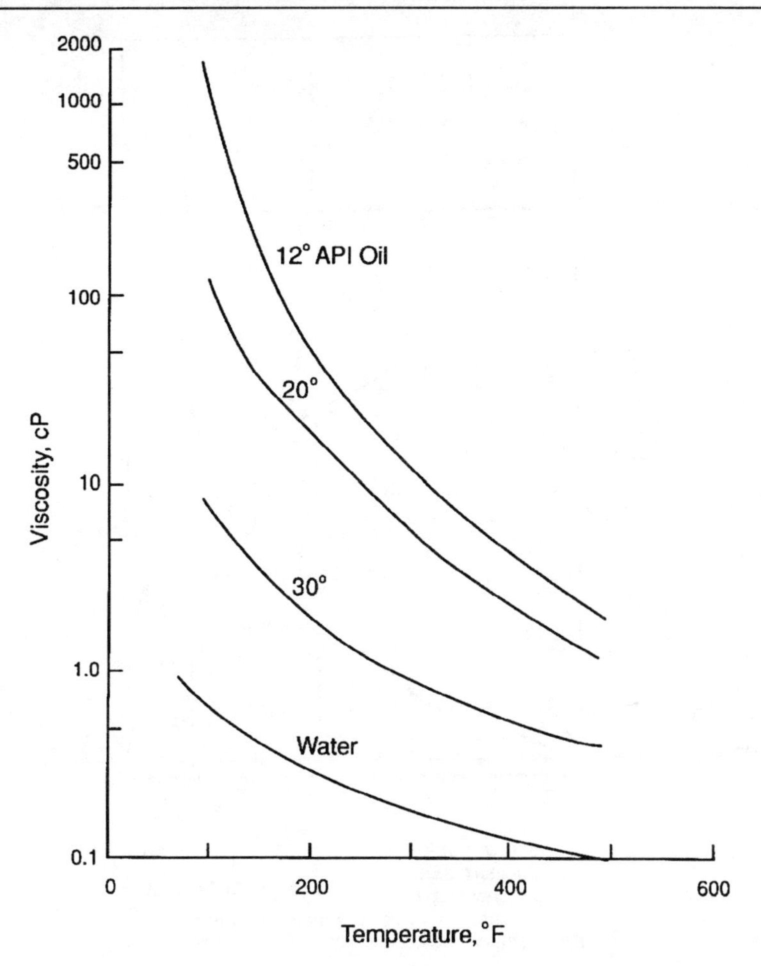

Normal steamflooding practice is to precede and accompany the steam drive by a cyclic steam stimulation of the producing wells.

Steamflooding is routinely used on a commercial basis. Including the steam stimulation process, there were 117 ongoing steam projects in the United States in 1992. About 454,000 BO/D or 60% of all EOR oil was produced by steam stimulation and steamflooding. In the United States, a majority of the field applications has occurred in California, where many of the shallow, high oil-saturated reservoirs are good candidates for thermal recovery. These reservoirs contain high-viscosity crude oils.

FIGURE 8–20. Regions of a Linear Steam Flood Process

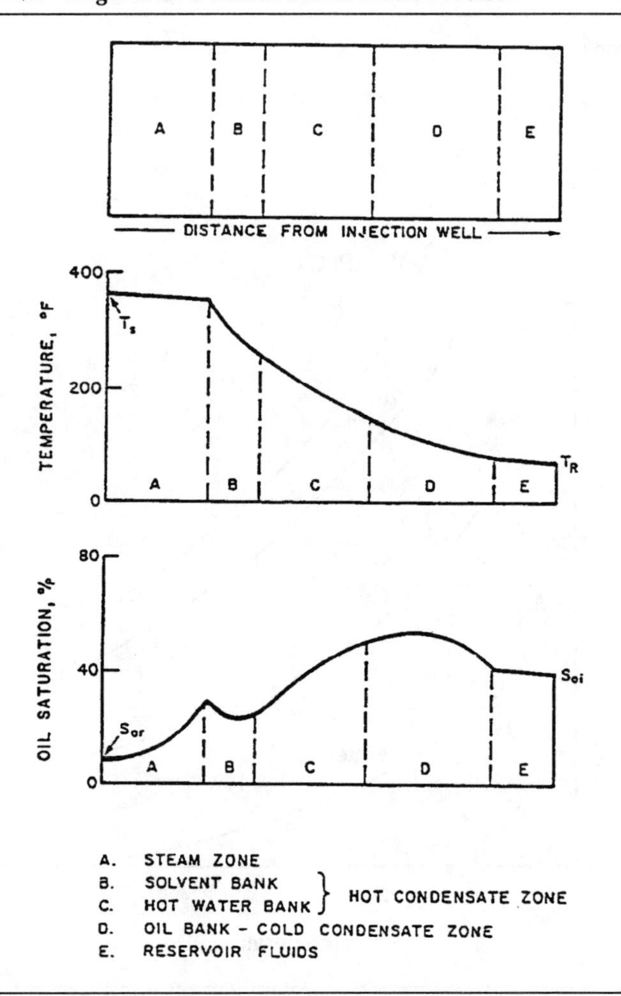

In-Situ Combustion

In-situ combustion or fireflooding involves starting a fire in the reservoir and injecting air to sustain the burning of some of the crude oil (see Table 8–7). The most common technique is forward combustion (Figure 8–21). One of the variations of this technique is a combination of forward combustion and waterflooding (COFCAW), and a second technique is reverse combustion.

In this process, which uses air and water, which are the world's cheapest and most plentiful fluids for injection, a significant amount of crude oil is burned (about 10% of the OOIP) to generate heat. The

FIGURE 8–21. Forward Combustion Process

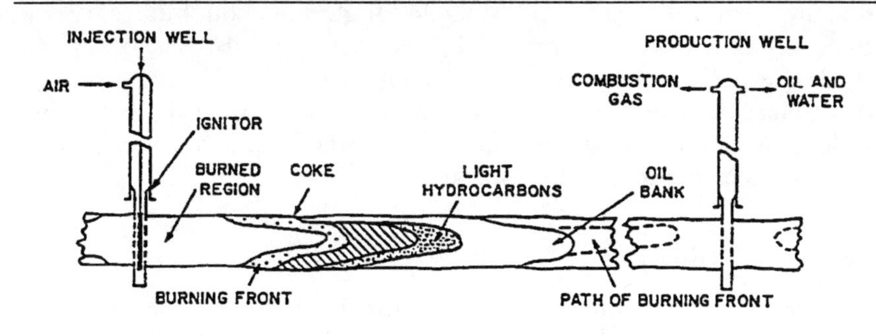

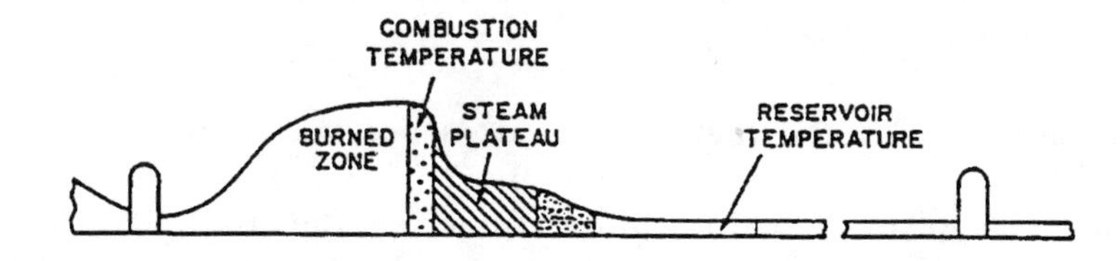

lighter ends of the oil are carried forward ahead of the burned zone upgrading the crude oil, whereas the heavy ends of the crude oil are burned. Heat is generated within a combustion zone at a very high temperature, about 600°C. As a result of burning the crude oil, large volumes of flue gas are produced. Thus corrosion is a major problem in this process.

Over 100 firefloods have been conducted in the world; however, there have not been many successes. There were eight ongoing combustion processes in the United States in 1992, producing about 4,700 BO/D. Seven of these appear to be successful or promising, whereas one is discouraging at this time.

CHEMICAL METHODS[14]

Chemical flooding processes produced only a very small amount (<1%) of EOR oil in the United States from the 26 active projects in 1992. These processes require conditions favorable to water injection because they are modifications of waterflooding. Chemical flooding is applicable to oils that are more viscous than oils suitable for gas injection, but less viscous than oils that can be recovered by thermal methods. Reservoirs with moderate permeability are desirable. The presence of a gas cap is not desirable, since there is the potential of resaturating the cap. Formations with high clay contents are undesirable, since the clays increase the adsorption of the injected chemicals.

Polymer Flooding

The objective of polymer flooding is to provide better displacement and volumetric sweep efficiencies during a waterflood (see Table 8–8).

Polymers improve recovery by:

- Increasing the viscosity of water.
- Decreasing the mobility of water.
- Contacting a larger volume of the reservoir.
- Reducing the injected fluid mobility to improve areal and vertical sweep efficiencies.

It should be noted that polymer does not lower the residual oil saturation.

Because polymer flooding inhibits fingering, the oil displacement is more efficient in the early stages as compared to a conventional waterflood. As a result, more oil will be produced in the early life of the flood, as indicated in Figure 8–22 at, for example, 1 pore volume. This is the primary economic advantage because it is generally accepted that ultimate recovery will be the same for polymer flooding as for waterflooding.

Many factors affect polymer flooding. These include polymer degradation due to salinity, temperatures, time, shear rates, and the presence of div-alent ions. Some polymers, like polysaccharides, are more resistant; however, they suffer from bacterial degradation problems and cause wellbore plugging. Also, polymers may be lost in the formation due to adsorption.

Polymers may be ineffective in a mature waterflood because of low mobile oil saturation. They show some promise in a reservoir with high vertical heterogeneity where the oil saturation may still be high and the vertical conformance poor. Some operators have been successful in treating injection wells (near-well treatment) with polymers to modify vertical profiles.

There were 23 polymer projects currently active in the United States in 1992; however, they produce only about 2,000 BO/D.

FIGURE 8–22. Water and Polymer Flood Recovery Comparison

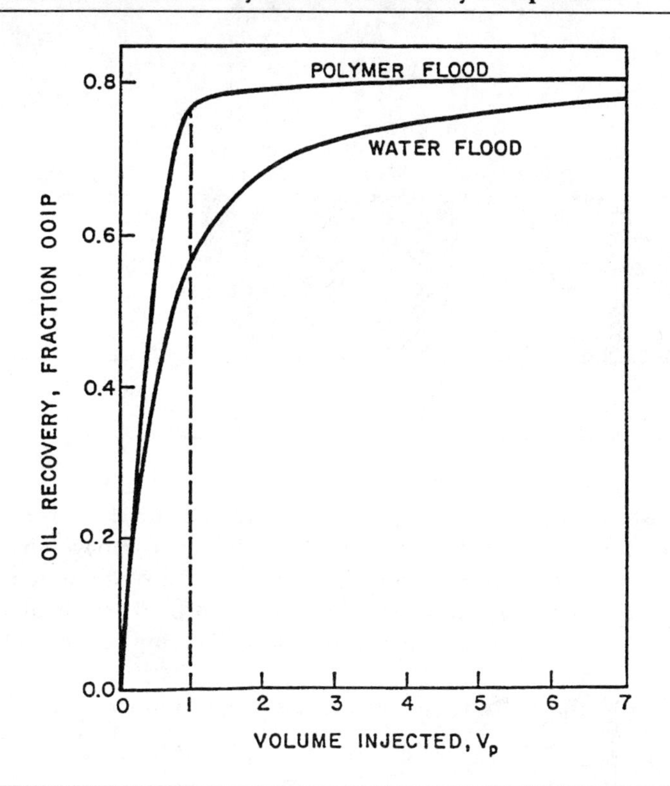

Surfactant/Polymer Flooding

Surfactant/polymer flooding, also called "micellar/polymer" or "microemulsion flooding" consists of injecting a slug that contains water, surfactant, electrolyte (salt), usually a cosolvent (alcohol), and possibly a hydrocarbon (oil) (see Figure 8–23 and Table 8–9).

Caustic Flooding

Caustic or alkaline flooding involves the injection of chemicals such as sodium hydroxide, sodium silicate, or sodium carbonate.

Oils in the API gravity range of 13–35° are normally the target for alkaline flooding. One of the desirable properties for the oils is to have enough organic acids so that they can react with the alkaline solution. Another desirable property is moderate oil gravity so that mobility control is not a problem.

FIGURE 8–23. Surfactant Flooding

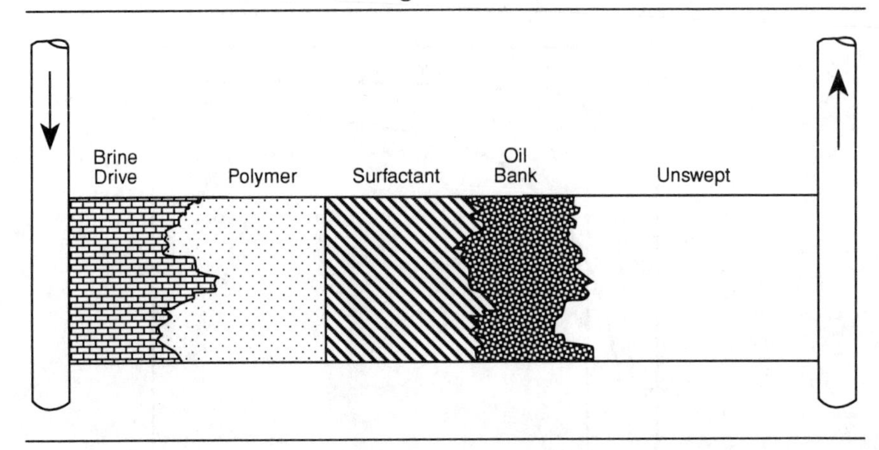

Sandstone reservoirs are generally preferred for this process, since carbonate formations often contain anhydride or gypsum which consume a large amount of alkaline chemicals. The alkali is also consumed by clays, minerals, or silica. In addition, the consumption is high at elevated temperatures. Another problem with caustic flooding is the scale formation in the producing wells.

Presently, there are no active caustic projects in the United States.

MISCIBLE METHODS[13,14,15,21–24]

Miscible flooding involves injecting a gas or solvent that is miscible with the oil. As a result, the interfacial tension between the two fluids (oil and solvent) is very low, and a very efficient microscopic displacement efficiency takes place. The displacing fluid may be hydrocarbon solvent that mixes with oil at first contact. For LPG slug or solvent flooding, enriched (condensing) gas drive and high pressure (vaporizing) gas drive, miscibility is achieved at different pressures.

Displacement of oil by propane or LPG is miscible at first contact in all proportions. With a high pressure gas (e.g., CO_2 or nitrogen), displacement of oil usually takes place by multiple contacts. In recent years the emphasis has been shifting to less valuable nonhydrocarbon gases such as CO_2, nitrogen, and flue gases. Although nitrogen and flue gases do not recover oil as well as the hydrocarbon gases (or liquids), the overall economics may be somewhat more favorable.

After thermal recovery, miscible flooding contributes the most among various EOR methods. About 40% of the total EOR production is by gas miscible/immiscible flooding. There were 89 active gas flooding projects in the United States in 1992.

Hydrocarbon Miscible Flooding

Hydrocarbon miscible flooding consists of injecting light hydrocarbons through the reservoir to form a miscible flood. Three different methods are used (see Table 8–11). Sometimes in the first LPG Slug Method, water is injected with the chase gas in a WAG (alternating water and gas) mode to improve mobility ratio between the solvent slug and the chase gas. In the second Enriched (Condensing) Gas Drive method, the enriching components are transferred from the gas to the oil. A miscible zone is formed between the injected gas and the reservoir oil, and this zone displaces the oil ahead. In the third High Pressure (Vaporizing) Gas Drive method, components from the crude oil being displaced result in multiple contact miscibility.

Carbon Dioxide Flooding

Carbon dioxide flooding is carried out by injecting large quantities of CO_2 (15% or more of the hydrocarbon PV) into the reservoir (Table 8–12).

Miscible displacement by carbon dioxide is similar to vaporizing gas drive. The only difference is a wider range of components; C_2 to C_{30} are ex-tracted. As a result, CO_2 flood process is applicable to a wider range of res-ervoirs at lower miscibility pressures than those for the vaporizing gas drive.

CO_2 is generally soluble in crude oils at reservoir pressures and temperatures. It swells the net volume of oil and reduces its viscosity even before miscibility is achieved by vaporizing gas drive mechanism. As miscibility is approached as a result of multiple contacts, both the oil phase and the CO_2 phase (containing intermediate oil components) can flow together because of the low interfacial tension. One of the requirements of the development of miscibility between the oil and CO_2 is the reservoir pressure.

CO_2 flooding should be used in moderately light-oil reservoirs (API gravity > 25), and the reservoir should be deep enough to have high enough pressure to achieve miscibility. CO_2 can dissolve in water; thus, it can lower the interfacial tension between oil and water. However, this process can also lead to more corrosion problems.

In this process, about 20–50% of the CO_2 slug is followed by chase water. Water is generally injected with CO_2 in a WAG mode to improve mobility ratio between the displacing phase and the oil.

CO_2 flooding is the fastest growing EOR method in the United States, and field projects continue to show good incremental oil recovery in response to CO_2 injection. The CO_2 flooding method works well as either a secondary or tertiary operation, but most large CO_2 floods are tertiary projects in mature reservoirs that have been waterflooded for many years.

There were 54 CO_2 floods currently active in the United States in 1992, out of which 52 were miscible. About 20% of the total EOR production was by CO_2 flooding.

Nitrogen and Flue Gas Flooding

Nitrogen and flue gas flooding are oil recovery methods that use these inexpensive nonhydrocarbon gases to displace oil in systems that may be either miscible or immiscible depending on the pressure and oil composition (see Table 8–13).

Both nitrogen or flue gas are inferior to hydrocarbon gases (and much inferior to CO_2) from an oil recovery point of view. Nitrogen has a lower viscosity, has poor solubility in oil, and requires a much higher pressure to generate or develop miscibility. The increase in the required pressure is significant compared to methane and very large (4–5 times) when compared to CO_2. Therefore, nitrogen will not reduce the displacement efficiency too much when used as a chase gas for methane, but it can cause a significant drop in the effectiveness of a CO_2 flood if the reservoir pressures are geared to the miscibility requirements for CO_2 displacements. Indeed, even methane counts as a desirable "light end" or "intermediate" in nitrogen flooding, but methane is quite deleterious to the achievement of miscibility in CO_2 flooding at modest pressures.

EOR SCREENING GUIDELINES

All of the processes described in this chapter have limitations in their application. These limitations have been derived partly from theory, partly from laboratory experiments, and partly from field experience. Prospect screening consists of (1) evaluating available information about the reservoir, oil, rock, water, geology, and previous performance, (2) supplementing available information with certain brief laboratory screening tests, and (3) selecting those processes that are potentially applicable and eliminating those that definitely are not. This is the first step in the enhanced recovery implementation sequence. The subsequent steps would be a further evaluation of candidate processes if more than one satisfies the screening criteria, pilot test design, pilot test implementation, pilot test evaluation/scale-up forecast, and commercial venture.

Table 8–14 presents screening guidelines for various EOR processes. A candidate reservoir for one or more EOR processes should not be

TABLE 8–14. Summary of Screening Criteria for Enhanced Recovery Methods *(Copyright © 1983, SPE, from paper 12069[14])*

	Oil Properties				Reservoir Characteristics				
	Gravity °API	Viscosity (cp)	Composition	Oil Saturation	Formation Type	Net Thickness (ft)	Average Permeability (md)	Depth (ft)	Temp. (°F)
Gas Injection Methods									
Hydrocarbon	> 35	< 10	High % of $C_2 - C_7$	> 30% PV	Sandstone or Carbonate	Thin unless dipping	N.C.	> 1000 (LPG) to > 5000 (H.P. Gas)	N.C.
Nitrogen & Flue Gas	> 24 > 35 for N_2	< 10	High % of $C_1 - C_7$	> 30% PV	Sandstone or Carbonate	Thin unless dipping	N.C.	> 4500	N.C.
Carbon Dioxide	> 26	< 15	High % of $C_5 - C_{12}$	> 20% PV	Sandstone or Carbonate	Thin unless dipping	N.C.	> 2000	N.C.
Chemical Flooding									
Surfactant/Polymer	> 25	< 30	Light intermediate desired	> 30% PV	Sandstone preferred	> 10	> 20	< 8000	< 175
Polymer	> 25	< 150	N.C.	> 10% PV Mobile oil	Sandstone preferred; Carbonate possible	N.C.	> 10 (normally)	< 9000	< 200
Alkaline	13–35	< 200	Some organic acids	Above waterflood residual	Sandstone preferred	N.C.	> 20	< 9000	< 200
Thermal									
Combustion	< 40 (10–25 normally)	< 1000	Some asphaltic components	> 40–50% PV	Sand or sandstone with high porosity	> 10	> 100*	> 500	> 150 preferred
Steamflooding	< 25	> 20	N.C.	> 40–50% PV	Sand or sandstone with high porosity	> 20	> 200**	300–5000	N.C.

N.C. = Not Critical
* Transmissibility > 20 md ft/cp
** Transmissibility > 100 md ft/cp

TABLE 8–15. Preferred Oil Viscosity Ranges for Enhanced Recovery Methods *(Copyright © 1983, SPE, from paper 12069[14])*

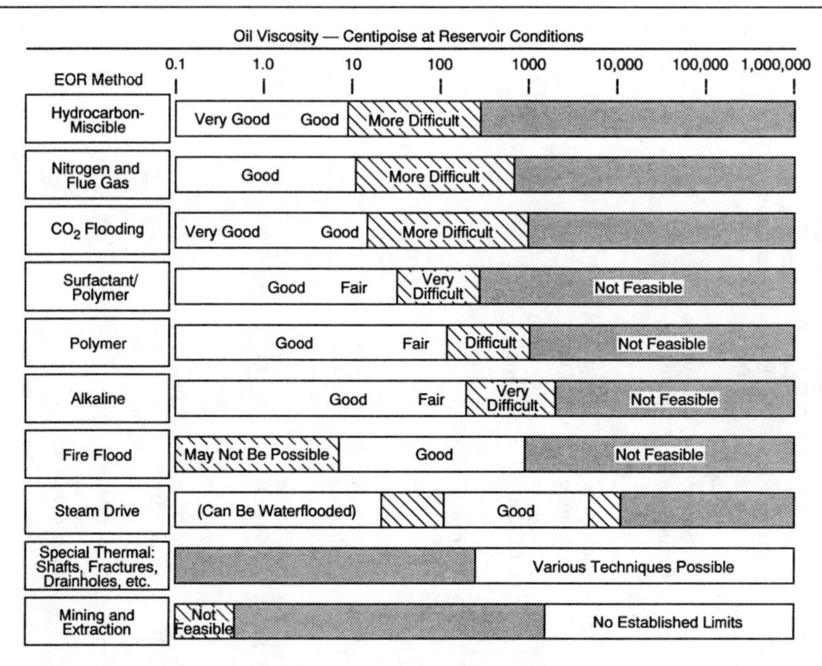

TABLE 8–16. Permeability Guides for Enhanced Recovery Methods *(Copyright © 1983, SPE, from paper 12069[14])*

Permeability (millidarcy): 1, 10, 100, 1000, 10,000

EOR Method	Permeability (millidarcy)
Hydrocarbon-Miscible	— Not Critical If Uniform —
Nitrogen and Flue Gas	— Not Critical If Uniform —
CO_2 Flooding	— High Enough For Good Injection Rates —
Surfactant/Polymer	Preferred Zone
Polymer	Possible; Preferred Zone
Alkaline	Preferred Zone
Fire Flood	Preferred Zone
Steam Drive	Preferred Zone

TABLE 8–17. Depth Limitation for Enhanced Oil Recovery Methods *(Copyright © 1983, SPE, from paper 12069[14])*

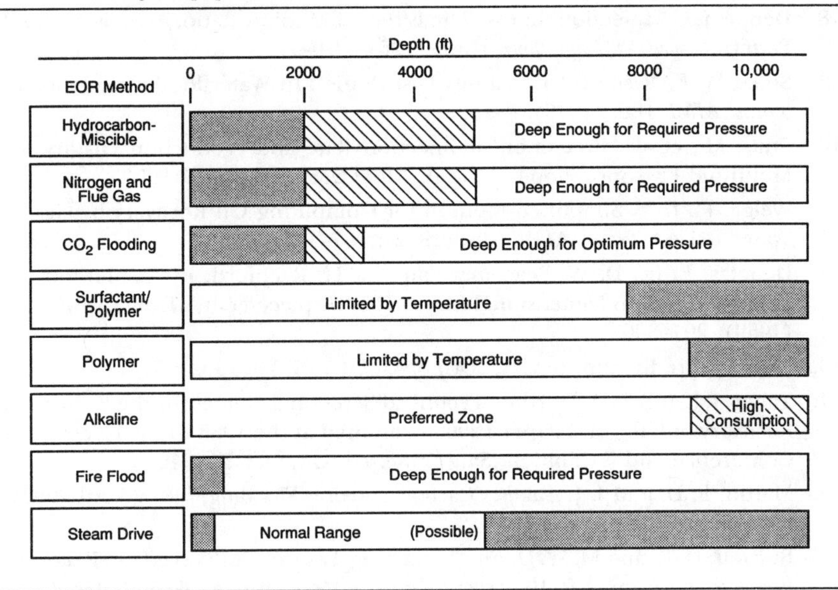

discarded because it does not satisfy one or two criteria. Each prospect should be evaluated on its own merits by analyzing the many reservoir, operational, and economic variables. The influences of oil viscosity, rock permeability, and reservoir depth on the technical feasibility of the various EOR methods are shown in Tables 8–15, 8–16 and 8–17, respectively.

REFERENCES

1. Craig, F. F., Jr. "The Reservoir Engineering Aspects of Waterflooding," SPE Monograph 3, Richardson, TX (1971).
2. Rose, S. C., J. F. Buckwalter, and R. J. Woodhall. The Design Engineering Aspects of Waterflooding, Monograph Series, SPE, Richardson, TX (1989): 11.
3. Dykstra, H. and R. L. Parsons. "The Prediction of Oil Recovery by Waterflooding," *Secondary Recovery of Oil in the United States*, 2nd ed., API (1950): 160–174.
4. Buckley, S. E. and M. C. Leverett. "Mechanisms of Fluid Displacement in Sands," *Trans. AIME* 146, (1942): 107–116.
5. Croes, G. A. and N. Schwarz. "Dimensionally Scaled Experiments and the Theories on the Water-Drive Process." *Trans. AIME* (1955): 204, 35–42.
6. Dyes, A. B., B. H. Caudle, and R. A. Erickson. "Oil Production After Breakthrough as Influenced by Mobility Ratio," *Trans. AIME* 201,(1954): 81–86.

7. Muskat, M. *Physical Principles of Oil Production.* New York: McGraw-Hill Book Co., Inc. 1949: 650–657.
8. Deppe, J. C. "Injection Rates—The Effect of Mobility Ratio, Area Swept, and Pattern," *Soc. Pet. Engr. Jour.* (June, 1961): 81–91.
9. Stiles, W. E. "Use of Permeability Distribution in Waterflood Calculations," *Trans. AIME* 186, (1949): 9–13.
10. Prats, M., et al. "Prediction of Injection Rate and Production History for Multifluid Five-Spot Floods," *Trans. AIME* 216, (1959): 98–105.
11. Welge, H. J. "A Simplified Method for Computing Oil Recovery by Gas or Water Drive," *Trans. AIME* 146, (1942): 107–116.
12. Douglas, J., Jr., D. W. Peaceman, and H. H. Rachford, Jr. "A Method for Calculating Multi-Dimensional Immiscible Displacement," *Trans. AIME* 216, (1959): 297–306.
13. "Annual Production Report," *OGJ Special,* (April 20, 1992): 51–79.
14. Taber, J. J. and F. D. Martin. "Technical Screening Guides for the Enhanced Recovery of Oil." SPE Paper 12069 presented at the 1983 Annual Technical Conference and Exhibition, San Francisco, CA, Oct. 5–8, 1983.
15. Martin, F. D. and J. J. Taber. "Carbon Dioxide Flooding," *JPT* (April 1992): 396–400.
16. Burnett, D. B. and M. W. Dann. "Screening Tests for Enhanced Oil Recovery Projects." SPE Paper 9710 presented at the 1981 Permian Basin Oil and Gas Recovery Symposium, Midland, TX, March 12–13, 1981.
17. Thermal Recovery Processes, SPE Reprint Series No. 7, 1985.
18. Thermal Recovery Techniques, SPE Reprint Series No. 10, 1972.
19. Prats, M.: Thermal Recovery, SPE Monograph 7, Richardson, Texas, 1982.
20. Boberg, T. C. *Thermal Methods of Oil Recovery,* New York: John Wiley & Sons, 1988.
21. Miscible Processes, SPE Reprint Series No. 8, 1971.
22. Miscible Processes 11, SPE Reprint Series No. 18, 1985.
23. Stalkup, F. I., Jr.: Miscible Displacement, SPE Monograph 8, 1983.
24. Klins, M. A. *Carbon Dioxide Flooding,* Boston: IHRDC, 1984.
25. Moore, T. F., and R. L. Slobod. "The Effect of Viscosity, and Capillarity on the Displacement of Oil by Water." *Producers Monthly* (Aug. 1956) 20–30.
26. Abrams, A. "The Influence of Viscosity, Interfacial Tension, and Flow Velocity on Residual Oil Saturation Left by Waterflood," SPE Reprint Series No. 24, vol. 1 (1988): 52–62.

CHAPTER 9

▼ ▼ ▼

Reservoir Management Case Studies

This chapter is devoted to reservoir management case studies. It presents brief reviews of several projects and an overall analysis of an EOR project. Specific contributions made by various team members of different case examples are shown in Table 9–1.

TABLE 9–1. Reservoir Management Case Studies

Case	Specific Contributions Made by Various Team Members
North Ward Estes Field	• Reservoir characterization, identification of the project area, and generation of a reservoir database—Geoscientists, reservoir engineers, and computer analysts (about 11 man years and $1.6 million of geoscientists' resources). • A comprehensive plan for a six-section CO_2 project, including project approval—Reservoir engineers and geoscientists, design and construction engineers, gas and chemical engineers, with assistance from drilling and production operations, research and service laboratory engineers/ geoscientists. • Environmental and regulatory aspects (approval for injection wells and estimating radius of exposure for H_2S/CO_2 leaks)—Land, legal and environmental staffs, gas and chemical engineers, design and construction engineers. • Identification of several hundred workover candidates and evaluation of many waterflood modification projects—Geoscientists, reservoir and production engineers, drilling and operations staff.

(continued)

TABLE 9–1. Reservoir Management Case Studies (continued)

	• CO_2 injection startup within 15 months of project initiation—Drilling and operations staff, gas and chemical engineers, design and construction engineers, and production and reservoir engineers. • Gas processing plant built and started within 18 months—Design and construction engineers, operations staff, outside contractors, and environmental staff.
McAllen Ranch Field	• 3-D seismic survey acquisition and interpretation and incorporate 3-D seismic into the planned drilling program—Geophysicist and geologist, reservoir engineer. • Adding reserves profitably and reducing non-contributing behind pipe reserves—Reservoir and production engineers, geologist, and field operations personnel. • Identify potential for commingling similar pay sands and obtain regulatory approval to commingle them—Reservoir engineer, production engineer, operations foreman, and land, permitting, and regulatory affairs staff. • Section 29 Federal Tight Gas Sand Credit—Tax department staff and attorney, and land, permitting, and regulatory affairs staff.
Brassey Oil Field	• Interpretation of seismic and offset VSP data to drill wells on the edge of the pool—Geophysicist, geologist, and reservoir engineers. • Defined pool edges by using seismic interpretations and well test data—Geologist, reservoir engineers, and geophysicist. • OOIP estimation—Reservoir engineer and geologist. • Management of production and injection rates to minimize breakthrough—Production operations staff, reservoir and production engineers, and geologist.
Means San Andres Unit	• Technique for estimating continuous and floodable pay to estimate potential additional recovery from infill drilling—Joint geology-engineering study. • A reservoir study conducted in 1981–82 that supported the applicability of CO_2 flooding—Reservoir engineers and geologists, with assistance from research and services organizations.

(continued)

TABLE 9–1. Reservoir Management Case Studies (continued)

	• A detailed surveillance program for CO_2 flooding—Production operations staff, production and reservoir engineers. • Infill wells—Drilling, geology and engineering staffs.
Teak Field	• Better utilization of a data set, management of mature field development operations, and detailed reservoir characterization and reservoir models—Geologist, geophysicist and reservoir engineer, working in close cooperation with operations personnel.
Esso Malaysia Field	• Data collection effort during pre-development and development phases—Operations, facilities, reservoir engineering, geology, and geophysics personnel. • Coordinated process to collect, analyze, validate, and integrate reservoir description and performance data into optimal development and depletion plans—Operations, facilities, reservoir engineering, geology, and geophysics personnel. • Monitor and review performance/ongoing surveillance activities—Operations and reservoir engineering personnel. • Revision of development plan and changes in operations—Operations, facilities, reservoir engineering, and geology staffs. • Identify problems and implement timely, innovative solutions—Operations, facilities, reservoir engineering, geology, and geophysics staffs.
Columbus Gray Lease	• Action plan—Production and reservoir engineers and geologist, in consultation with surface facilities engineer and field foreman. • Modification of OOIP—Geologist and reservoir engineer. • Lease performance improvement as a result of well workovers, installation of larger pumps, and increasing injection rate—Production engineers, reservoir engineers, facilities engineer and field foreman, in consultation with drilling engineer.

NORTH WARD ESTES FIELD AND COLUMBUS GRAY LEASE

North Ward Estes field study illustrates the application of a comprehensive management approach for large reservoirs, and Columbus Gray lease study discusses the problem-solving approach for small reservoirs.[1] These projects have been discussed in Chapter 3. Although the two approaches are philosophically quite different, both approaches resulted in positive results.

To illustrate the importance and value of team effort, the design and implementation of an EOR project and the improvement of existing waterfloods, the North Ward Estes field, a mature field located in Ward and Winkler Counties, Texas, has been considered (see Appendix E for details about this project[2]). A study team, as shown in Figure 3–8, was formed. All team members were asked to focus on the previously mentioned objectives. During the design phase, a reservoir engineer led the team, whereas during the implementation phase involving construction of a $40 million plant, a team leader with an operational background was chosen.

All of the team members had a full-time assignment on this project, except as specified below:

Design Phase

Reservoir Engineer	1 (Team Leader)
Additional Reservoir Engineers	5
Production Engineers	2
Design and Construction Engineer	1
Gas and Chemical Engineer	1
Drilling	1 (mostly full-time)
Production Operations	2 (part-time)
Geologists	4
Land, Legal and Environmental	2 to 4

Implementation Phase

Production Operations Superintendent	1 (Team Leader)
Additional Production Operations Personnel	12 to 15
Production Engineers	3
Design and Construction Engineers	4 (+ many contractors)
Gas and Chemical Engineers	2
Drilling	3 to 5 (+ many contractors)
Research and Service Laboratory Engineers	1

Geologists	4
Reservoir Engineers	4
Land, Legal and Environment	2

As a result of the design phase, a comprehensive plan for a six-section CO_2 project was completed. In addition, hundreds of workover candidates were identified, and several waterflood modification projects were evaluated.

As mentioned in Chapter 3, CO_2 injection was started in the six-section area within 15 months of project initiation. Moreover, within 1½ years, the gas-processing plant with a capital expenditure of $40 million was built and started. The teams' goal for every aspect of the project—from well workovers, reservoir studies, CO_2 injection, and gathering system construction to start up—was accomplished in a short time without sacrificing quality.

In summary, the previous case study illustrates that a full-blown team effort and teamwork across the function lines has resulted in a successful design and the implementation of many projects in the North Ward Estes field.

Attributes of the successful team effort followed for the North Ward Estes Field are given below:

- Empowerment.
- Accountability.
- Cross-functional.
- Clear and common objectives.
- Preparation of action plan, involving all functions.
- Defined deadlines.
- Flexible plans.
- Management commitment.
- Knowledgeable team members, requiring little or no supervision.
- Periodic project reviews with participation from all functions.

Difficulties encountered by the North Ward Estes Field Team are listed below:

- Sometimes difficult to reach consensus.
- Cost and work schedule estimate repeated.
- No team training.
- People not used to teamwork initially had a tendency to go to their functional managers.
- Occasional interference from functional managers.

There is no doubt that this case study is a rigorous illustration of team effort, involving environmental, legal, management, land, research and service laboratories, as well as geology, reservoir engineering, production engineering, design and construction engineering, gas and chemical engineering, and production operations. However, the question is "Can we afford to carry on a full-blown team effort for every project?"

The obvious answer to this question is "no;" every team effort should be screened while keeping the cost-benefit analysis in mind. To illustrate a problem-solving approach to reservoir management, a case study involving Columbus Gray lease is considered.

As a result of a problem-solving session involving two (production and reservoir) engineers and a geologist over a period of two weeks, a preliminary action plan for the lease was prepared. The results of the study were then discussed in a one-half day session with the field foreman and surface facilities engineer. (Chapter 3 describes the results in detail.)

A simple team make-up with a total study time of two weeks is illustrated below:

Reservoir Engineer	1
Production Engineer	1
Geologist	1
Field Foreman	short-term involvement
Facilities (Design & Construction) Engineer	short-term involvement
Drilling Engineer	none (utilized the knowledge of the production engineer as he had considerable prior drilling experience)
Environmental, Legal, Land	none (geologist and production engineer shared their knowledge)
Research and Service Laboratories	none (except contact by a few phone calls)

As we can see, the reservoir management approach followed was very simple because the lease production rate was only a few hundred barrels a day. Also, based upon reservoir heterogeneity and past performance, the expected increase in production was not considered high. Thus, a decision was made to design and implement a cost-effective reservoir management program.

At this time, a question is "Should we have done something differently?" The answer to this question is "no." Although one could have designed the team differently, the overall objective and the approach would have been similar.

McALLEN RANCH FIELD

Durrani et al.[3] described the process of managing redevelopment of the 30-year-old geopressured McAllen Ranch gas field by a cross-functional team. The make-up of the team is given below:

Engineers (reservoir, production, petrophysics, and facilities)
Geologists
Geophysicists
Field Operations Personnel
Land, Permitting, Business/Regulatory Affairs Staff
Tax and Legal Staff

The concept of cross-functional team management in this case became necessary, just like in the North Ward Estes case described earlier, because of declining production and concerns about substantial noncontributing reserves.

At the initiation of a study in early 1989, the engineering function was organized along specialty lines that followed separate drilling, facilities, petrophysical, production, and reservoir activities. A group of engineers of similar discipline (e.g., drilling) formed a section and reported to a division engineer. The various division engineers did not always report to the same manager and each had their own specific goals in mind. For example, the reservoir engineers targeted reserve additions, whereas the production engineers concentrated on production increase through recompletion and reconditioning, and drilling engineers concentrated on reduced trouble costs. As a result, incompatible goals of various specialties developed. A lack of team effort was realized and, therefore, a cross-functional team effort was initiated.

The cross-functional team was assigned to a specific field, the McAllen Ranch, and involved all functions. It was made sure early on that all team members had compatible and consistent targets. This team approach had somewhat of a "matrix organization" (i.e., each member of the team remained in their own specialty and had the technical backup and review available but their actual targets were concentrated towards the McAllen Ranch Asset Team).

Shell Western E&P Inc. also followed the asset team approach for several major South Texas fields. Note that the functions represented on the asset management team were determined by the activities and objectives of each field. Not all functions were required on all teams, and the team membership expanded or contracted based on the objectives.

A major focus of the McAllen Ranch Asset Management Team included a 3-D seismic survey acquisition and interpretation, development drilling, commingling (including regulatory approval), and field-producing operations, and remedial well work. Specifically, the following goals were identified: (1) develop a technique to allow accelerated production of existing reserves, (2) add reserves profitably, (3) incorporate 3-D seismic into the planned drilling program, and (4) identify potential for commingling similar pay sands and obtain regulatory approval to commingle them. Team members focused on these overall goals rather than on those segments that were specifically applicable to their functions. The team membership included:

Team member	Number	Area
Geophysicist	1	3-D seismic information
Drilling Engineer	1	
Drilling Foreman	1	
Geological Engineer	1	High level of drilling activity
Petrophysical Engineer	1	
Landman	1	
Permitting Staff	1	
Reservoir Engineer	1	Commingling effort and remedial work
Production Engineer	1	
Operations Foreman	1	Development and commingling activities
Facilities Engineer	1	
Regulatory Staff	1	Commingling approval of the Texas Railroad Commission
Tax Department Staff	1	Section 29 Federal Tight Gas Sand Credit
Attorney	1	
Team Leader		Chosen from amongst the team members

Based upon the overall goals defined, the team developed the following specific targets to be achieved in a two-year (1990–91) time frame.

- Increase total field gas production rate above 100 MMcf/D. Reduce noncontributing behind pipe reserves by 50%.
- Complete 3-D seismic interpretation and mapping, and identify at least 10 new drilling locations.
- Reduce drilling costs by at least 10%.
- Commingle production from various zones.

At the end of this time period, the team achieved the following:

- Field gas production rate increased from 50 MMcf/D to over 130 MMcf/D.
- Application of fine-grid 3-D seismic and synergistic subsurface work resulted in a success rate of 93% for drilling 15 new wells. These wells added over 100 Bcf of gas reserves.
- Time to drill new wells was cut in half, and overall drilling costs were reduced by 25% (about $7 million).
- Average new well initial production rate was 7 MMcf/D, compared to previous rates of 1–3 MMcf/D.

- The concept of commingling several massive hydraulic fractured intervals proved successful.
- Behind pipe noncontributing reserves were reduced from 40 to 20% by commingling.

The McAllen Ranch Team effort, as documented by Durrani et al., is a well-described example of team excellence. This case study shows that the effective cross-functional teamwork can indeed produce exemplary results for complex oil and gas operations. It documents that conventional organizational structure did not function properly and was producing poor results because of conflicting functional goals and objectives.

Attributes of the successful team effort followed for the McAllen Ranch Field are listed below:

- Empowerment.
- Innovation and risk taking.
- Reduced routine supervision.
- Minimum individual technical reviews by functional heads in favor of joint reviews with other heads.
- Focusing on a common team goal (rather than separate objectives for each specialty).
- Development of a consensus plan.
- Cross-functional team.
- Defining team's targets and goals.
- Management "buy-in" and commitment.
- Informal communication and clarification of priorities.
- Quick approval process.
- Functional heads' roles more like advisors than bosses.
- Periodic reviews (formal and informal).
- Decision making by all team members.
- Technology transfer between various teams (learning from other teams).
- Well-trained, highly motivated individuals (with a mix of experienced and inexperienced).
- State-of-the-art technical tools, sufficient resources, and funding without a strict budget.
- Rewarding team excellence.

BRASSEY OIL FIELD

Woofer and MacGillivary[4] and Anderson et al[5] presented a case study of how an aggressive engineering/geoscience team approach provided the development plan for the Brassey oil field in British Columbia. (See

Appendix B for a complete description of this case study.) Miscible flooding of the field began only 2 years after discovery. The geological and geophysical technical contributions to field development were noteworthy. The interpretation of seismic and offset VSP data provided the development team with the confidence to drill wells on the edge of the pool. A team approach combined the knowledge from a geological model, seismically defined pool edges, continuity information from well-test data, and material-balance calculations to predict reservoir volume, areal extent, and continuity on the basis of an integrated reservoir model.

The benefits of the team effort followed in this field study are tremendous and are summarized below:

- Development of a reservoir model
  Geophysicist— Defined pool edges by seismic interpretations.
  Engineers— Described pools by using well-test pressures and pulse-test responses to determine well continuity.
  Geologist— Mapped reservoir properties utilizing excellent core controls throughout the field.
  (The integration of the reservoir description, volumetrics, seismic delineation, and well-test data provided the basics for determining reservoir continuity and size.)
- OOIP Calculation
  Engineering— Material-balance analysis
  Geology— Hand-contoured volume
  (Agreement of independently calculated volumes gave a high level of confidence to proceed with the field development plan.)
- Reservoir pressure was maintained at or near target pressure and never fell into the vicinity of minimum miscibility pressure.
- The results of tracer breakthrough monitoring—coupled with pressure, voidage, gas compositional analysis, and GOR—provided management of production and injection rates in the field to minimize breakthrough. This led to the maximum utilization of existing injection compressure facilities.

In summary, a team approach to reservoir management showed the value of open communication between office and field, and between different technical disciplines.

Attributes of the successful reservoir management effort followed for the Brassey Oil Field are listed below:

- Setting goals and objectives.
- Creating a plan to achieve these objectives.
- Monitoring operations.

- Establishing good communication between disciplines and verifying that objectives have been achieved.
  —Creating a reservoir management bulletin board, mounted in a prominent location in the office and updated monthly
  —Spur-of-the-moment hallway meetings around the bulletin board and generation of new ideas
  —Frequent office staff visits to the field and creation of interests regarding the reservoir performance among the field operators and supervisors.

MEANS SAN ANDRES UNIT

Stiles[6] discussed reservoir management in the San Andres Unit in West Texas from primary to secondary to tertiary operations. (See Appendix F for description of this case study.) The reservoir management in this field started in 1935, just one year after discovery. Reservoir management has evolved from relatively simple primary operations to elaborate techniques as the reservoir has been produced by primary, secondary, and tertiary methods. In addition, the reservoir management programs included a team effort from several groups and had management support. The author concluded that reservoir management must be dynamic and sensitive to changes in performance, technology, and economic conditions.

Stiles documented a comprehensive reservoir management program used at the Means San Andres Unit. A detailed surveillance program was developed and implemented in 1975. A detailed engineering and geologic study was conducted during 1968–1969 to determine a new depletion plan. This cooperative study provided the basis for a secondary surveillance program and later for a design and implementation of the CO_2 tertiary project.

As a result of a joint geology-engineering study, a technique for estimating continuous and floodable pay was developed. This technique assisted in estimating potential additional recovery from infill drilling with pattern densification. About 140 infill wells were drilled through 1981 that would recover over 15 million barrels of incremental oil.

A reservoir study was conducted in 1981–1982 that supported the applicability of CO_2 flooding in this unit. The reservoir description conducted in 1968–1969 provided the basis for planning the CO_2 tertiary project. Although this reservoir description was the building block for the project, it was continuously updated during the planning and implementing phases of the CO_2 project as more data became available. The CO_2 project plan consisted of 167 patterns on about 6,700 acres. Project implementation began in late 1983. In less than two years, 205 infill producers and 158 infill injectors were drilled.

Attributes of the successful reservoir management effort for the Means San Andres Unit follow:

- Early start (just one year after discovery).
- Reservoir management evolved from simple to elaborate techniques.
- Team effort from several groups.
- Management support.
- Joint geology-engineering reservoir description.
- Extensive data acquisition and management effort.
- Comprehensive surveillance program.
- Economic considerations.
- Dynamic reservoir management and sensitivity to changes in performance, technology, and economics.
- Continuous goal of obtaining better reservoir description to better understand the reservoir processes. The application of seismic sequence stratigraphic concepts has yielded significant insights into the reservoir complexities.

TEAK FIELD

Lantz and Ali[7] discussed development of the Teak field in Trinidad. This mature giant offshore oil field responded favorably to a deliberate and systematic team approach to continuing development, halting, and even reversing a steep natural decline. The impact of the multidisciplinary team approach has been to better utilize an ever expanding data set and to provide added precision to mature field development operations.

The multidisciplinary team approach utilized a geologist, geophysicist, and reservoir engineer in a study group working in close cooperation with operations personnel. Detailed structure mapping and reservoir characterization, accurate reservoir models, and good teamwork were responsible for the success of the two 1989 development wells, which were completed at a cumulative rate of 9,000 BOPD, about 25% of the total field rate. The team also developed workover, recompletion, and waterflood prospects, in addition to optimizing waterflood patterns and preparing long-term depletion plans.

The interaction of disciplines followed in this case study is described in Figure 9–1.

ESSO MALAYSIA FIELDS

A well-integrated, multidisciplinary team approach is used in many of Esso Malaysia fields.[8] The team includes personnel from functional groups (operations, surveillance, facilities, reservoir geology, and geo-

FIGURE 9–1. Interaction of Different Disciplines—Teak Field

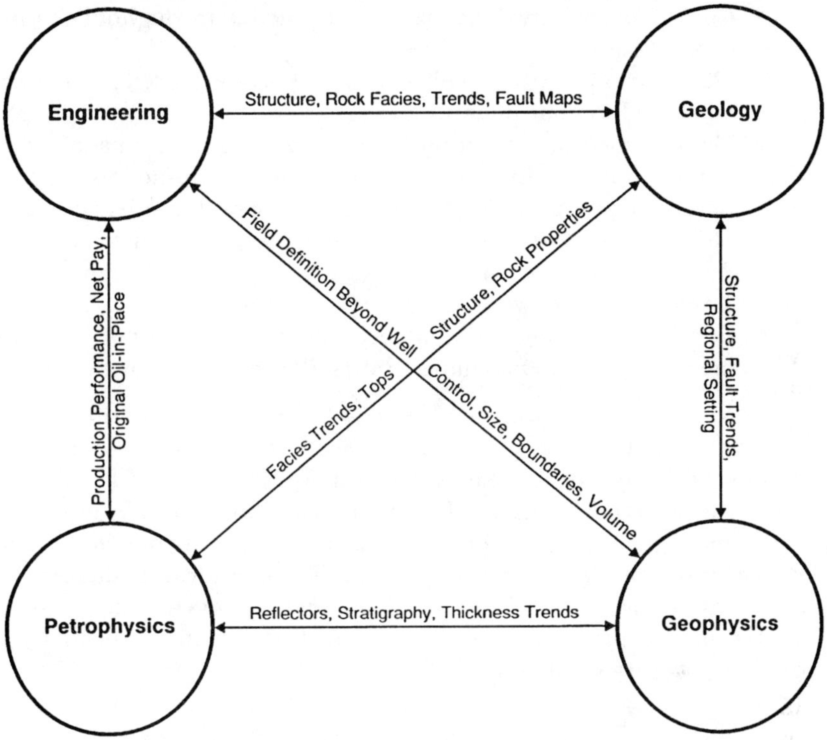

physics) on an as-needed basis. Group members have specific field and platform area assignments and work together to ensure that proper reservoir description data are acquired. The team approach allows real-time synergism of needed technical expertise and ensures that various operating scenarios are considered in developing and revising plans.

Attributes of the successful reservoir management effort for the Esso Malaysia Fields follow:

- Clearly defined and endorsed plans provide direction and control of the reservoir management program.
- Coordinated process to collect, analyze, validate, and integrate reservoir description and performance data into optimal development and depletion plans.
- Use of well-integrated, multidisciplinary team to identify problems and to implement timely, innovative solutions.
- Regular reports to management and frequent discussions among functional groups.

- Reservoir management planning begins during predevelopment phases.
- Continuous reservoir management practice throughout the life of the reservoir.
- Multidisciplinary data collection effort during predevelopment and development phases.
- Maintenance of a comprehensive and integrated database supporting both geologic and reservoir simulation models.
- Reservoir operating guidelines developed with model results.
- Monitor and review performance—ongoing surveillance activities.
- Revision of development plan and changes in operations.
- To improve teamwork, hold team-building sessions.
- Continue to integrate new technology into reservoir management program to maximize profitability and economic recovery.

The above six cases illustrate the full-blown team effort and demonstrate the advantages of team effort. Most of these case studies list attributes of the successful team effort followed.

Several case studies outline the difficulties encountered by not working as a management team (i.e., noncompatible and inconsistent goals and objectives of different team members). A critical evaluation of various team efforts clearly indicates that effective cross-functional teamwork can indeed produce exemplary results for maximizing profits and economic recovery of oil and gas.

REFERENCES

1. Thakur, G. C. "Implementation of a Reservoir Management Program." SPE Paper 20748 presented at the SPE Annual Technical Conference and Exhibition, New Orleans, Sept. 23–26, 1990.
2. Winzinger, R. et al. "Design of a Major CO_2 Flood—North Ward Estes Field, Ward County, Texas," *SPERE* (February 1991): 11–16.
3. Durrani, A. J. et al. "The Rejuvenation of the 30-year-old McAllen Ranch Field: An Application of Cross-Functional Team Management." SPE 24872 presented at the SPE Annual Technical Conference and Exhibition, Washington, DC, October 4–7, 1992.
4. Woofer, D. M. and J. MacGillivary. "Brassey Oil Field, British Columbia: Development of an Aeolian Sand—A Team Approach," *SPERE* (May 1992): 165–72.
5. Anderson, J. H. et al. "Brassey Field Miscible Flood Management Program Features Innovative Tracer Injection." SPE Paper 24874 presented at the SPE Annual Technical Conference and Exhibition, Washington, DC, October 4–7, 1992.

6. Stiles, L. H. "Reservoir Management in the Means San Andres Unit." SPE Paper 20751, presented at the Annual Technical Conference and Exhibition, New Orleans, Sept. 23–26, 1990.
7. Lantz, J. R. and N. Ali. "Development of a Mature Giant Offshore Oil Field, Teak Field, Trinidad." OTC paper 6237, 22nd Annual OTC Meeting, Houston, TX, May 7–9, 1990.
8. Trice, M. L. and B. A. Dawe. "Reservoir Management Practices," *JPT* (December 1992): 1296–1303 & 1349.

CHAPTER 10

▼ ▼ ▼

Reservoir Management Plans

This chapter presents several reservoir management plans to illustrate the following reservoir management concepts and processes:

- Newly discovered field.
- Secondary and EOR operated field.

NEWLY DISCOVERED FIELD[1]

Offshore Wizard Field, analogous to many Gulf Coast reservoirs, was recently discovered by rank wildcat Well No. 1 and Wells 2, 3, 4, and 5 were drilled to delineate the field (see Figures 10–1 and 10–2). A drill stem test was performed at the discovery well and the results indicated two productive zones, 4,000 ft and 4,500 ft sands, with 875 STBOPD and 1,456 STBOPD productions, respectively. Well No. 4 was a dry hole, and Well No. 3 penetrated the oil-water contact (see structure maps in Figures 10–1 and 10–2). The available data indicated that this field had a significant amount of potential reserves. It was essential to form a multidisciplinary, integrated team that was charged by the management to develop an economically viable plan for the reservoir within 60 days. The team consisted of the following professionals:

- A geologist responsible for geological and petrophysical works.
- A reservoir engineer responsible for providing production and reserves forecasts and economic evaluation.
- A drilling and a completion engineer responsible for drilling and completing wells, respectively.
- An equipment engineer responsible for designing surface-processing facilities.
- A structural engineer responsible for designing platforms and production decks.
- Other professionals, if needed, such as pipeline engineer, land manager, etc.

FIGURE 10–1. Wizard Field, 4,000' Sand Data *(Copyright © 1992, SPE, from paper 22350[1])*

Gross Area, A (acres)	2664
Initial Pressure, P_i psia	1930
*Initial Temperature, Ti (°F)	132
Net Thickness, h_{avg} (ft)	34
Initial Oil Saturation, S_i (%)	80
Permeability, K_{avg} (md)	345
Porosity, ϕ (fraction)	0.321
*Oil gravity, γ_o °API	37.2
*Gas Gravity, γ_g (Air=1.0)	0.673
*Bubble Point Pressure P_b (psia)	1616
*Solution Gas-Oil Ratio @ P_b, R_s (SCF/STB)	530
Oil Formation Volume Factor, P_b, B_o (RB/STB)	1.295
Oil Viscosity @ P_b, μ_o (cp)	0.819
Water Viscosity μ_w (cp)	0.547
Oil Compressibility, C_o (psi^{-1})	18.2E-6
Formation Compressibility, C_f (psi^{-1})	3.2E-6
Original Oil-In-Place, N_i (MMSTB)	55.477

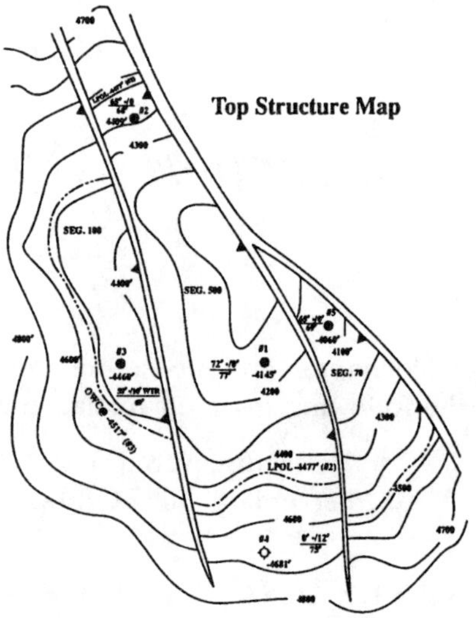

Top Structure Map

***Parameters used for PVT Data Correlations**

FIGURE 10–2. Wizard Field, 4,500' Sand Data *(Copyright © 1992, SPE, from paper 22350[1])*

Gross Area, A (acres)	2010
Initial Pressure, P_i psia	2205
*Initial Temperature, Ti (°F)	141
Net Thickness, h_{avg} (ft)	57
Initial Oil Saturation, S_i (%)	78
Permeability, K_{avg} (md)	304
Porosity, ϕ (fraction)	0.315
*Oil gravity, γ_o °API	35.5
*Gas Gravity, γ_g (Air=1.0)	0.664
*Bubble Point Pressure P_b (psia)	1745
*Solution Gas-Oil Ratio @ P_b, R_s (SCF/STB)	580
Oil Formation Volume Factor, P_b, B_o (RB/STB)	1.322
Oil Viscosity @ P_b, μ_o (cp)	0.776
Water Viscosity μ_w (cp)	0.502
Oil Compressibility, C_o (psi^{-1})	18.2E-6
Formation Compressibility, C_f (psi^{-1})	3.2E-6
Original Oil-In-Place, N_i (MMSTB)	101.988

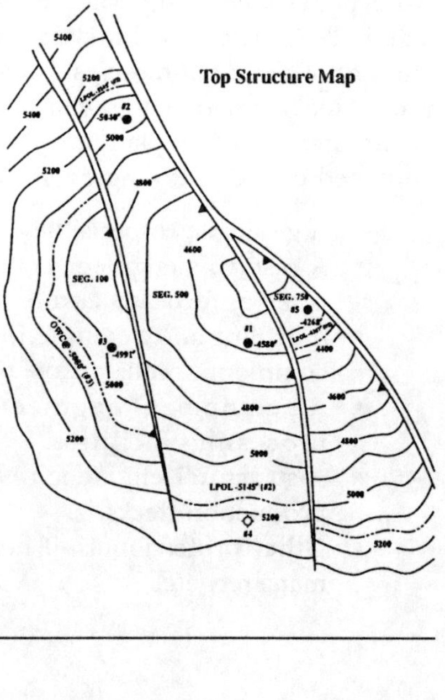

Top Structure Map

***Parameters used for PVT Data Correlations**

The reservoir engineer with an overall knowledge of reservoir management was elected as team leader by the team members.

This case illustrates the application of the reservoir management process/methodology described in Chapter 3.

Development And Depletion Strategy

The input of all disciplines, mutual understanding, and interdiscipline communication was the key to developing a successful optimum plan. The team needed to address the following main questions in order to come up with an economically viable development and depletion strategy:

1. Recovery scheme—natural depletion or natural depletion augmented by fluid (water or gas) injection?
2. Well spacing—number of wells, platforms, reserves, and economics?

Preliminary data indicated that the reservoir was undersaturated and had the initial pressure several hundred psi above the bubble point. The regional geological data and production experience in this area suggested moderate natural water drive as a potential recovery mechanism in addition to rock and fluid expansion and solution gas drive. However, the possibility of secondary gas cap drive may exist because of relatively thick pays with high porosity and permeability.

Production from the reservoir by primary depletion, as well as waterflooding, was considered. It was also decided that all the wells would be completed in the lower sands with the plug back potential in the upper sands when the lower sands were depleted. It was recognized that selective perforation intervals in both sands would maximize the oil recovery and prevent early high GOR productions.

In order to realistically forecast oil production rates and reserves, a full field reservoir simulation study was necessary. Considering several well spacings for the field development, the simulated production performance results were used to economically optimize the number of wells and platforms.

Reservoir Data

The Wizard Field is characterized by faulted structural traps (see structure maps in Figures 10–1 and 10–2). It consists of two main oil-bearing, highly porous and permeable sandstone formations. The upper 4,000 ft sand is separated by some 500 ft of shale from the lower 4,500 ft sand and both the formations are intersected by sealing faults. The formations consist of interbedded shales and sands; however, the shales do not appear to be continuous. Based upon permeability variation, the upper and lower sands could be subdivided into two and three layers, respectively. The reservoir properties and the data sources are given in Figures 10–1 and 10–2 and Tables 10–1 and 10–2. These data were stored in a comprehensive geosciences-engineering database for future use.

TABLE 10–1. Data Sources *(Copyright © 1992, SPE, from paper 22350[1])*

Data	Source
Structure and Isopach Maps	Seismic surveys, revised with well log information
Reservoir Pressure and Temperature	Drill stem test on Well #1
Porosity	Well logs (Sonic, FDC-CNL, GR, ILD, etc.) from Wells 1, 2, 3, 4, and 5, and conventional cores from Well #2
Permeability	Conventional cores from Well #2
Fluid Saturations	Well logs from Wells 1, 2, 3, 4, and 5
PVT Properties	Correlation parameters (see Figures 10–1 and 10–2)
Relative Permeability	Correlation parameters (see Table 10–2)

TABLE 10–2. Layer Data *(Copyright © 1992, SPE, from paper 22350[1])*

Sand Tops, Sub-Surface Feet

		Well #				
Reservoir	**Layer**	**1**	**2**	**3**	**4**	**5**
4000'	A	4145	4409	4460	4681	4060
Sand	B	4180	4442	4491	4714	4095
4500'	A	4580	5040	4991	5230	4268
Sand	B	4616	5071	5026	5261	4300
	C	4679	5120	5093	5320	4344

Net Sand Thickness, Feet

		Well #					
Reservoir	**Layer**	**1**	**2**	**3**	**4**	**5**	**Average**
4000'	A	14	11	11	13	12	12
Sand	B	22	22	22	23	20	22
	Average	36	33	33	36	32	34
4500'	A	14	11	14	13	12	13
Sand	B	27	21	27	26	18	24
	C	22	19	25	16	19	20
	Average	63	51	66	55	49	57

TABLE 10–2. Layer Data (continued)

Porosity, Fraction

Reservoir	Layer	Well # 1	2	3	4	5	Average
4000'	A	0.324	0.314	0.325	0.318	0.328	0.322
Sand	B	0.321	0.311	0.327	0.320	0.324	0.321
	Average	0.323	0.313	0.326	0.319	0.326	0.321
4500'	A	0.322	0.316	0.317	0.318	0.312	0.317
Sand	B	0.310	0.317	0.312	0.325	0.312	0.315
	C	0.308	0.317	0.314	0.320	0.310	0.314
	Average	0.313	0.317	0.314	0.321	0.311	0.315

Reservoir Permeability (md)

Reservoir	Layer	Well # 1	2	3	4	5	Average
4000'	A	397	315	208	245	323	298
Sand	B	502	451	290	306	417	393
	Average	449	383	249	276	370	345
4500'	A	130	224	257	191	305	221
Sand	B	239	168	395	213	426	288
	C	513	267	421	311	503	403
	Average	294	220	358	238	411	304

Relative Permeability—End Point Properties

Parameters	4000' Sand	4500' Sand
S_{oi}, %	20	22
S_{orw}, %	24	25
S_{org}, %	35	34
S_{gc}, %	3	3
k_{row} @ S_{wir}, Frac.	0.65	0.61
k_{rw} @ S_{or}, Frac.	0.29	0.25
k_{rog} @ S_{gc}, Frac.	0.61	0.61
k_{rg} @ S_{org}, Frac.	0.35	0.35
n_w	2.2	2.2
n_{ow}	1.9	1.9
n_{og}	2.3	2.1
n_g	2.1	2.1

Reservoir Modeling

While considering the development of the field by using 40, 80, 120, and 160-acre well spacings, a full-field reservoir simulation model was constructed to predict depletion drive performance. A commercial black-oil simulator was used with 30 columns and 23 rows oriented along the faults. The input data for both sands were obtained from Table 10–2. The reservoir layers were considered continuous and homogeneous throughout the field. The vertical-to-horizontal permeability ratio was chosen to be 1 to 10. PVT and relative permeability data were based upon correlations using the appropriate parameter values given in Figures 10–1, 10–2 and Table 10–2.

Most wells would be initially completed in the lower sands except for the few wells that would encounter oil-bearing column only in the upper sands. The bottom two layers in the lower sand and the bottom layer in the top sand would be perforated. The optimum production rates were derived from a NODAL analysis. Table 10–3 shows the wells scheduled for production for the various well spacings along with the initial production rates for each sand. The model used the following well production limitations:

- Economic rate of 30 STBOPD per well.
- Flowing bottom hole pressure of 600 psia.
- Gas-oil ratio of 20,000 SCF/STB.
- Water cut of 95%.

Production Rates and Reserves Forecasts

A 30 × 23 × 6 grid model was used to predict field-wide primary reservoir performance for 40, 80, 120, and 160-acre well spacings. In addition to the five layers, an impermeable layer was added to isolate the two sands. The simulation runs were made with an active aquifer, which is ten times the reservoir volume. The results show that the larger the spacing, the longer the life with less oil recovery (see Table 10–4 and Figure 10–3).

TABLE 10–3. Drilling, Completion and Production Schedule *(Copyright © 1992, SPE, from paper 22350[1])*

	Primary Depletion					
	Initial Prod. Rate STBOPD		No. of	No. of Wells		Drilling & Completion Time
Case	4000' Sd	4500' Sd	Platforms	Pre-Prod.	During Prod.	Days/Well
40-Ac	1000	1750	2	20	35	39
80-Ac	1250	2000	1	10	20	41
120-Ac	1500	2500	1	8	12	45
160-Ac	1500	2500	1	8	7	45

TABLE 10–4. Economic Evaluation *(Copyright © 1992, SPE, from paper 22350[1])*

	Primary Development				Primary Followed by Waterflood
Parameters	**40-acre**	**80-acre**	**120-acre**	**160-acre**	**80-acre**
Investment, $MM	325	222	202	162	220
Reserves, MMSTBO	40.3	40.2	38.7	38.0	81.3
Economic Life, yrs.	9	11	15	15	22
Payout, yrs.	5.1	4.8	4.7	4.7	4.9
Disc. Cash Flow Return on Inv. (DCFROI), %	29.0	38.8	35.8	40.4	42.7
Net Present Value (NPV), $MM	112	161	144	157	309
Present Worth Index (PWI)	1.63	2.31	2.15	2.49	3.64
Development Costs, $/BO	5.95	3.91	3.62	2.87	2.18

FIGURE 10–3. Wizard Field (Effect of Well Spacing on Recovery) *(Copyright © 1992, SPE, from paper 22350[1])*

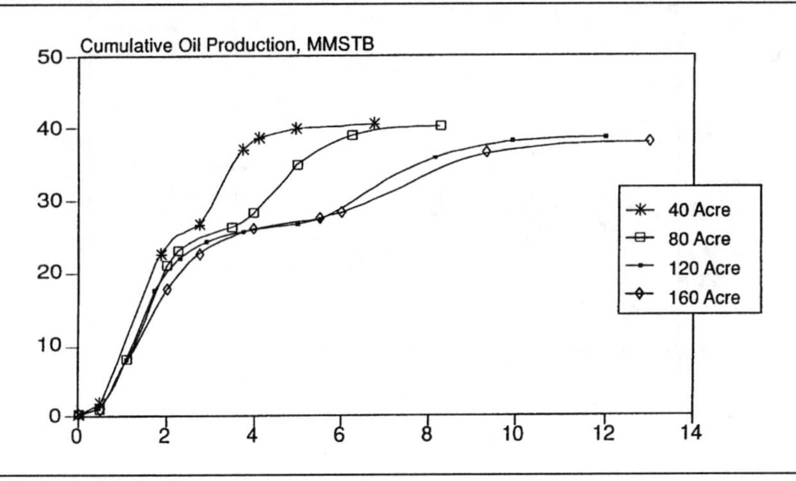

The production performances are shown in Figures 10–4 and 10–5 for the 160-acre well spacing, which turned out to be the optimum spacing case for developing the field.

A sensitivity analysis of the aquifer size on the recovery was made using the 160-acre well spacing. Results, which are shown in Figure 10–6,

FIGURE 10–4. Pressure and Cumulative Production vs. Time (160 Acre Spacing) *(Copyright © 1992, SPE, from paper 22350[1])*

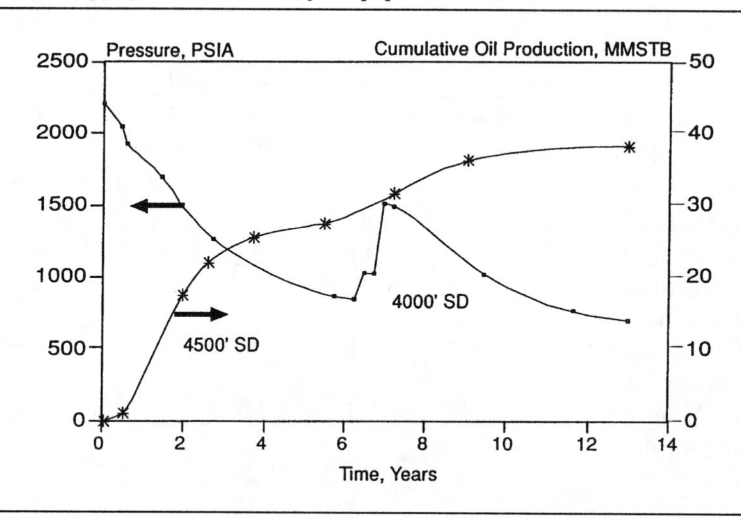

FIGURE 10–5. Water Cut and GOR vs. Time (160 Acre Spacing) *(Copyright © 1992, SPE, from paper 22350[1])*

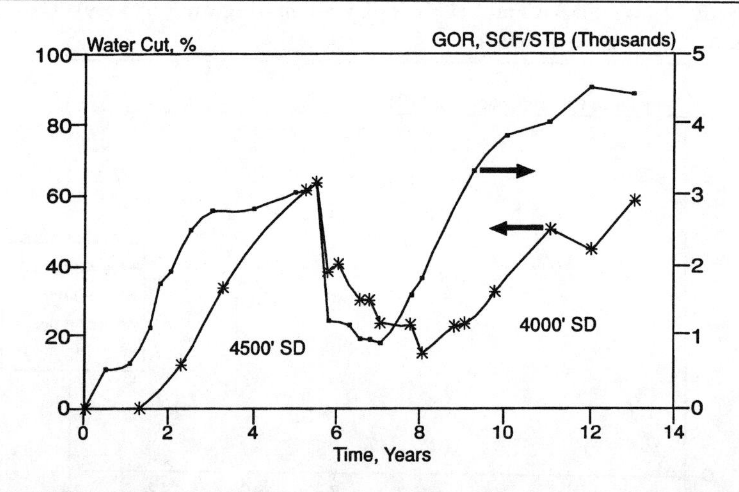

indicate significant recovery without any aquifer support. The oil recovery is 33.816 MM STBO (21.5% OOIP) for no aquifer influx as compared to 38.029 MM STBO (24.2% OOIP) with a 10:1 aquifer size.

An additional run was made using 160-acre well spacing for initiating primary depletion followed by 80-acre, 5-spot infill waterflooding after two years. In this case, 12 water injection, 18 production, and 3 water

FIGURE 10–6. Wizard Field (Effect on Aquifer Size on Rate & Cumulative Production) *(Copyright © 1992, SPE, from paper 22350[1])*

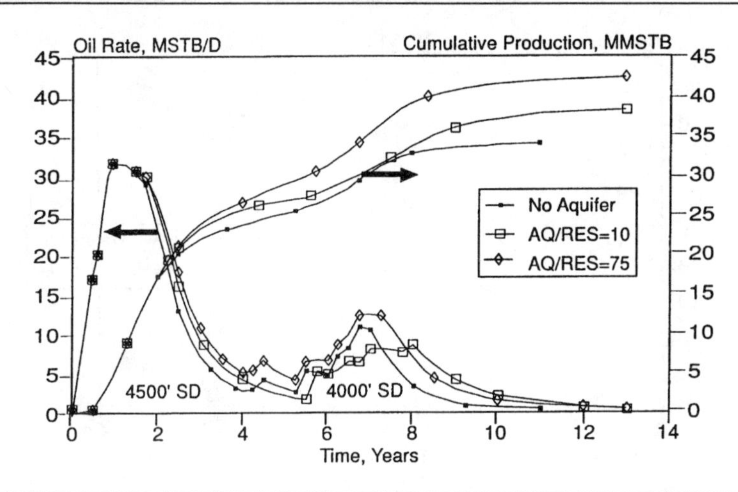

source wells would be needed. Water injection would be initiated at the 4,500 ft sand and recompleted to 4,000 ft sand as all production wells plugged back to this sand. The production performance of this case is compared in Figure 10–7 with the 160-acre depletion case with 10:1 aquifer size. The oil recovery from the waterflood operation was computed to be 81.305 MMSTBO (51.7% OOIP), which more than doubled the primary recovery.

Facilities Planning

The simulated production performance results were used to size platforms, production decks, and surface facilities, etc. Also, drilling, well completions, and production practices requirements were established. Subsequently, estimates of capital requirements and operating expenses were made for economic analyses.

Economic Optimization

Using estimated production, capital, operating expenses, and other financial data (see Table 10–5), economic analyses of the primary development plans with 40, 80, 120, and 160-acre well spacings were made. Table 10–4 shows the evaluation results for these cases. The 160-acre well spacing case with the lowest capital investment, development cost, and payout time and the highest PWI, DCFROI, and next highest NPV offered the economically optimum primary development plan. Even though the 80-acre case yielded the highest NPV ($161 million), the additional

FIGURE 10–7. Wizard Field (Cumulative Production vs. Time) *(Copyright © 1992, SPE, from paper 22350[1])*

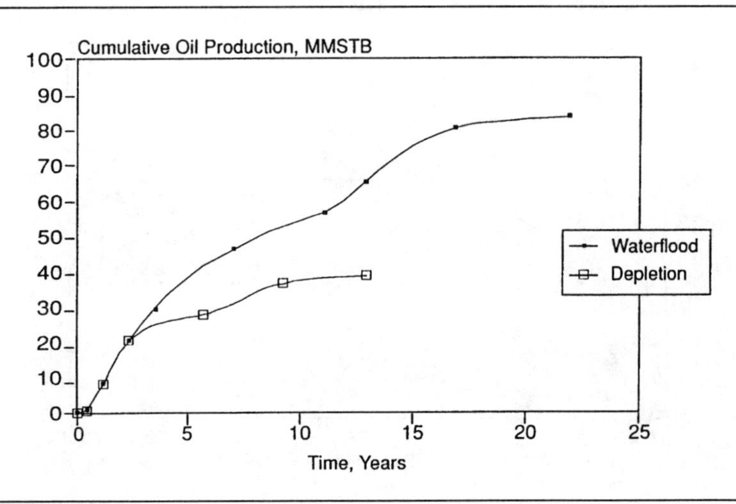

TABLE 10–5. Economic Parameters *(Copyright © 1992, SPE, from paper 22350[1])*

	Primary Development				Primary Followed by Waterflood
Parameters	**40-acre**	**80-acre**	**120-acre**	**160-acre**	**80-acre**
Design Production Rate, BOPD	55,000	50,000	45,000	35,000	40,000
Maximum Water Handling Capacity, BWPD	20,000	10,000	5,000	5,000	80,000
Number of Production Wells	55	30	20	15	18
Number of Injection Wells	–	–	–	–	12
Number of Water Source Wells	–	–	–	–	3
Average Well Depth, Ft.	6,049	6,464	7,670	7,670	6,464
Avg. Drilling & Completion Time, Days/Well	39	41	45	45	41

(continued)

TABLE 10–5. Economic Parameters (continued)

	Primary Development				Primary Followed by Waterflood
Parameters	**40-acre**	**80-acre**	**120-acre**	**160-acre**	**80-acre**
Average Cost per Well, $ Million	2.446	2.552	3.111	3.411	2.248
Production Platform Cost, $ Million	38.019	37.456	36.279	22.937	37.861
Template Structure Cost, $ Million	7.826	4.200	3.484	3.052	4.200
Satellite Well Platform Cost, $ Million	12.941	0	0	0	0
Process Facility Cost, $ Million	19.069	15.148	14.175	12.191	15.026
Engineering/Design/ Project Mgmt., $ Million	19.042	13.011	12.556	9.546	12.434
Base Annual Operating Cost, $ Million	17.088	12.832	11.732	11.810	16.671
Abandonment Cost, $ Million	12.695	8.674	8.371	6.364	8.289
Oil Price, $/STB	20	20	20	20	20
Discount Rate, %	11.5	11.5	11.5	11.5	11.5
Escalation Rate, %	8.0	8.0	8.0	8.0	8.0

capital investment of $60 million over the 160-acre case ($222 million) only gave an incremental NPV return of $4 million.

The 160-acre development case without any aquifer support showed project life of 13 years, DCFROI of 39%, NPV of $140 million, and development costs per barrel of $3.12. Therefore, the 160-acre primary development still looked very attractive.

Results of the economic analysis of the waterflood case (Table 10–4) show the highest oil reserves, DCFROI, NPV, and PWI and the lowest development costs per barrel of oil. Therefore, the early waterflood offers the most economic means to exploit this field. The platform needs to be designed so that the water injection facilities could be installed later (i.e., some deck space would be left for future water injection equipment).

Based on the economic evaluation results, the team recommended to its management the initial 160-acre primary development followed by 80-acre, 5-spot infill waterflooding after two years.

Implementation

After management approval of the project, the next major assignment would be to implement the development plan in order to get the production on stream as soon as possible. A project manager with full authority would be needed to manage the various activities as follows:

1. Design, fabricate, and install production platform and surface facilities. This is usually a critical path for the whole project that requires tremendous efforts and experience to preplan, monitor, and complete the project on time.
2. Develop a drilling program allowing predrilling while the platform is being fabricated.
3. Develop a completion program that includes the tie back of predrilled wells after production platform and process facilities have been installed.
4. Acquire and analyze necessary logging, coring, and initial well test data from the development wells to better define reservoir characterization.
5. Upgrade the reservoir data base and make simulation runs using the latest data to update the depletion strategy and predict reservoir performance.
6. Drill and complete additional production, water injection, and water source wells as warranted.

Monitoring, Surveillance, and Evaluation

An integrated and comprehensive monitoring and surveillance program needs to be initiated at the start of production from the field. Dedicated and coordinated efforts of the various functional groups working on the project are essential. The performance of the reservoir would be monitored as follows:

1. Accounting for daily oil, gas and water productions and water injections by well.
2. Tracing the water-oil contacts at the wells in the vicinity of the aquifer.
3. Systematic and periodic bottomhole pressure testing of selected wells.
4. Periodic measurements of static and flowing casing and tubing pressures.
5. Recording of workovers and results.

The reservoir performance needs to be reviewed periodically to ensure that the plan is working. This can be done effectively by comparing the actual well/reservoir production and pressure behavior with the simulated performance.

SECONDARY AND EOR OPERATED FIELD

Reservoir management plans for a secondary and EOR operated field, the North Ward Estes field, are described in detail in Appendix E[2]. In addition, another secondary and EOR operated field, the Means San Andres Unit, is presented in Appendix F[3]. The first case demonstrates all team members' roles, whereas the second concentrates more on the interaction of geoscientists and engineers.

REFERENCES

1. Satter, A., J. E. Varnon, and M. T. Hoang. "Reservoir Management: Technical Perspective." SPE Paper 22350, SPE International Meeting on Petroleum Engineering, Beijing, China, March 24–27, 1992.
2. Thakur, G. C. "Implementation of a Reservoir Management Program." SPE 20748, SPE Annual Tech. Conf. and Exhib., New Orleans, LA, Sept. 23–24, 1990.
3. Stiles, L. H. and J. B. Magruder. "Reservoir Management in the Means San Andres Unit," *Jour. Pet. Tech.* (April 1992): 469–475.

CHAPTER 11

▼ ▼ ▼

Conclusions

We are not claiming to reinvent the wheel; rather we have packaged the available information in a manner that will benefit geoscientists, engineers, and others involved in reservoir management.

We hope that the objective of this book—to provide readers with a broader and better understanding of the practical approach to reservoir management utilizing multidisciplinary, integrated teams—has been fulfilled. Based upon this knowledge, they will have gained insight into approaching the reservoir management task from a new point of view. However, the real test to the success of this book will be whether the readers do their job differently as a result of this book.

Our further thoughts and perceptions concerning reservoir management follow.

THE STATE OF THE ART

Reservoir management has advanced tremendously during the past 30 years. The techniques and tools are better, reservoir characterization has improved, and automation using mainframe and personal computers has helped data processing and management.

The synergism provided by the interaction between geoscience and engineering has been quite successful, and the reservoir management team concept involving relevant functions is becoming more popular. Team members are beginning to work more like a "well-coordinated basketball team" rather than "a relay team."

It is believed that using an integrated approach to reservoir management along with the latest technological advances will allow companies to extract the maximum economic recovery during the life of an oil or a gas field. It can add years of recovery to the life of a reservoir.

Because production in all fields declines over the years, innovations that prolong cost-effective recovery should be of global interest.

IMPORTANCE OF INTEGRATIVE RESERVOIR MANAGEMENT

As previously described, the modern reservoir management process involves goal setting, planning, implementing, monitoring, evaluating, and revising plans. Setting a reservoir management strategy requires knowledge of the business, political, and environmental climate. Formulating a comprehensive management plan involves depletion and development strategies, data acquisition and analyses, geological and numerical model studies, production and reserves forecasts, facilities requirements, economic optimization, and management approval. Implementing the plan requires management support, field personnel commitment, and multidisciplinary, integrated teamwork. Success of the project depends upon careful monitoring/surveillance and thorough, ongoing evaluation of its performance. If the actual behavior of the project does not agree with the expected performance, the original plan needs to be revised, and the cycle (implementing, monitoring, and evaluating) should be reactivated.

It is a well-known fact that reservoir studies are more effective when geoscientists and engineers work together, and reservoir management plans are more productive if all functional groups are involved. Thus, it is essential to have an integrated reservoir management approach for maximizing the economic recovery from a reservoir.

CURRENT CHALLENGES AND AREAS OF FURTHER WORK

The current challenge primarily concentrates on the need (A) for improved definition of reservoir characteristics, (B) to track the movement of fluids through the reservoir, and (c) to control that movement. The problem is not only in our ability to displace oil, but also to contact a major portion of the oil. This clearly implies that we have to develop technology further in order to improve volumetric sweep efficiency in the reservoir.

OUTLOOK AND THE NEXT STEP

In the years ahead, even more attention will be given to integrated reservoir management. The areas that will play an increasing role are:

- *Team Effort*
  Economic recovery from a reservoir can be maximized by an integrated group effort. Decisions pertaining to a field will not be

made by a single person but by a team that will consider the entire field (i.e., reservoir, wellbores and surface facilities, in addition to economics). A team effort involving people from various functional areas will become necessary for the development and implementation of a successful reservoir management program.

- *Role of Geophysics in Reservoir Management*
  Any reservoir model must provide a description of the reservoir that correctly accounts for spatial variation and continuity of porosity, permeability, and fluid saturations. An integrated geoscience and engineering model provides information about the likely fluid flow paths and an information management system for better surveillance and monitoring of a reservoir.

  Today, geophysics is beginning to play a key role in reservoir development, production, and EOR projects. People are beginning to realize the value of geophysics for reservoir description and monitoring of projects.

  Core analyses and well logs provide the detailed data about the immediate vicinity of the well; however, the question of how to interpolate these data between wells always arises. Reservoir geophysics (seismic reflection) can provide this information, but it requires closer interaction between engineers, geologists, and geophysicists.

  It is anticipated that geophysicists will be involved throughout the life of a reservoir. During the initial phase of development, they will play a key role in identifying key reservoir features and developing geologic models. During secondary and EOR projects, they will help monitor the fluid movement through the reservoir.

- *Storing and Retrieval of Data*
  Storing and retrieval of data during reservoir life cycle poses a major challenge today. The industry is poised to develop technology to create a seamless flow of geoscience and engineering data from heterogeneous hardware and database systems. The goal is to provide user access to multidisciplinary information from a common platform through a common user interface.

- *Integrated Software*
  A major breakthrough in reservoir modeling has occurred with the advent of integrated geoscience and engineering software. However, the challenge is now to affirm the effectiveness of the software in real life situations utilizing multidisciplinary groups working together.

APPENDIX A

▼ ▼ ▼

Waterflood Surveillance Techniques—A Reservoir Management Approach

This appendix describes waterflood surveillance techniques in the light of the reservoir management approach.[1] This approach considers a system consisting of reservoir characterization, fluids and their behavior in the reservoir, creation and operation of wells, and surface processing of the fluids. These are interrelated parts of a unified system. The function of reservoir management in waterflood surveillance is to provide facts, information, and knowledge necessary to control operations and to obtain the maximum possible economic recovery from a reservoir.

Guidelines for waterflood management should include information on (1) reservoir characterization, (2) estimation of pay areas containing recoverable oil, (3) analysis of pattern performance, (4) data gathering, (5) well testing and reservoir pressure monitoring, and (6) well information database.

Today, sufficient performance history is available that surveillance techniques can be documented in detail. This appendix highlights waterflooding in light of practical reservoir management practices. Case studies that illustrate the best surveillance practices are referenced.

KEY FACTORS IN WATERFLOODING SURVEILLANCE

Key monitoring points in the traditional waterflood cycle[1] are described in Figure A–1. In the past, attention was focused mainly on reservoir performance. However, with the application of the reservoir management approach, it has become industry practice to include wells, facilities, water system, and operating conditions in surveillance programs.

FIGURE A–1. Waterflood Cycle *(Copyright © 1974, SPE, from* JPT, *December 1988[1])*

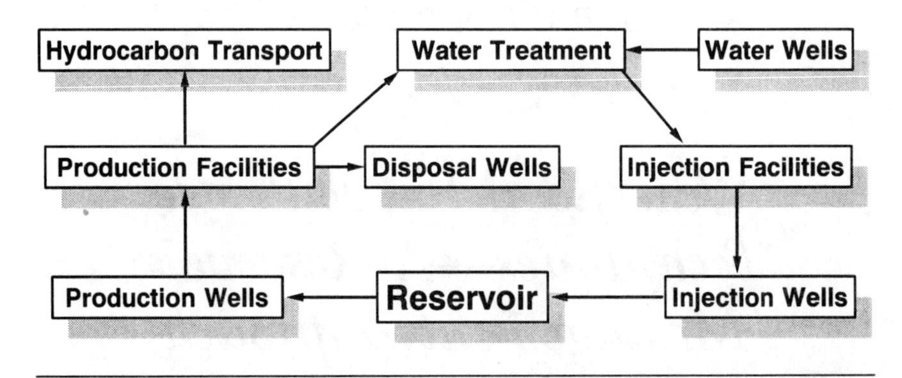

It is important to consider the following items in the design and implementation of a comprehensive waterflood surveillance program (see Table A–1).[2]

- Accurate and detailed reservoir description.
- Reservoir performance and ways to estimate sweep efficiency and oil recovery at various stages of depletion.
- Injection/production wells and their rates, pressures, and fluid profiles.
- Water quality and treating.
- Maintenance and performance of facilities.
- Monthly comparison of actual and theoretical performance to monitor waterflood behavior and effectiveness.
- Reservoir-management information system and performance control (accurate per-well performance data).
- Diagnosis of existing/potential problems and their solutions.
- Economic surveillance.

RESERVOIR CHARACTERIZATION AND PERFORMANCE MONITORING

- *Physical characteristics of the reservoir*
  Reservoir characteristics must be defined: permeability, porosity, thickness, areal and vertical variations, areal and vertical distributions of oil saturation, gas/oil and oil/water contacts, anisotropy (oriented fracture system or directional permeability), in-situ stress, reservoir continuity, vertical flow conductivity, and portion of pay containing the bulk of recoverable oil. To manage a

TABLE A–1. Waterflood Surveillance *(Copyright © 1991, SPE, from* JPT, *October 1991*[2]*)*

Reservoir	Wells	Facilities	Water System
Pressure	Perforations	Production/ injection	Water quality
Rates	Production/injection logging	Monitoring equipment	Presence of corrosive dissolved gases, minerals, bacterial growth, dissolved solids, and suspended solids
Volumes	Injected water in target zone		Ion analysis
Cuts	Tracer		pH
Fluid samples	Tagging fill		Corrosivity
Hall plots	Cement integrity		Oil content
Fluid drift	Downhole equipment		Iron sulfide
Pattern balancing	Wellbore fractures		On-site or laboratory analysis
Pattern realignment	Formation damage		Data gathering on source and injection wells and injection system
	Perforation plugging		
	Pumped-off condition		

FIGURE A–2. Geologic Concepts *(Copyright © 1974, SPE, from* JPT, *June 1974*[3]*)*

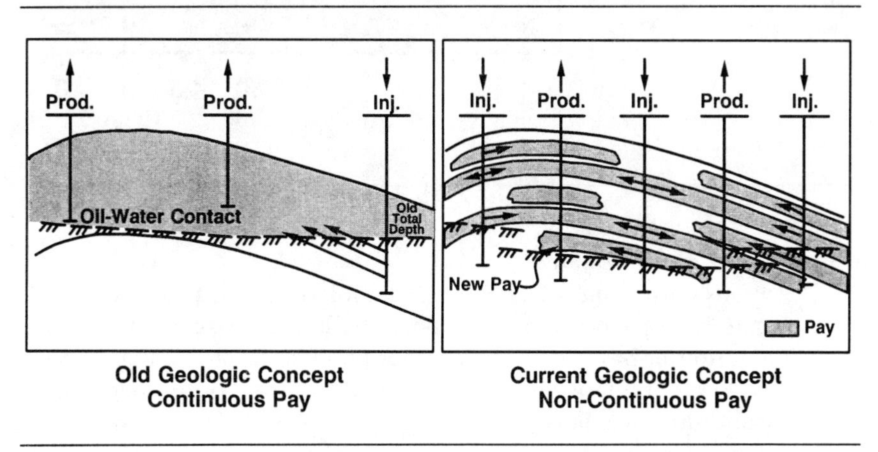

waterflood accurately, detailed knowledge of the reservoir architecture also is necessary. Figures A–2 and A–3 show some examples of geological characterization, involving changing geological concepts and zonation.[3,4]

FIGURE A–3. Type Log for North Ward Estes Field *(Copyright © 1990, SPE, from paper 20748[4])*

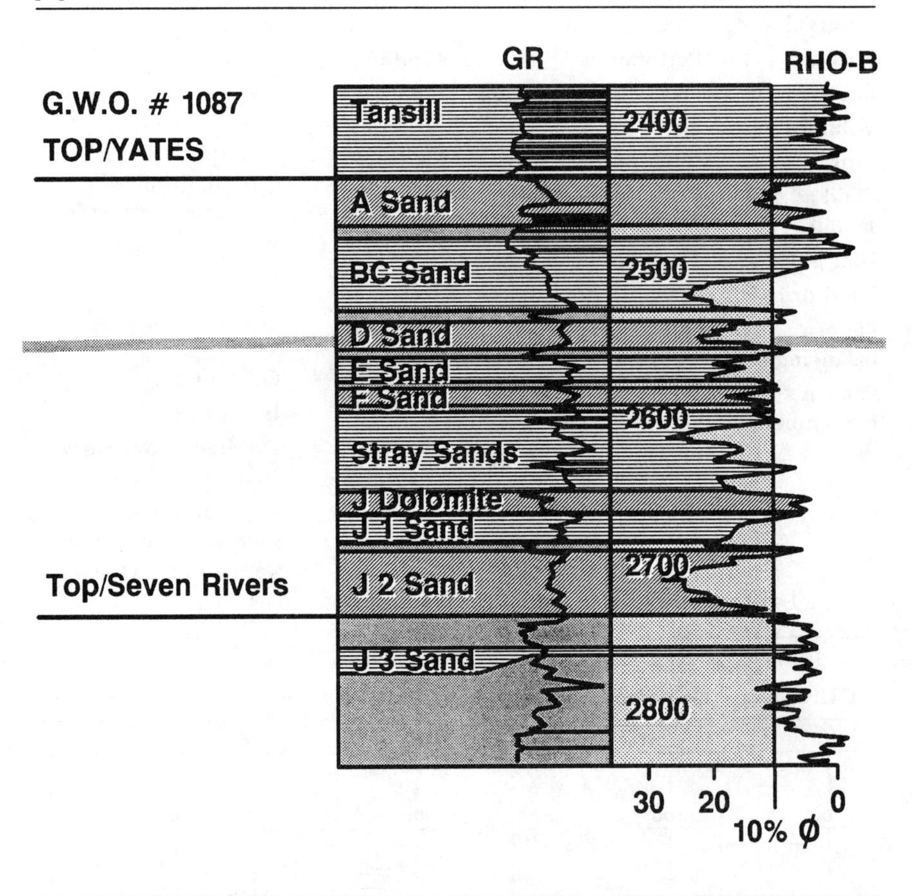

- *Primary performance*
  Wells indicating relatively high cumulative production may indicate high permeability and porosity, higher pay-zone thickness, or another pay zone. On the other hand, wells indicating relatively low cumulative production may indicate poor mechanical condition, wellbore skin damage, or isolated pay intervals.
- *Production curves*
  Percent oil cut in the produced stream (log scale) vs. cumulative recovery during secondary performance may result in an estimate of future recovery or may indicate improvement in the waterflood performance as a result of more uniform injection profile. Figure A–4 illustrates the performance curve of a typical successful waterflood, and Figure A–5 illustrates various examples of waterfloods.[2]

FIGURE A–4. Typical Successful Waterflood Performance *(Copyright © 1991, SPE, from* JPT, *October 1991*[2]*)*

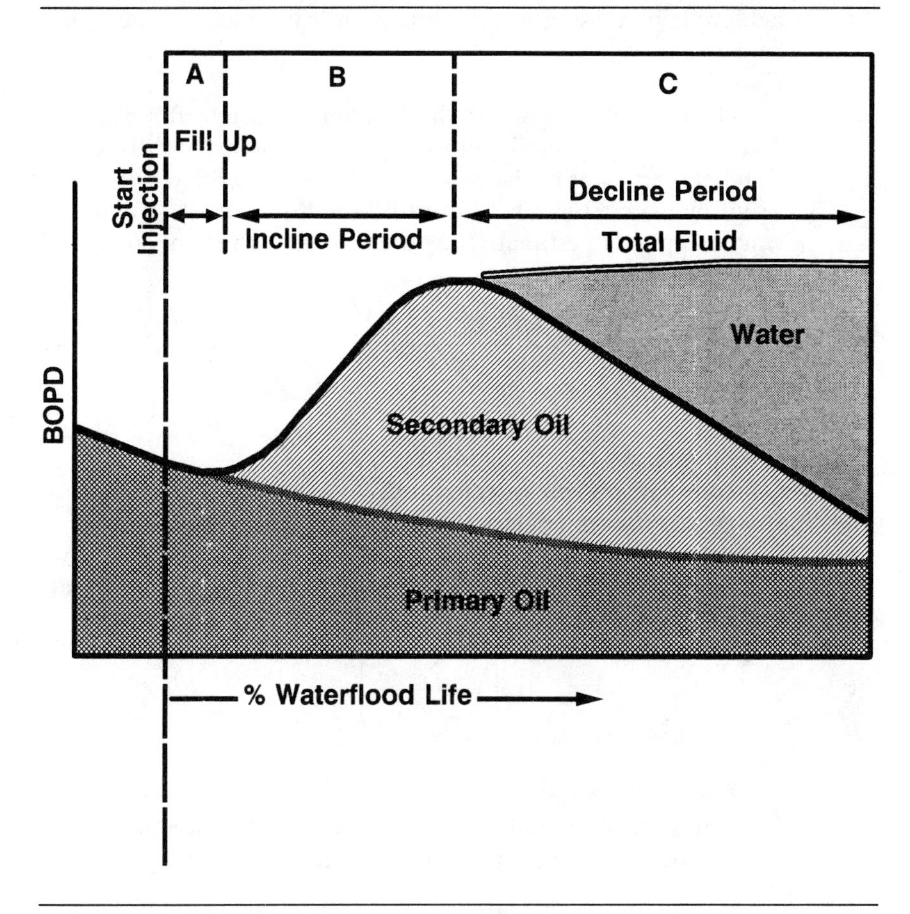

FIGURE A–5. Cumulative Injection vs. Cumulative Total Fluid and Cumulative Oil *(Copyright © 1991, SPE, from* JPT, *October 1991*[2]*)*

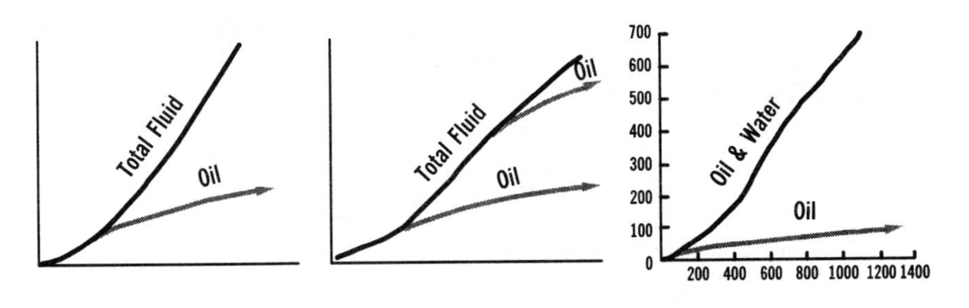

- *Gas-oil-ratio*
  Decreasing gas-oil-ratio's indicate that fluid fill-up is being achieved. Increasing gas-oil-ratio's indicate that voidage is not replaced by injection.
- *Flood front map*
  This pictorial display shows the location of various flood fronts. The maps, often called "bubble maps," allow visual differentiation between areas of the reservoirs that have and have not been swept by injected water.[5] Before fill-up, Equations A–1 and A–2 can be applied to estimate the outer radius of the banked oil and the water-bank radius.

$$r_{ob} = \left(\frac{5.615\, I_{cw} E}{\pi \phi h S_g} \right)^{1/2} \quad \text{(A–1)}$$

where
r_{ob} = outer radius of the banked oil, ft
I_{cw} = cumulative water injected, bbl
S_g = gas saturation at start of injection, fraction
E = layer injection efficiency (fraction of water volume that enters the layer where effective waterflood is taking place)
h = thickness, ft

$$r_{wb} = r_{ob} \left(\frac{S_g}{\overline{S}_{wbt} - S_{iw}} \right)^{1/2} \quad \text{(A–2)}$$

where:
r_{wb} = water bank radius, ft
$\overline{S}_{wbt}$ = average water saturation behind front, fraction
S_{iw} = connate water saturation, fraction

If zones are correlative from well to well and if limited vertical communications exist, then the bubble map can be drawn for each zone. The bubble map can be used to identify areas that are not flooded and areas with infill drilling opportunities.

- *X-plot*
  Because extrapolation of past performance on the graph of water cut vs. cumulative oil is often complicated, a method was devised to plot recovery factor vs. X that yielded a straight line.[6,7] X was defined as

$$X = \ln\left(\frac{1}{f_w} - 1 \right) - \frac{1}{f_w} \quad \text{(A–3)}$$

where f_w = fractional water cut.

This method is more general than the conventional plot of water cut vs. cumulative oil. Both of these methods are more applicable when the water cut exceeds 0.75.

- *Hall plot*[8]

This technique, used to analyze injection-well data, is based on a plot of cumulative pressure vs. cumulative injection. It can provide a wealth of information regarding the characteristics of an injection well, as shown in Figure A–6.

Early in the life of an injection well, the water-zone radius will increase with time, causing the slope to concave upward, as shown by Segment ab in Figure A–6. After fillup, Line bA indicates stable or normal injection. An increasing slope that is concave upward generally indicates a positive skin or poor water quality (Line D). Similar slopes may occur if a well treatment is designed to improve effective volumetric sweep. In this case, however, the slope will first increase and then stay constant. Line B indicates a decreasing slope, which indicates negative skin or injection above parting pressure. The injection under the latter condition can be verified by running step-rate tests. A very low slope value, as shown by Line bC, is an indication of possible channeling or out-of-zone injection.

FIGURE A–6. Typical Hall Plot for Various Conditions

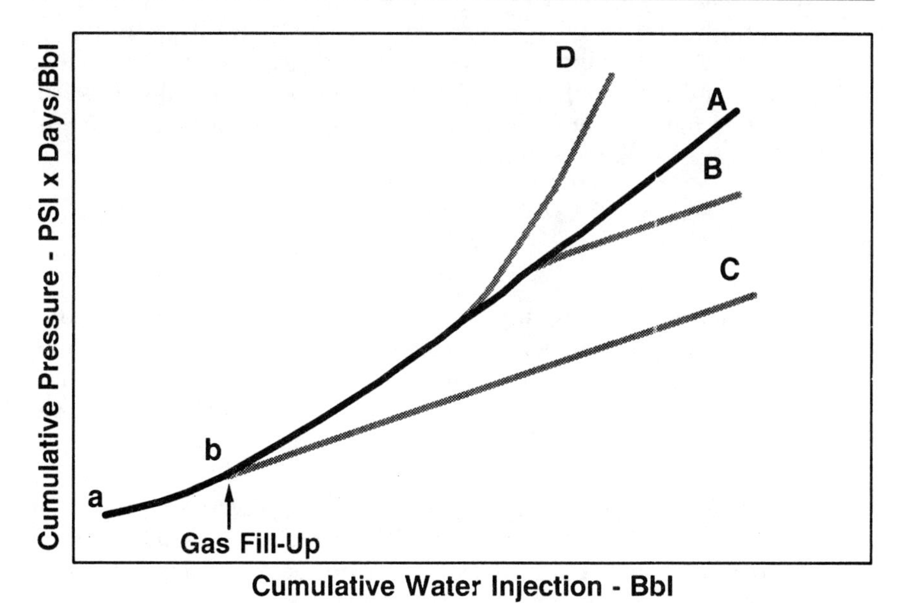

- *Controlled waterflood*
  Maximum profit and recovery would be realized if all wells reached the flood-out point simultaneously. This means producing the largest oil volumes from the wells draining the largest pore volumes. This scenario will result in minimum life with minimum operating expense while realizing maximum oil recovery. Note that if there is a large variation in pore volumes, this task is difficult because each well is allocated a production/injection rate on the basis of pore volume fractions.
- *Pattern balancing*
  Minimizing oil migration across pattern boundaries improves the capture of the mobilized oil and reduces the volume of recycled water. Pattern balancing generally increases sweep efficiency. In addition, realignment of flood patterns in conjunction with pattern balancing provides more opportunities to increase oil recovery. Simple reservoir modeling work can be helpful. For example, the modeling work can identify an unswept area and what kind of improvement in sweep can be obtained by changing producer/injector configuration.
- *Produced water analysis*
  Injected-water breakthrough can be detected by monitoring the chloride content of the produced water if there is a significant difference in the salinities.
- *Injection profile surveys*
  Periodic surveys of injection-well fluid entry profiles can detect formation plugging, injection out of the target zone, thief zones, and underinjected zones. Allocation of injection volumes with data obtained from the profile surveys allows tracking of waterflood histories of each zone.

Wells

- *Problem areas*
  Formation plugging, injection out of the target zone, and nonuniform injection profile caused by stratification are all problem areas. They cause major problems in waterflood operations and low vertical sweep efficiency. Thin, high-permeability layers serve as highly conductive streaks for the injected water.
- *Well completion*
  Condition of the casing and/or cement bond plays an important role in waterflood surveillance. Because of poor cement, water flow can occur behind the casing. Also, openhole injectors and producers and fractured wells with large volume treatments are

not generally desirable. The latter condition may sometimes have a significant negative effect on sweep efficiency. Note that these conditions do not preclude a successful waterflood, but they require more concentrated efforts in surveillance.

- *Injection well testing*
  These tests are conducted to optimize waterflood performance by maximizing pressure differential, minimizing skin damage, ensuring proper distribution of water, and monitoring the extent of fracturing.
- *Quality of producers*
  If producers are converted to injectors, care should be taken to avoid conversion of all poor producers because generally poor producers make poor injectors.
- *Converting producers*
  Producers are converted and high gas producers are shut in to accelerate fill-up time.
- *Backpressure*
  If the producing wells are not pumped off, a backpressure is applied to cause crossflow. As a result, the low-pressure zones may not produce.
- *Changing injection profiles*
  This can be done with selective injection equipment, selective perforating, low-pressure squeeze cementing, acidizing, and thief zone blockage through polymer treatments.
- *Regular cleanouts*
  Regular cleanouts of injectors are necessary, especially if they become plugged with time because of unfiltered water. A Hall-plot analysis may provide some guidance regarding the well cleanout necessities. Regular cleanouts not only increase injectivity, but they also improve the injection profile because the low permeability zones are the first ones to plug, which causes the profile to become less than desirable. Of course, the decision on the frequency of well cleanouts should be made based upon economics.
- *Completion and workover techniques*
  Selection of completion and workover fluids, perforating, and perforation cleaning should be carefully made. This may maximize the completion efficiency of the injection and production wells.
- *Flow regulation*
  Proper utilization of surface and downhole regulator and single/dual-string injector should be made.
- *Profile control*
  Polymer, cementing, chemical, and microbial methods may assist in controlling profiles for improving vertical sweep.

Facilities/Operations

The literature on waterflood surveillance is generally aimed at reservoir performance. Overall project success, however, is often critically affected by daily field operations. While reservoir engineers and geologists play a very important role in reservoir performance and waterflood optimization, facilities/operations staff are concerned with daily management of field operations, information collection, and diagnosis of potential or existing problems (mechanical, electrical, or chemical).

Surface equipment considerations should include a surface gathering and storage system, injection pumps, water distribution systems, metering, water treatment and filtering system, oil/water separation, corrosion and scale, plant and equipment sizing, and handling of separated waste products.

WATER-QUALITY MAINTENANCE

If water quality is not maintained, higher injection pressures are required to sustain the desired injection rates. Also, corrosion problems increase with time when lower-quality water is used. It is important to protect the injection system against corrosion to preserve its physical integrity and to prevent the generation of corrosion products.

Ideally, the water quality should be such that the reservoir does not plug and injectivity is not lost during the life of the flood. However, cost considerations often prohibit the use of such high-quality water. The expense of obtaining and preserving good-quality water must be balanced against the loss of income incurred as a result of decreased oil recovery and increased workover and remedial operations requirements.

Questions are often asked about the determination of acceptable water quality. Tighter formations require better quality water. Sometimes poor quality water can be injected above parting pressures, but injection through fractures could reduce sweep efficiency.

Although it is impossible to predict quantitatively, the minimum water quality required for injection water into a given formation, some attempts have been made and documented in the literature to define injection water-quality requirements from on-site testing.[9] Table A–2 and Figure A–7 describe other considerations regarding water systems.[2]

It is interesting to note that incompatible barium and sulfate waters were injected into the Baylor County Waterflood Unit No. 1.[10] Produced and makeup waters were not mixed; instead, they were injected through two separate systems and into separate wells. No problems were encountered through mixing and precipitation in the reservoir, nor were any problems in the producing system experienced.

TABLE A–2. Water System *(Copyright © 1991, SPE, from* JPT, *October 1991*[2]*)*

Water source (produced, source well, separated)

Water-quality requirements:

- source water—produced-water compatibility
- injection water—reservoir/rock interaction (clay swelling)
- dispersed oil
- corrosion
- scale
- bacteria (sulfate reducing, oxidize soluble iron in water, produce organic acids)
- marine organisms
- pH control
- corrosive dissolved gases (CO_2, H_2S, O_2)
- total dissolved and suspended solids (iron content, barium sulfate)
- corrosion inhibitors (not sufficiently soluble)
- scale inhibition
- closed vs. open injection facilities
- treatment program to ensure acceptable water for formation and to minimize corrosion

Other important considerations:

- oil/water separation
- filtration (gathering station, treatment plan, types of filters, wellhead filters and strainers)
- waste treatment
- water-supply wells (solids, corrosion products, bacteria)
- surface water (oxygen, bacteria, marine organisms, suspended inorganic solids)

MONITORING

Reservoir

- Pressure (portable test equipment, fluid-level testing; repeat formation, buildup/fall-off, and step-rate tests; fieldwide pressure surveys to determine pressure gradient for use on balancing injection/production rates).
- Rate (oil, water, gas, water-cut, GOR, well testing—production rates).
- Pattern balancing (voidage control, areal/vertical sweep efficiency using stream-tube models).
- Waterflood pattern realignment.
- Observation/monitoring wells.

FIGURE A–7. Water System *(Copyright © 1991, SPE, from* JPT, *October 1991*[2]*)*

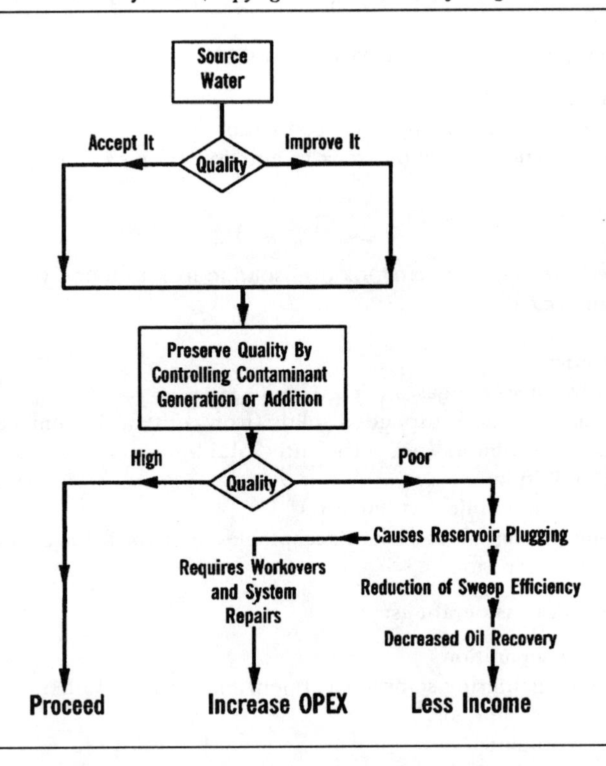

- Reservoir sweep and bypassed oil.
- Fracture communications.
- Thief zones and channels.
- PV injected.
- Gravity underriding and fingering/coning.

Wells

- Production/injection logging (openhole/cased hole, temperature/spinner/tracer).
- Injected water in target zone.
- Hall plots (well plugging/stimulation).
- Tracer (single well/interwell).
- Tagging fill.
- Cement integrity.
- Downhole equipment.
- Surface equipment.

- Wellbore fractures.
- Formation damage/perforation plugging.
- Pumped-off condition.
- Corrosion/scale-inhibition residuals.

Facilities

- Production/injection.
- Monitoring equipment and maintenance.

Water System

- Presence of corrosive dissolved gases (CO_2, H_2S, 0_2); minerals; bacterial growth; dissolved solids; suspended solids, concentration and composition; ion analysis; pH.
- Corrosivity (corrosion coupons and corrosion rate monitoring), oil content (dispersed or emulsified oil in water), and iron sulfide.
- On-site or laboratory analysis.
- Data gathering at the water-source well, water-injection wells, and several points in the injection system.

CASE HISTORIES

Means San Andres Unit

A comprehensive surveillance program used at the Means San Andres Unit was documented by Stiles.[11] A detailed surveillance program was developed and implemented in 1975. It included monitoring production (oil, gas, and water) and water injection, controlling injection pressures with step-rate tests, pattern balancing with computer balance program, running injection profiles to ensure optimum distribution, selecting specific production profiles, and choosing fluid levels to ensure pump-off of producing wells.

The following were implemented during tertiary recovery (water-alternating-gas injection), but they also apply to waterflood surveillance.

- Areal flood balancing (optimizing the arrival of flood fronts at producers) performed by annual pressure-falloff tests in each injection and computer balancing programs.
- Production/injection monitoring.
- Data acquisition and monitoring.
- Pattern performance monitoring to maximize oil recovery and flood efficiency by evaluating and optimizing the performance of each pattern.

- Optimization (it must be dynamic and sensitive to changes in performance, technology, and economics).
- Vertical conformance monitoring to optimize vertical sweep efficiency while minimizing out-of-zone injection. Several cross-sections were constructed for each pattern to ensure completions in all the floodable pay. Annual profiles were run on all injection wells. For each profile, casing or packer leaks were identified, out-of-zone injection was identified, and zonal injection from profile was compared with porosity-feet profiles.

The main objective of an injection survey is to provide a means of monitoring the injection water so that efforts can be made to ensure that injection rates conform with zonal porosity-thickness. These efforts have paid substantial dividends in increased vertical sweep and ultimate recovery.

South Hobbs Unit

Production at the South Hobbs Unit increased almost 100% within a year.[12] The reason for boosted performance was an aggressive program of well surveillance, general record keeping, and remedial action. Five operational efforts had positive effects on production.

- Lift capacity of a number of wells was increased so that a pumped-off condition could be maintained.
- Operating pressures in the satellite battery separators were reduced, thereby reducing backpressure through the flowlines back to the well.
- Adverse effects of scale accumulation were decreased by remedial and preventive measures.
- Injection pressures just under the parting pressure were maintained.
- Tracer surveys were run to ensure that fluid is entering into the proper zones in the right amounts.

West Yellow Creek Field

This case study describes the importance of a thorough, well-organized reservoir surveillance for the West Yellow Creek field.[13] This effort involved many activities, including pressure-falloff tests, a computerized flood balancing program, and a produced-water sampling program.

Ventura Field

Schneider described the role of geological factors on the design and surveillance of waterfloods in the structurally complex reservoirs in the

Ventura field, California.[14] Geologic factors strongly influenced the profiles of injection wells and the responses of producing wells. The waterflood was monitored to determine the cause of injection anomalies and to predict their effect on waterflood response.

JAY/LITTLE ESCAMBIA CREEK FIELD

The application of reservoir management techniques was key to the success of this waterflood.[15–17] Surveillance information and reservoir description data provided new insights into water movement and zonal depletion. Operating decisions based on these data proved to be highly profitable.

Surveillance was used for both the vertical and horizontal conformances. Cased-hole logging, pressure-buildup and production tests, and permeability data from core analysis were used for the vertical conformance surveillance; radioactive tracers, reservoir pressure data, and interference tests were used for the areal surveillance.

To achieve vertical conformance, injection wells were acid-fracture-treated in multiple stages to create connecting vertical fracture systems. Temperature surveys, noise logs, and flowmeters were used for the vertical conformance surveillance. The entire section in the producing wells was opened without acid fracturing to maintain the flexibility of future water production. Flowmeter/gradiomanometer surveys, pressure buildups along with core analysis data, noise logs, and gamma ray logs were used for monitoring. In addition, pulsed-neutron-capture logs tracked edgewater encroachment.

Radioactive tracer data provided a means of determining the source of water breakthrough, which was later confirmed by the interference test performed between the producer and the suspect injection well. On the basis of these results, injection rates were adjusted to minimize trapped oil behind the water fronts.

Wasson Denver Unit

Ghauri et al.[3,18,19] described several innovative techniques to increase this unit's production rates and reserves, including novel geological concepts (Figure A–2), major modifications in flood design, infill drilling, and careful surveillance. The waterflood surveillance incorporated such common techniques as computer-generated analyses of production/injection data, water-bank radii or bubble maps, pressure contour maps, artificial-lift monitoring, and specific items like careful monitoring of the relationship between reservoir withdrawals and the water-injection rate. The latter was monitored on both a unit and individual battery basis.

Based upon an energy balance of injection and withdrawals,

$$B_w i_w = B_o q_o + B_g (R - R_s) q_o + B_w q_w, \tag{A–4}$$

where

B_w = water FVF, RB/STB
i_w = water injection rate, STB/D
B_o = oil FVF, RB/STB
q_o = oil production rate, STB/D
B_g = gas FVF, RB/scf
R = producing GOR, scf/STB
R_s = solution GOR, scf/STB
q_w = water production rate, STB/D

or,

$$q_o = \frac{B_w (i_w - q_w)}{B_o + B_g (R - R_s)} \tag{A–5}$$

With 800-psi PVT data and injection and production rates of 416,000 and 70,000 STB/D water, the oil rate for producing GOR'S's of 700 and 750 scf/STB are 148,000 and 138,000 STB/D, respectively (B_o = 1.213 RB/STB, B_g = 0.003125 RB/scf). The rate model was history matched with actual performance in individual battery areas and then used to investigate the effects of changes in operating policy.

A significant effort was also made to improve the vertical sweep efficiency in both existing and new water-injection wells. Cemented liners were installed in openhole producers that were converted to injection, and the zones to be flooded were selectively perforated. All new producers and injectors were cased through the zones of interest and selectively perforated rather than completed openhole, which had been practiced before. Treating pressures during acid stimulation jobs were kept under formation fracturing pressures to maintain zonal isolation, and injection rates below fracturing pressure were maintained.

REFERENCES

1. Talash, A. W. "An Overview of Waterflood Surveillance and Monitoring," *JPT* (December 1988): 1539–43.
2. Thakur, G. C. "Waterflood Surveillance Techniques—A Reservoir Management Approach," *JPT* (October 1991): 1180–1188.
3. Ghauri, W. K., A. F. Osborne, and W. L. Magnuson. "Changing Concepts in Carbonate Waterflooding—West Texas Denver Unit Project—An Illustrative Example," *JPT* (June 1974): 595–606.

4. Thakur, G. C. "Implementation of a Reservoir Management Program." Paper SPE 20748 presented at the 1990 SPE Annual Technical Conference and Exhibition, New Orleans, September 23–26.
5. Staggs, H. M. "An Objective Approach to Analyzing Waterflood Performance." Paper presented at the Southwestern Petroleum Short Course, Lubbock, Texas, 1980.
6. Ershaghi, I. and O. Omoregie. "A Method for Extrapolation of Cut vs. Recovery Curves," *JPT* (February 1978): 203–204.
7. Ershaghi, I. and D. Abdassah. "A Prediction Technique for Immiscible Processes Using Field Performance Data," *JPT* (April 1984): 664–670.
8. Hall, H. N. "How to Analyze Waterflood Injection Well Performance," *World Oil* (October 1963): 128–130.
9. McCune, C. C. "On-Site Testing to Define Injection-Water Quality Requirements," *JPT* (January 1977): 17–24.
10. Roebuck, I.F., Jr. and L. L. Crain. "Water Flooding a High Water-Cut Strawn Sand Reservoir," *JPT* (August 1964): 845–50.
11. Stiles, L. H. "Reservoir Management in the Means San Andres Unit." Paper SPE 20751 presented at the SPE Annual Technical Conference and Exhibition, New Orleans, Sept. 23–26, 1990.
12. Sloat, B. F. "Measuring Engineering Oil Recovery," *JPT* (January 1991): 8–13.
13. Gordon, S. P. and O. K. Owen. "Surveillance and Performance of an Existing Polymer Flood: A Case History of West Yellow Creek." Paper SPE 8202 presented at the SPE Annual Technical Conference and Exhibition, Las Vegas, Sept. 23–26, 1979.
14. Schneider, J. J. "Geologic Factors in the Design and Surveillance of Waterfloods in the Thick Structurally Complex Reservoirs in the Ventura Field, California." Paper SPE 4049 presented at the SPE Annual Meeting, San Antonio, Oct. 8–11, 1972.
15. Langston, E. P. and J. A. Shirer. "Performance of Jay/LEC Fields Unit Under Mature Waterflood and Early Tertiary Operations," *JPT* (February 1985): 261–68.
16. Langston, E. P., J. A. Shirer, and D. E. Nelson. "Innovative Reservoir Management—Key to Highly Successful Jay/LEC Waterflood," *JPT* (May 1981): 783–91.
17. Shirer, J. A. "Jay-LEC Waterflood Pattern Performs Successfully." Paper SPE 5534 presented at the SPE Annual Technical Conference and Exhibition, Dallas, Sept. 28–Oct. 1, 1975.
18. Ghauri, W. K. "Innovative Engineering Boosts Wasson Denver Unit Reserves," *Pet. Eng.* (December 1974): 26–34.
19. Ghauri, W. K. "Production Technology Experience in a Large Carbonate Waterflood, Denver Unit, Wasson San Andres Field," *JPT* (September 1980): 1493–1502.

APPENDIX B

▼ ▼ ▼

Brassey Oil Field

This case study presents how an aggressive engineering/geoscience team approach provided the development plan for the Brassey oil field.[1] Miscible flooding of the field began only two years after discovery. The team approach combined the knowledge from a geological model, seismically defined pool edges, continuity information from well-test data, and material-balance calculations to predict reservoir volume, areal extent, and continuity on the basis of an integrated reservoir model.

The Brassey oil field, located in northeast British Columbia, produces from the Artex member of the Upper Triassic Charlie Lake formation (see Figures B–1 and B–2). The Artex member is an aeolian sand at a depth of 10,000 ft, and it has an average porosity of 16%, a permeability of 152 md (see Figure B–3), and well capabilities of 2,000 BOPD. The 57° API, volatile, undersaturated oil makes it a good candidate for a miscible flood. The oil viscosity is 0.09 cp at 5,000 psi, with a formation volume factor and a gas-oil ratio of 1.8 RB/STB and 1,433 SCF/STB, respectively.

The Artex member is a porous aeolian sandstone encased by dolomitic and argillaceous siltstones (see Figure B–4). The maximum thickness of the Artex is about 15 ft. Oil within the Artex is stratigraphically trapped. The water saturation is determined as less than 2%. The oil-wet nature of the reservoir has been confirmed by the Amott wettability tests.

Geophysical involvement in the Brassey field began soon after the completion of Well C-34-F (see Figure B–5). This well provided the sonic log necessary to begin a geophysical modeling project. The project's objective was to determine if the porous Artex sand could be observed on seismic data and what effects changing the thickness and porosity of the Artex sand would have on the seismic signature (see Figures B–6 and B–7). Results of the study indicated that the Artex sand could be observed if high-frequency seismic data (above 80-cycles/second or 80-Hz) were recorded and the sand was greater than 7 ft thick.

FIGURE B–1. Brassey Field Location Map *(Copyright © 1992, SPE, from* SPE Reservoir Engineering, *May 1992*[1]*)*

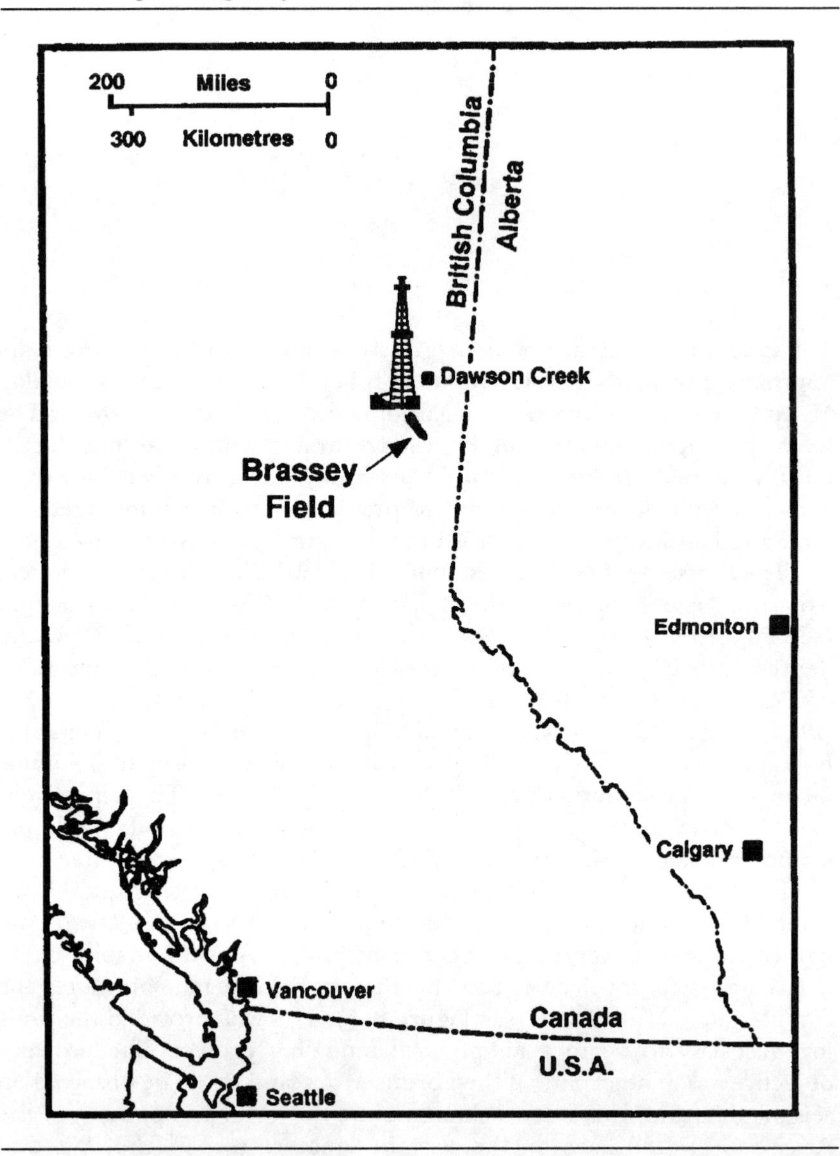

FIGURE B–2. General Stratigraphic Succession in Brassey Field Vicinity *(Copyright © 1992, SPE, from* SPE Reservoir Engineering, *May 1992*[1]*)*

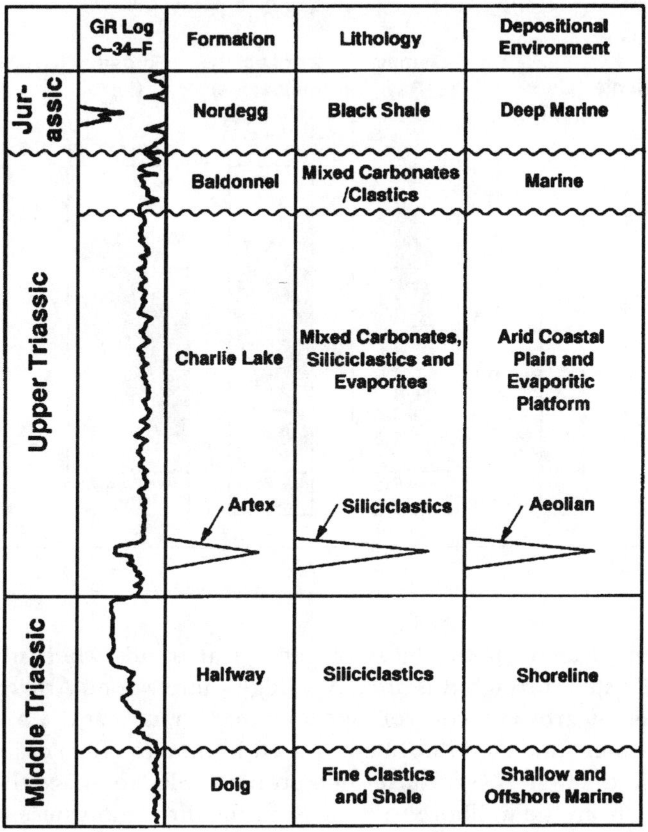

	GR Log c–34–F	Formation	Lithology	Depositional Environment
Jurassic		Nordegg	Black Shale	Deep Marine
Upper Triassic		Baldonnel	Mixed Carbonates /Clastics	Marine
Upper Triassic		Charlie Lake	Mixed Carbonates, Siliciclastics and Evaporites	Arid Coastal Plain and Evaporitic Platform
Upper Triassic		Artex	Siliciclastics	Aeolian
Middle Triassic		Halfway	Siliciclastics	Shoreline
Middle Triassic		Doig	Fine Clastics and Shale	Shallow and Offshore Marine

Based upon the successful completion of a test seismic line tying a productive well with 11 ft thick sand to a dry hole with zero feet of porous sand (Wells c-34-F and d-33-F), a detailed program of three lines (Phase 1) was conducted at the north end of the field (see Figure B–5). The program was successful in tying Well 6-1 which had 8.5 ft of porous Artex sand. As defined by seismic, the 7-ft edge of the pool was mapped.

The important contribution of the first seismic program was realized when a well was whipstocked 722 ft to the east (Well dA-46-F/93-P-10, see Figure B–5) of Well d-46-F/93-P-10 with no reservoir thickness. This well

FIGURE B–3. Artex Member Type Section, Reservoir Parameters and Oriented-Core Data *(Copyright © 1992, SPE, from* SPE Reservoir Engineering, *May 1992*[1]*)*

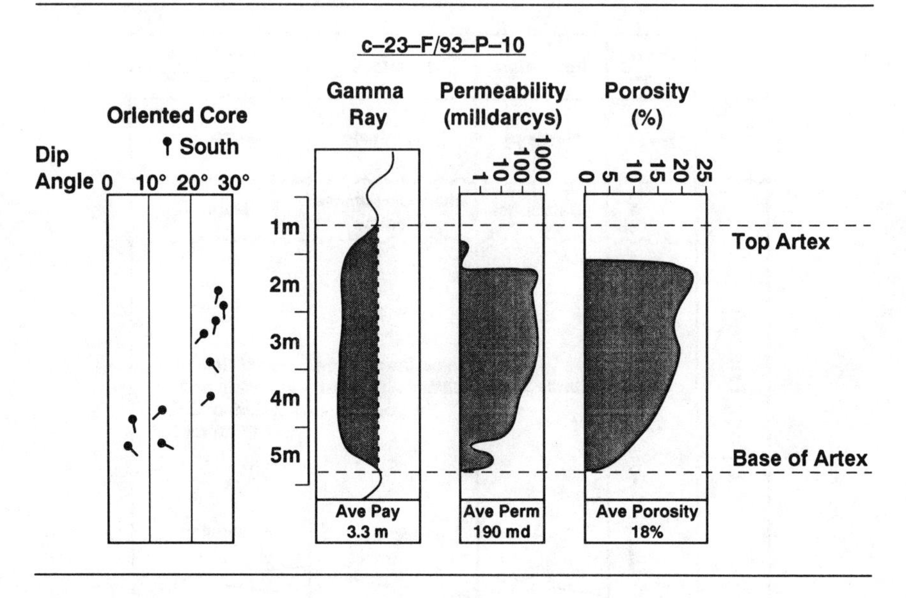

had been located on the basis of geological trends when no seismic control existed through this area. Although there was no Artex reservoir in the well, nearby well control indicated that it was near the Artex pool. Also, a seismic line could not be acquired in time to effect a decision to whipstock the well. A vertical seismic profile (VSP), to image the subsurface away from the wellbore to the east in the direction where the Artex was known to be present, was shot (see Figure B–8).

The offset VSP interpretation showed that the porous Artex sand was about 492 ft to the east (see Figure B–9). The well was then whipstocked 722 ft to the east, where about 8 ft of the porous sand was encountered. The well was later completed with initial production rates of 2,500 BOPD.

Although a seismic line is preferred over an offset VSP because of the amount of the subsurface imaged, the value of the VSP technique is that it can be acquired much faster than with a seismic line (see Figure B–10).

Because Brassey became a candidate for miscible flooding very early in the development program, the wells were located precisely to maximize sweep by drilling at the pool edges (see Figure B–11). Seismic data helped achieve this objective and resulted in the drilling of successful wells.

FIGURE B–4. Artex Member Type Section, Lithology, Sedimentology, and Gamma Ray Profile *(Copyright © 1992, SPE, from* SPE Reservoir Engineering, *May 1992*[1]*)*

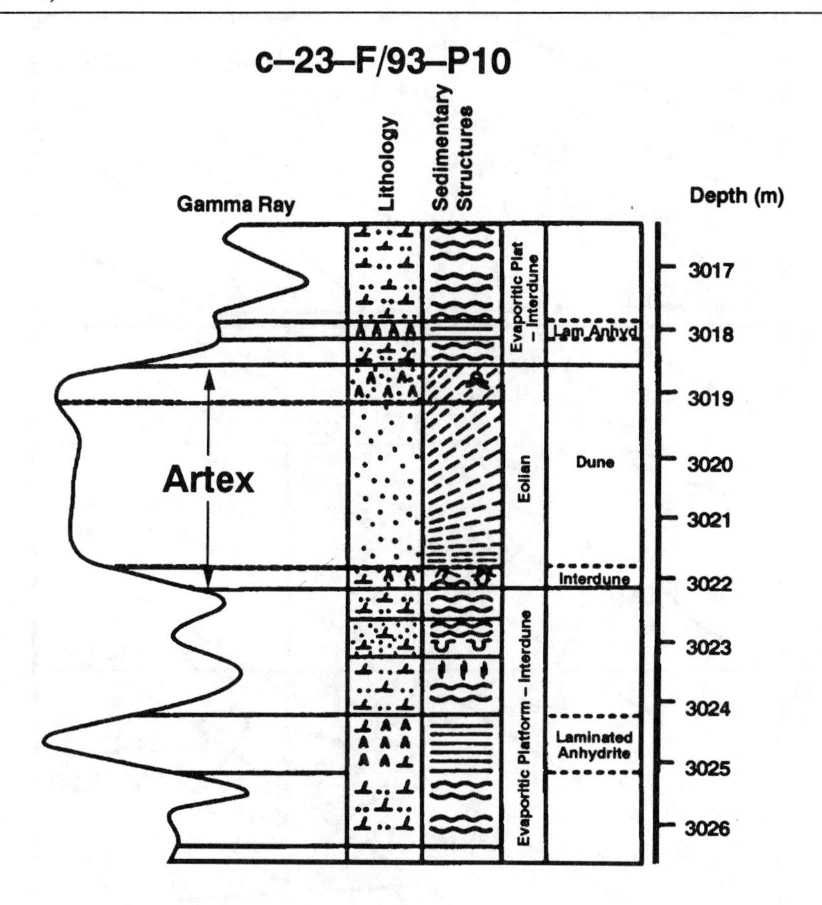

FIGURE B–5. Brassey Field Map Showing Well, Seismic, and Pool Distribution
(Copyright © 1992, SPE, from SPE Reservoir Engineering, *May 1992*[1]*)*

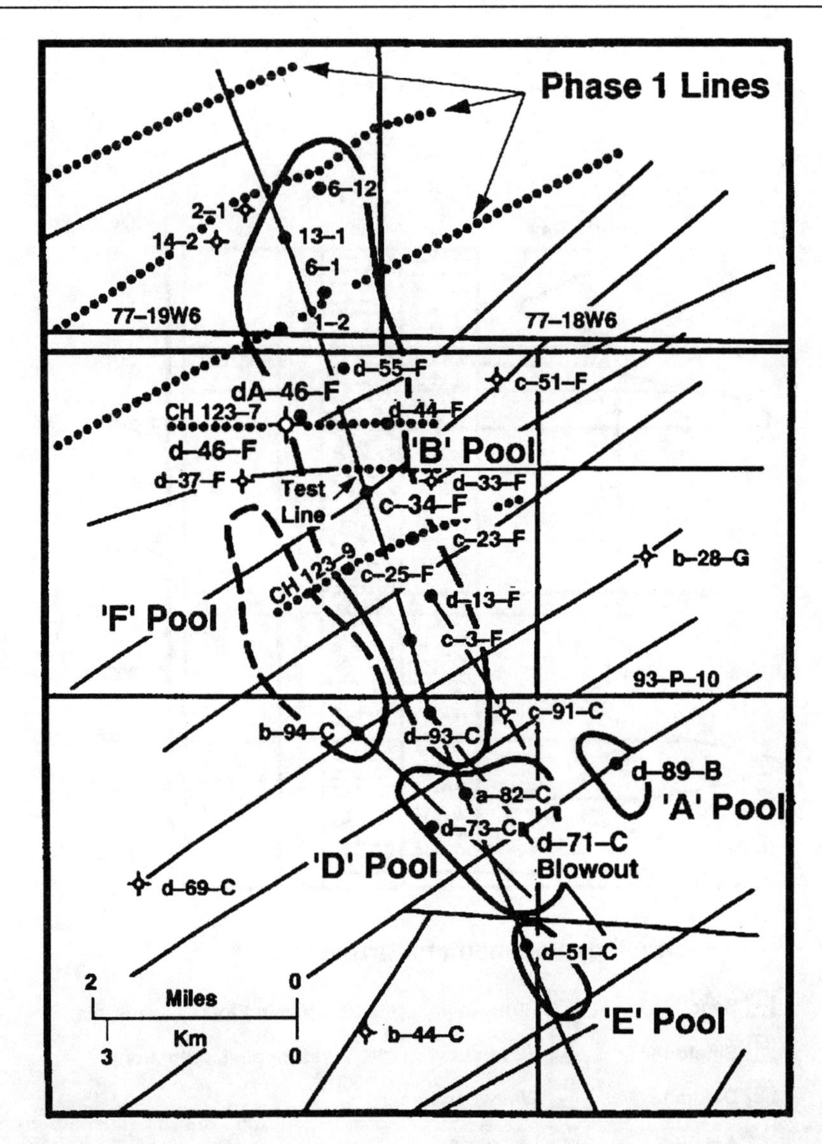

FIGURE B–6. Artex Sand Thickness Model (The 5-m Artex Zone in Well c-34-F Thinned from 5 to 0 m. The Traces Show No Change Until 2 m of Porous Sand is Present. As the Thickness of Porous Sand Increases, the Amplitude Increases.) *(Copyright © 1992, SPE, from SPE Reservoir Engineering, May 1992[1])*

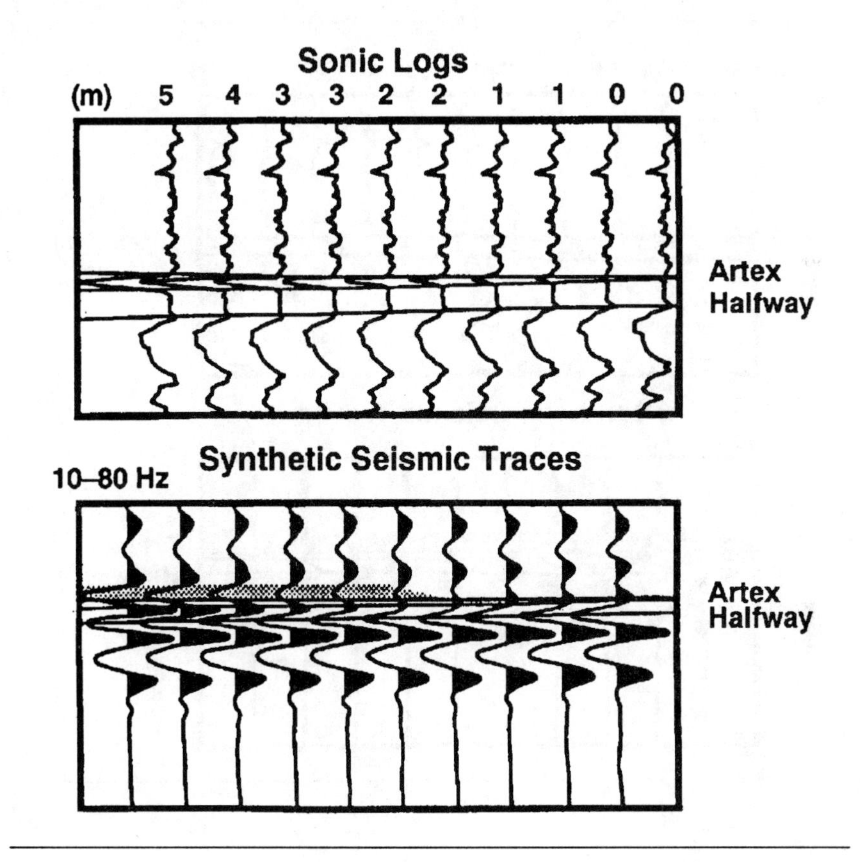

achieve this objective and resulted in the drilling of successful wells.

A geological model was developed before the design of the miscible flood (see Figure B–12). Good core control throughout the field aided in detailed sedimentologic and petrographic interpretations, but the wide well spacing of 0.5 mile provided limited interwell correlations. Figure B–13 shows the Artex member, which consists of an aeolian layer (Layer 1) underlain by an interdune facies (Layer 2). A tight sand formed as a result

FIGURE B–7. Artex Sand Porosity Model (The 5-m Artex Zone from Well c-34-F with Porosity Varied from 3 to 26%. The Resulting Traces Show a Subtle Change Between 3 and 6% Porosity. The Amplitude of the Response Increases as the Porosity Increases) *(Copyright © 1992, SPE, from* SPE Reservoir Engineering, *May 1992*[1]*)*

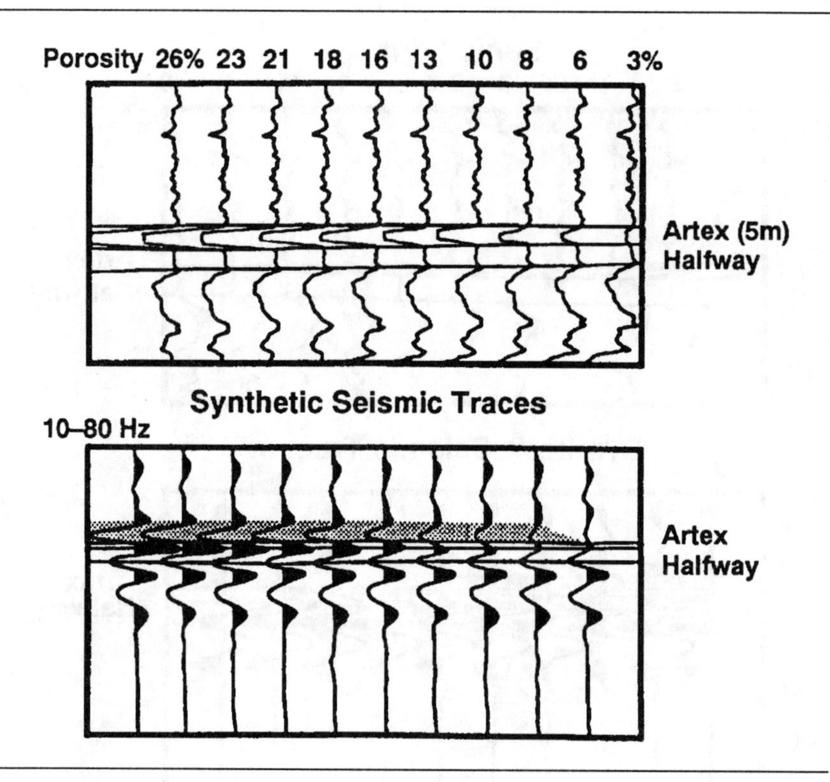

of anhydrite cementation caps Layer 1.

Figure B–12 provides cross-plots of permeability-porosity data for Layers 1 and 2. The interdune layer (No.2) has much lower permeability and porosity because it is subaqueously deposited, aeolian-derived sand. This layer is not considered detrimental to sweep efficiency during EOR because it occurs only at the base of the reservoir and accounts for less than 5% of the total oil volume.

Layer 1 depicts an excellent porosity-permeability relationship. The permeability variation within Layer 1 results from the presence of anhydrite cement. The anhydrite ranges in volume from 2 to 15% in Layer 1, with 25% in the most extreme case capping the reservoir.

In the Artex porous sand, the magnitude of preferred permeability

FIGURE B–8. Offset VSP Schematic (As the Downhole Detector Moves Up the Borehole, Different Subsurface Points Are Imaged) *(Copyright © 1992, SPE, from* SPE Reservoir Engineering, *May 1992[1])*

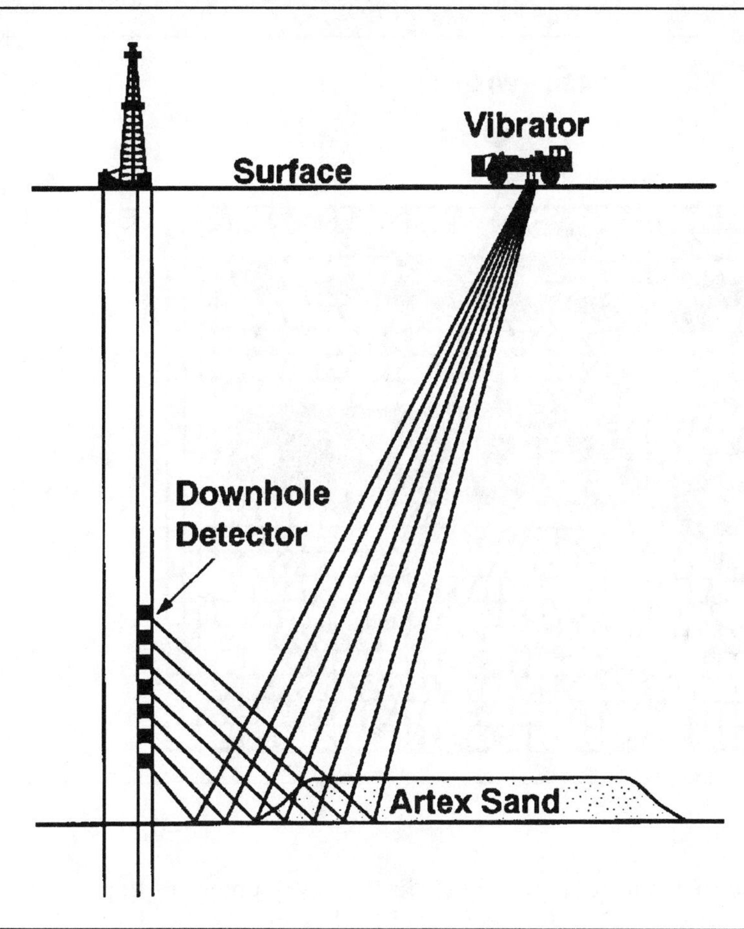

absence of directional permeability is also confirmed by the pulse test data.

Reservoir simulation models were prepared using the geological model. Although the well spacing and thinness of the reservoir limited interwell correlations to two layers, additional model sensitivities were run with more layers on the basis of core permeability variations. The model runs showed that under miscible displacement over 60% of the original oil-in-place would be recovered.

FIGURE B–9. Offset VSP (Well d-46-F/93-P-10 Character Associated with Porous Artex Is First Observed 150 m from Wellbore. Well was Whipstocked 220 m from the Wellbore to Encounter 2.4 m of Porous Artex Sand) *(Copyright © 1992, SPE, from* SPE Reservoir Engineering, *May 1992*[1]*)*

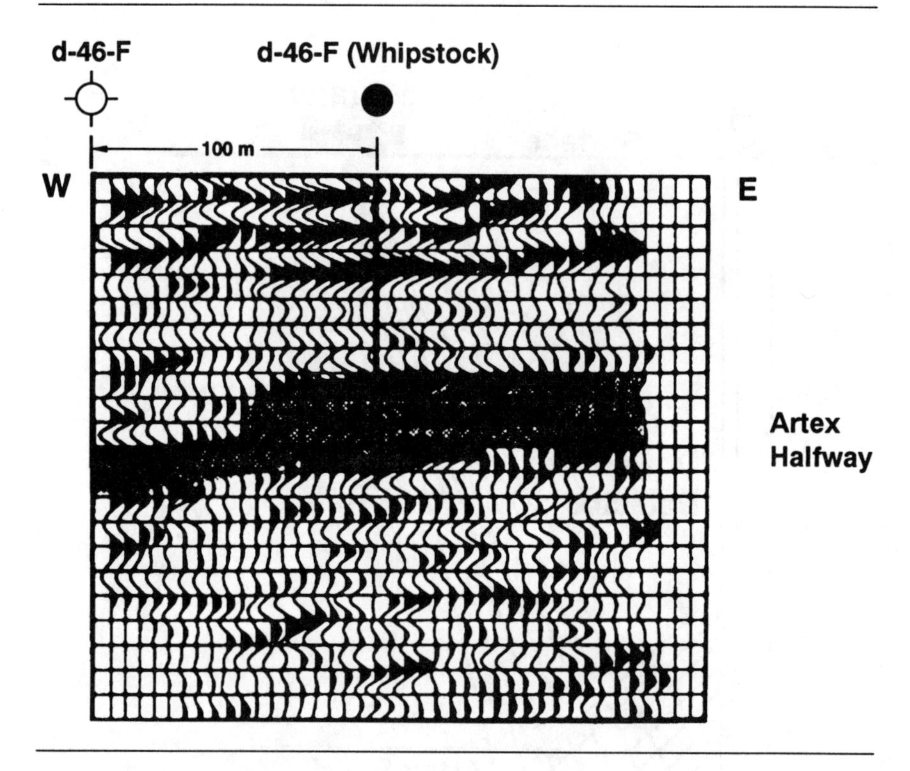

Teamwork played a critical role in the development of the reservoir model. For example, the geophysicists defined the pool edges, the engineers described pools by utilizing well-test pressures and pulse test responses, and the geologists mapped the reservoir properties (accounting for dune morphology, sedimentological and lithological layers, and reservoir geometry).

Original oil-in-place was derived from material-balance analysis and calculated from hand-contoured volumes. The material-balance calculations show 22 to 27 MMSTB, compared with 24 MMSTB from the mapped volumes.

Field production started August 1989, and pressure maintenance began from day one utilizing a hydrocarbon miscible flood.

FIGURE B–10. Seismic Line CH-123-7 (Line Aquired after the Whipstock of Well d-46-F/93-p-10 Showing Excellent Agreement with the Offset VSP) *(Copyright © 1992, SPE, from* SPE Reservoir Engineering, *May 1992[1])*

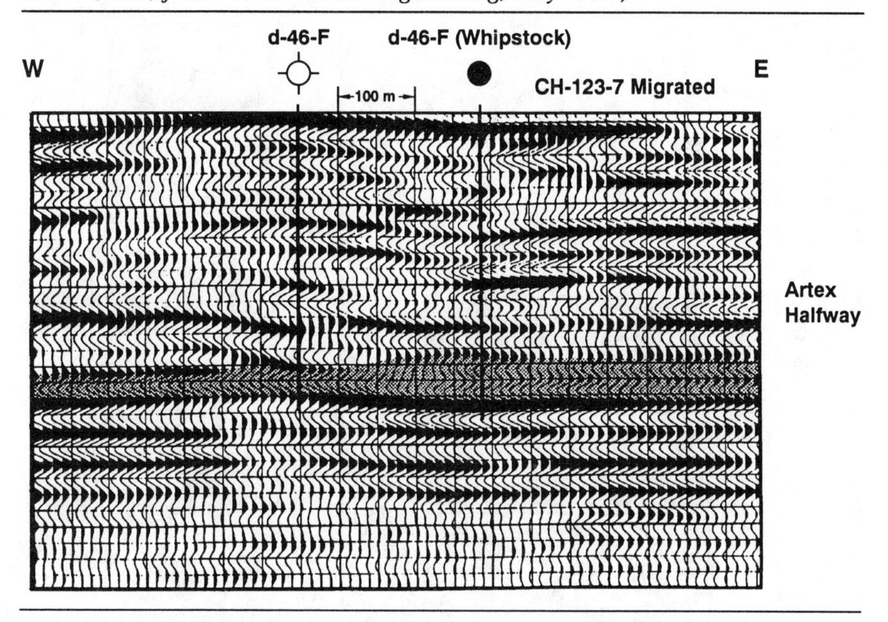

FIGURE B–11. Seismic Line CH-123-9 Designed to Locate Pool Edge and Two Development Wells (Porous Sand Is Observed to Extend Only a Short Distance Beyond the Wells. The Loss of Porous Character Indicated the Pool Boundary.) *(Copyright © 1992, SPE, from* SPE Reservoir Engineering, *May 1992[1])*

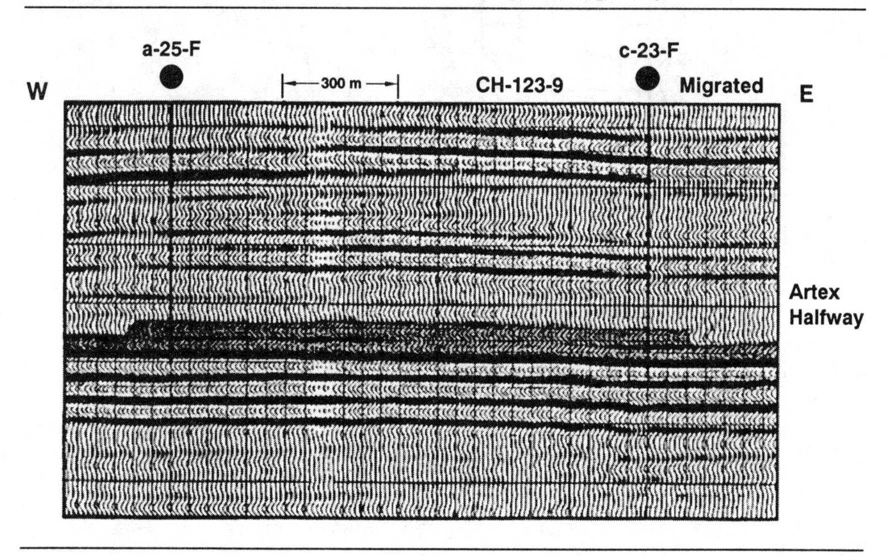

FIGURE B–12. Brassey B Pool Net-Pay Isopach and Reservoir Model Schematic Illustrated with Core Permeability (Correlations Were Based on Sedimentologic and Lithologic Boundaries from Core) *(Copyright © 1992, SPE, from* SPE Reservoir Engineering, *May 1992*[1])

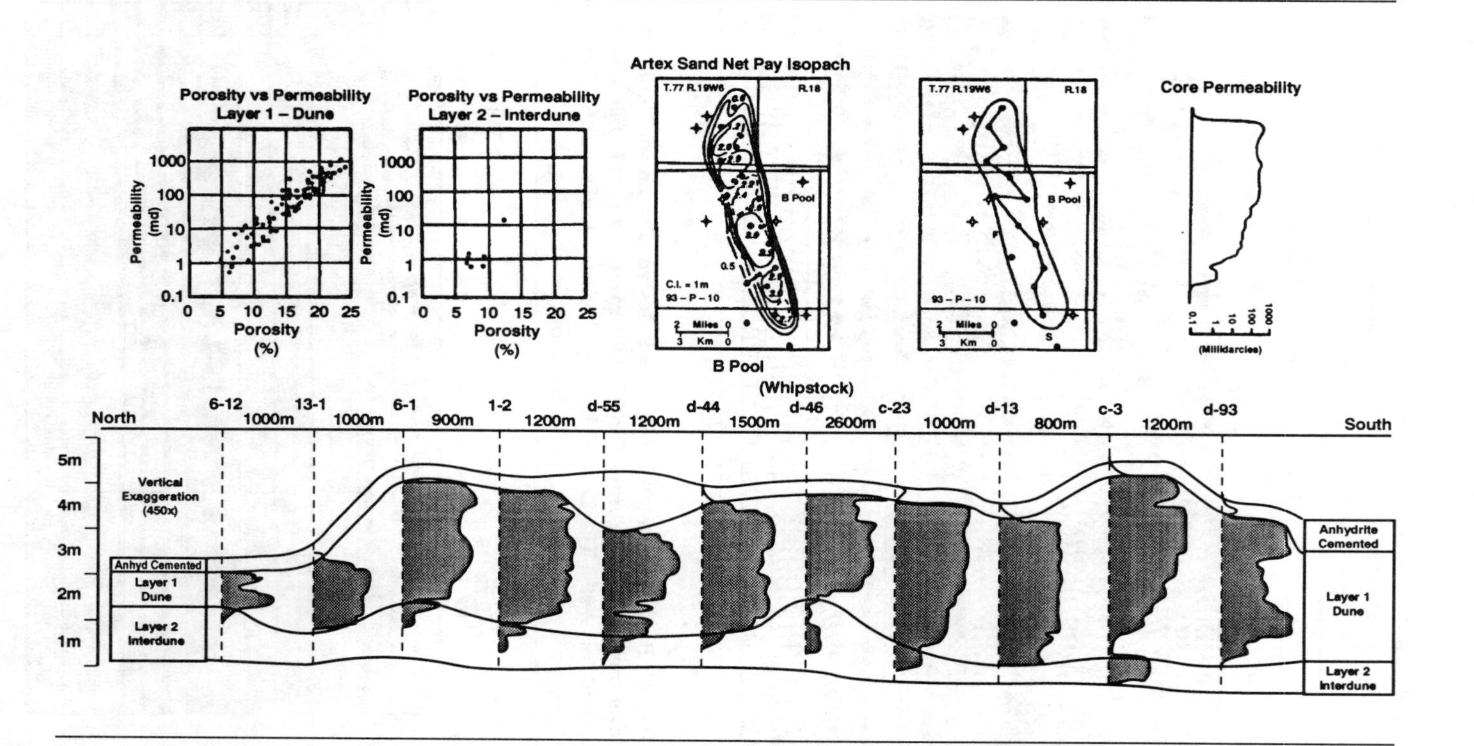

FIGURE B–13. Artex Sand Core Crossplot of k_{max} (Maximum Permeability) Versus k_{90} (Permeability Intermediate Measured Perpendicular to k_{max}) *(Copyright © 1992, SPE, from* SPE Reservoir Engineering, *May 1992[1])*

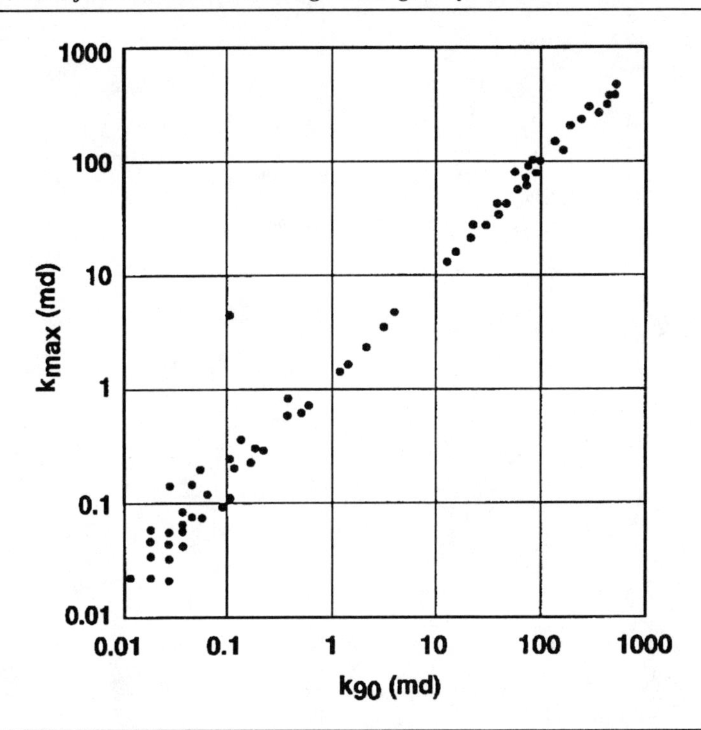

REFERENCE

1. Woofter, D. M. and J. MacGillivray. "Brassey Oil Field, British Columbia: Development of an Aeolian Sand—A Team Approach." SPE Reservoir Engineering (May 1992): 165–172.

APPENDIX C

▼ ▼ ▼

Reserves Estimation Techniques

This appendix presents details of reserves estimation techniques:

1. Volumetric.
2. Decline Curves.
3. Material Balance.
4. Mathematical Simulation.

VOLUMETRIC METHOD

Original Hydrocarbon In Place

The original-oil-in-place can be estimated as follows:

$$N = \frac{7758\ Ah\phi S_{oi}}{B_{oi}} \qquad \text{(C–1)}$$

where:

$$7758 = \frac{43{,}560\ \text{ft}^2/\text{Ac}}{5.614\ \text{ft}^3/\text{bbl}}$$

N = OOIP, STB
A = area, acre
h = average thickness, ft (oil interval)
ϕ = average porosity, fraction
S_{oi} = initial oil saturation, fraction
B_{oi} = initial oil formation volume factor, RB/STB
RB = reservoir barrel
STB = stock tank barrel

The original-solution gas-in-place, OGIP, is given by:

$$G_s = NR_{si} \qquad \text{(C–2)}$$

where:

G_s = OSGIP, scf
N = OOIP, STB
R_{si} = initial solution gas-oil ratio, scf/STBO

The original-free gas-in-place in the gas cap, if present in the reservoir, is given by:

$$G = \frac{7758\ Ah\phi S_{gi}}{B_{gi}} \tag{C–3}$$

where:

G = original free gas in place, scf
S_{gi} = initial gas saturation, fraction
B_{gi} = initial gas formation volume factor, RB/scf
h = average thickness, ft (gas interval)

API Correlations[1,2]

API correlation for recovery efficiency for solution gas drive reservoirs (sands, sandstones, and carbonate rocks) is given by

$$E_R = 41.815 \left[\frac{\phi(1 - S_{wi})}{B_{ob}} \right]^{0.1611} \times \left(\frac{k}{\mu_{ob}} \right)^{0.0979} \times (S_{wi})^{0.3722} \times \left(\frac{p_b}{p_a} \right)^{0.1741} \tag{C–4}$$

where:

E_R = recovery efficiency, % OOIP at bubble point
ϕ = porosity, fraction of bulk volume
S_{wi} = interstitial water saturation, fraction of pore space
B_{ob} = oil formation volume factor at bubble point, RB/STB
k = absolute permeability, darcy
μ_{ob} = viscosity of oil at bubble point, cp
p_b = bubble point pressure, psia
p_a = abandonment pressure, psia

Recovery efficiency for water drive reservoirs (sands and sandstones):

$$E_R = 54.898 \left[\frac{\phi(1 - S_{wi})}{B_{oi}} \right]^{0.0422} \times \left(\frac{k\mu_{wi}}{\mu_{oi}} \right)^{0.0770} \times (S_{wi})^{-0.1903} \times \left(\frac{p_i}{p_a} \right)^{-0.2159} \tag{C–5}$$

where:

E_R = recovery efficiency, % OOIP
B_{oi} = initial oil formation volume factor, RB/STB
μ_{wi} = initial water viscosity, cp
μ_{oi} = initial oil viscosity, cp
p_i = initial reservoir pressure, psia

DECLINE CURVE EQUATIONS

A general mathematical expression for the rate of decline, D, can be expressed as[3,4]

$$D = -\frac{dq/dt}{q} = Kq^n \tag{C–6}$$

where:

q = production rate, barrels per day, month or year
t = time, day, month or year
K = constant
n = exponent

The decline rate in Equation (C–6) can be constant or variable with time yielding three basic types of production decline as follows:

1. Exponential or constant decline

$$D = -\frac{dq/dt}{q} = K = -\frac{\ln\left(\frac{q_t}{q_i}\right)}{t} \tag{C–7}$$

when $n = 0$, K = constant
q_i = initial production rate
q_t = production rate at time t

The rate-time and rate-cumulative relationships are given by

$$q_t = q_i \cdot e^{-Dt} \tag{C–8}$$

$$Q_t = \frac{q_i - q_t}{D} \tag{C–9}$$

where:

Q_t = cumulative production at time t

A familiar rate constant for exponential decline is as follows:

$$D' = \frac{\Delta q}{q_i} \tag{C–10}$$

where Δq is the rate change in the first year. In this case, the relationship between D and D' is given below:

$$D = -\ln\left(1 - \frac{\Delta q}{q_i}\right) = -\ln(1 - D') \tag{C–11}$$

2. Hyperbolic decline

$$D = -\frac{dq/dt}{q} = Kq^n (0 < n < 1) \tag{C–12}$$

Note that this is the same equation as the general decline rate equation (Equation C–6) except for the constraint on n.

For initial condition

$$K = \frac{D_i}{q_i^n}$$

The rate-time and rate-cumulative relationships are given by:

$$q_t = q_i(1 + nD_i t)^{-\frac{1}{n}} \tag{C–13}$$

$$Q_t = \frac{q_i^n\left(q_i^{1-n} - q_t^{1-n}\right)}{(1-n)D_i} \tag{C–14}$$

where D_i = initial decline rate

3. Harmonic decline

$$D = -\frac{dq/dt}{q} = Kq \tag{C–15}$$

when $n = 1$

For initial condition

$$K = \frac{D_i}{q_i}$$

The rate-time and rate-cumulative relationships are given by:

$$q_t = \frac{q_i}{\left(1 + D_i t\right)} \quad \text{(C–16)}$$

$$Q_t = \frac{q_i}{D_i} \ln \frac{q_i}{q_t} \quad \text{(C–17)}$$

Both exponential and harmonic declines are special cases of the hyperbolic decline.

MATERIAL BALANCE METHOD

Oil Reservoirs[5-8]

The basis for material balance is the law of conservation of mass (i.e., mass is neither created nor destroyed). Therefore, the general material-balance equation can be expressed as follows:

Underground withdrawal = Expansion of oil
+ Original dissolved gas
+ Expansion of gas caps
+ Reduction in hydrocarbon pore volume due to connate water expansion and decrease in the pore volume
+ Natural water influx

The material balance as an equation of straight line[5,6] is given by

$$F = N\left(E_o + mE_g + E_{fw}\right) + W_e \quad \text{(C–18)}$$

where:

F = underground withdrawal, RB

$$= N_p\left[B_o + \left(R_p - R_s\right)B_g\right] + W_p B_w - W_i B_w - G_i B_g \quad \text{(C–19)}$$

$$= N_p\left[B_t + \left(R_p - R_{si}\right)B_g\right] + W_p B_w - W_i B_w - G_i B_g \quad \text{(C–20)}$$

and,

N_p = cumulative oil production, STB
B_o = oil formation volume factor, RB/STB
R_s = gas in solution in oil, SCF/STBO
B_g = gas formation volume factor, RB/SCF

W_p = cumulative water production, STB
W_i = cumulative water injection, STB
G_i = cumulative gas injection, SCF
R_p = cumulative gas oil ratio
= cum gas production/cum oil production, SCF/STB
N = original oil in place, STB
E_o = expansion of oil and original gas in solution, RB/STB

$$= B_t - B_{ti} \tag{C–21}$$

$$= (B_o - B_{oi}) + (R_{si} - R_s)B_g \tag{C–22}$$

$$B_t = B_o + (R_{si} - R_s)B_g \tag{C–23}$$

m = initial gas cap volume fraction

$$= \frac{\text{initial hydrocarbon volume of the gascap}}{\text{initial hydrocarbon volume of the oil zone}}, \; RB/RB$$

E_g = expansion of gas cap gas, RB/STB

$$= B_{oi}\left(\frac{B_g}{B_{gi}} - 1\right) \tag{C–24}$$

E_{fw} = expansion of the connate water and reduction in the pore volume, RB/STB

$$= (1+m)B_o\left(\frac{C_w S_{wi} + C_f}{1 - S_{wi}}\right)\Delta P \tag{C–25}$$

c_w, c_f = water and formation compressibilities, respectively, psia
S_{wi} = initial water saturation, fraction
Δp = pressure drop, psi
W_e = cumulative natural water influx, RB

$$= US(p,t) \tag{C–26}$$

and

U = aquifer constant, RB/psi
$S(p,t)$ = aquifer function.

Wang and Teasdale list theoretical aquifer constants and aquifer functions for various aquifers.[7]

The material-balance equations can be used to estimate the original-oil-in-place by history matching the past production performance, and to predict future performance.

For a solution gas-drive reservoir, where there is no initial gas cap (m = 0), no gas and water injection (G_i = 0), W_i = 0), and no natural water influx ($W_e = 0$), the general material-balance Equation C–18 can be reduced to

$$N_p\left[B_t + \left(R_p - R_{si}\right)B_g\right] + W_p B_w = N\left[\left(B_t - B_{t_i}\right) + B_{oi}\left(\frac{C_w S_{wi} + C_f}{1 - S_{wi}}\right)\Delta p\right] \tag{C–27}$$

Above the bubble point pressure (under saturated oil), $R_p = R_s = R_{si}$, $B_{ti} = B_{oi}$ and $B_t = B_o$. Neglecting water production, then Equation C–27 is reduced to

$$\frac{N_p}{N} = \left(\frac{B_{oi}}{B_o}\right)C_e \Delta p \tag{C–28}$$

where

$$C_e = \frac{C_o S_o + C_w S_w + C_f}{\left(1 - S_{wi}\right)} \tag{C–29}$$

S_o = oil saturation, fraction
C_o = oil compressibility, psi^{-1}

If original-oil-in-place is known, Equation C–28 can be used to calculate future production for sequential pressure drops starting from the initial pressure.

Below the bubble point pressure, neglecting water production and rock compressibility, Equation C–27 is reduced to

$$\frac{N_p}{N} = \frac{B_t - B_{t_i}}{B_t + \left(R_p - R_{si}\right)B_g} \tag{C–30}$$

Calculation of future production requires not only the solution of Equation C–30, but also of the subsidiary equations for liquid saturation, produced gas-oil ratio, and cumulative gas production given below:

$$S_o = \left(1 - \frac{N_p}{N}\right)\left(\frac{B_o}{B_{oi}}\right)\left(1 - S_{wi}\right) \tag{C–31}$$

$$R = R_s + \left(\frac{B_o}{B_g}\right)\left(\frac{\mu_o}{\mu_g}\right)\left(\frac{k_{rg}}{k_{ro}}\right) \tag{C–32}$$

$$R_p = \frac{G_p}{N_p} = \frac{\int_o^t R dN_p}{N_p} \approx \frac{\sum R \Delta N_p}{N_p} \tag{C–33}$$

where

μ_o, μ_g = oil and gas viscosities, respectively, cp
k_{rg}, k_{ro} = gas and oil relative permeabilities, respectively, fraction

Simultaneous solutions of the material balance and the subsidiary equations are required based upon time steps or corresponding pressure steps.

Gas Reservoir[5–8]

The general material balance as an equation of straight line is given below:

$$F = G\left(E_g + E_{fw}\right) + W_e \tag{C–34}$$

where:

F = underground withdrawal, RB

$$= G_{wgp} B_g + W_p B_w \tag{C–35}$$

G_{wgp} = cumulative wet gas production, SCF

$$= G_p + N_{pc} F_c \tag{C–36}$$

G_p = cumulative dry gas production, SCF
N_{pc} = cumulative condensate production, STB
F_c = condensate conversion factor, SCF/STB

$$= 132.79\, \gamma_c / M_c \tag{C–37}$$

γ_c = specific gravity of condensate (water = 1.0)

$$= \frac{141.5}{131.5 + °API} \tag{C–38}$$

M_c = molecular weight of condensate

$$= \frac{6084}{°API - 5.9} \tag{C–39}$$

G = wet gas in place, SCF
E_g = expansion of gas, RB/SCF

$$= B_g - B_{gi} \tag{C–40}$$

E_{fw} = expansion of connate water and reduction in the pore volume, RB/SCF

$$= B_{gi} C_e \left(P_i - P\right)$$

$$= B_{gi}\left(\frac{C_w S_{wi} + C_f}{1 - S_{wi}}\right)\Delta p \tag{C–41}$$

Δp = initial reservoir pressure – average reservoir pressure, psi

$$= p_i - p \tag{C–42}$$

W_e = cumulative water influx, RB

$$= U\,S(p,t) \tag{C–43}$$

U = aquifer constant, RB/psi
$S(p,t)$ = aquifer function, psi

For depletion drive reservoir, W_e and E_{fw} can be neglected. Then considering no water and condensate production, general material-balance Equation C–34 can be reduced to a more popular form as follows:

$$\frac{p}{z} = \frac{p_i}{z_i}\left(1 - \frac{G_p}{G}\right) \tag{C–44}$$

A plot of p/z vs. G_p should yield a straight line. The original-gas-in-place (G) can be obtained by extrapolating the straight line to 0. Equation C–44 can be used directly to calculate future gas production corresponding to a given pressure. If the abandonment pressure is known, this equation can be used to estimate the ultimate gas production.

MATHEMATICAL SIMULATION

The reservoir is divided into many small tanks, cells, or blocks to take into account reservoir heterogeneity. Computations using material balance and fluid flow equations are carried out for oil, gas, and water phases for each cell at discrete time steps, starting with the initial time. This section provides an introduction to numerical reservoir simulation concepts and applications.[8]

Spatial and Time Discretization

The discretization of the reservoir into blocks will depend upon the size and complexity of the reservoir, quality and quantity of the reservoir data, objective of the simulation study, and the accuracy of the solution needed. In practice, the number of blocks will be limited principally by the cost of calculations and the time available to prepare input data and to interpret results. The model should contain enough blocks and dimensions to represent the reservoir and to simulate its performance adequately. Figure 6–27 illustrates few of the many types of models used in reservoir simulation.

The life of the reservoir must also be discretized or divided into time increments. Starting at the initial time, pressure, saturations, etc. computations are carried out for all phases at each block—over each of many finite time increments. In general, the accuracy with which reservoir behavior can be calculated will be influenced by the size of the time steps and the number of grid blocks.

Approximation of Complex Flow Equations

In simulating conventional oil and gas reservoirs, the same material-balance principles are used as in the classical approach. The difference lies in applying the rules of conservation of mass and Darcy's fluid flow law for each fluid phase (i.e., oil, gas and water) to each block, rather than to the reservoir as a whole. This procedure, in turn, leads to a set of partial differential equations, involving saturations, pressures, rock and fluid properties, and time. Analytical solutions of these equations for saturations and pressure distributions as functions of time are impossible to obtain. Luckily, however, approximate solutions of the complex partial differential equations can be obtained by using finite difference schemes.

Computing Power

A reservoir simulator depends upon complex matrix solutions of the finite difference approximations of multidimensional, multiphase partial differential fluid flow equations. Simulation has become a reality because of the computing power of the mainframe high-speed computers. Even a personal computer today, such as Compaq DeskPro 386 PC, is powerful enough with its storage capacity and computing speed to handle small scale reservoir simulation. Technological advances in computational techniques, data handling, report writing, and graphics have also made reservoir simulation more practical and widely used.

Differential Equations

A mathematical description of fluid flow in a porous medium can be obtained from the following physical principles:

1. The Law of Conservation of Mass, or Material Balance.
2. Fluid Flow Law, such as Darcy's Law.
3. PVT behavior of fluids.

For simplicity, let us consider 1-dimensional flow of oil, gas and water in the x-direction through an arbitrary volume element (Figure C–1).

FIGURE C–1. Volume Element

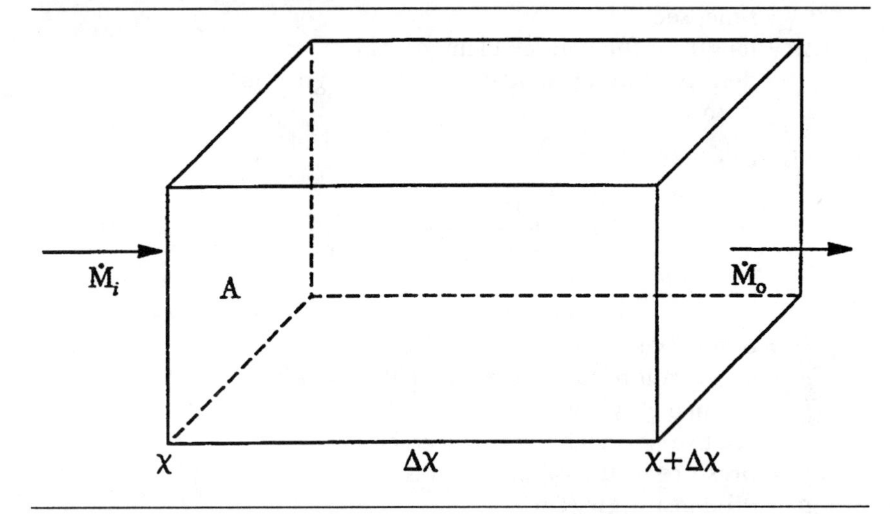

According to the law of conservation of mass:

Mass Rate In – Mass Rate Out = Mass Rate of accumulation.

Then, considering the flow of oil, material balance equation gives:

$$\frac{\partial M_o}{\partial t} = \dot{M}_{o_i} - \dot{M}_{o0} \tag{C–45}$$

where:

$$M_o = \rho_{os} A \Delta x \phi S_o / B_o \tag{C–46}$$

$$\dot{M}_{o_i} = \rho_{os}\left(\frac{q_o}{B_o}\right)_x \tag{C–47}$$

$$\dot{M}_{o0} = \rho_{os}\left(\frac{q_o}{B_o}\right)_{x+\Delta x} \tag{C–48}$$

where:

A = cross-sectional area of the element in the direction of flow, sq. cm.
B_o = oil formation volume factor, std. cm^3/res. cm^3
M_o = mass of oil in the volume element, gm
M_{o_i} = mass rate of oil flowing into the volume element, gm/sec
M_{o_0} = mass rate of oil flowing out of the volume element, gm/sec
q_o = oil flow rate, cm^3/sec

S_o = oil saturation in the volume element, fraction of pore volume
t = time, sec
Δx = length of the volume element, cm
ρ_{os} = density of oil at standard condition, gm/cm^3
ϕ = porosity

Using Darcy's law:

$$q_o = -\frac{Akk_{ro}}{\mu_o}\left(\frac{\partial p_o}{\partial x} - \rho_o g \frac{\partial D}{\partial x}\right) \tag{C–49}$$

where:

D = depth, cm
g = acceleration due to gravity (980.7), cm/sec^2
k = permeability, darcy
k_{ro} = relative permeability to oil, fraction
p_o = pressure in the oil phase, atm
ρ_o = oil density, gm/cm^3
x = direction of flow, cm
μ_o = oil viscosity, cp

Substituting Equations (C–46), (C–47), (C–48), and (C–49) in Equation (C–45), the differential equation for oil flow in one dimension is obtained as follows:

$$\frac{\partial}{\partial x}\left[\frac{Akk_{ro}}{B_o\mu_o}\left(\frac{\partial p_o}{\partial x} - p_o g \frac{\partial D}{\partial x}\right)\right] = A\frac{\partial}{\partial t}\left(\frac{\phi S_o}{B_o}\right) \tag{C–50}$$

Similarly, water and gas (including not only gas in the gas phase, but also the gas dissolved in the oil phase) flow equations can be obtained as given below:

$$\frac{\partial}{\partial x}\left[\frac{Akk_{rw}}{B_w\mu_w}\left(\frac{\partial p_w}{\partial x} - p_w g \frac{\partial D}{\partial x}\right)\right] = A\frac{\partial}{\partial t}\left(\frac{\phi S_w}{B_w}\right) \tag{C–51}$$

$$\frac{\partial}{\partial x}\left[\frac{Akk_{rg}}{B_g\mu_g}\left(\frac{\partial p_g}{\partial x} - p_g g \frac{\partial D}{\partial x}\right)\right] + \frac{AR_s kk_{ro}}{B_o\mu_o}\left(\frac{\partial p_o}{\partial x} - p_o g \frac{\partial D}{\partial x}\right) = A\frac{\partial}{\partial t}\left[\phi\left(\frac{S_g}{B_g} + \frac{S_o R_s}{B_o}\right)\right] \tag{C–52}$$

where:

R_s = gas in solution, std. cm^3/res. cm^3
g, o, w = subscripts referring to gas, oil, and water, respectively

Equations (C–50), (C–51), and (C–52) relate the saturations to the pressure in the phases (p_o, p_w, p_g) and the rock (ϕ, k, k_r) and fluid properties (B, μ, R_s, ρ).

In addition to the partial differential equations, certain auxiliary relationships must be satisfied to solve these equations. First, the sum of the volumes of the oil, gas, and water must always be equal to the pore volume at any point in the system. Therefore,

$$S_o + S_g + S_w = 1 \tag{C–53}$$

If the rock and fluid properties are assumed to be known functions of pressure, then there are four equations and the six unknowns (S_o, S_g, S_w, p_o, p_g, p_w).

The capillary pressures at any position can be taken to be functions of saturations alone as shown below:

$$P_{c_{ow}} = P_o - P_w = P_{c_{ow}}\left(S_o, S_w\right) \tag{C–54}$$

and

$$P_{c_{go}} = P_g - P_o = P_{c_{go}}\left(S_o, S_g\right) \tag{C–55}$$

where:

P_c = capillary pressure, atm
ow,go = subscripts referring to oil-water and gas-oil, respectively

Now, there are six equations (i.e., Equations (C–50), (C–51), (C–52), (C–53), (C–54), (C–55), involving six unknowns (i.e., three saturations and three pressures). With appropriate boundary and initial conditions, the system of six equations can be solved for saturation and pressure distributions in the reservoir.

Finite Difference Approximations

The numerical solution of the partial differential equations by finite differences involves replacing the partial derivatives by finite difference quotients. Then, instead of obtaining a continuous solution, an approximate solution is obtained at a discrete set of grid blocks or points at discrete times. A general treatment of finite difference schemes are beyond the scope of this book. However, certain basic concepts will be exemplified.

Finite difference approximations of partial differential equations require spatial and time discretizations. Figure C–2 shows the discretizations of a 1-dimensional flow system. Note that the length of the system is divided into a discrete set of grid blocks of size Δx. The subscript, i, is used to identify the grid blocks in the x direction. Time is divided into a set of discrete time

FIGURE C–2. Discretization of a 1-D System

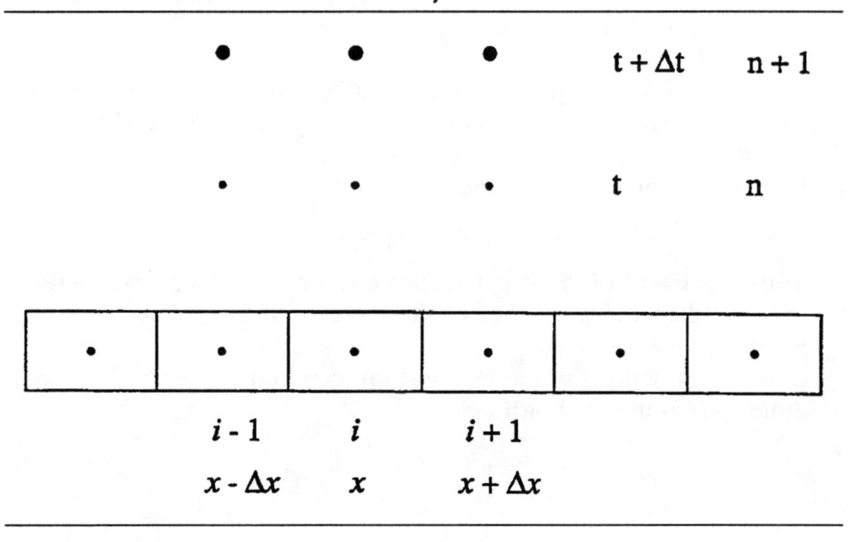

intervals, Δt. The subscript, *n*, is used to identify the time level. The partial derivatives can be evaluated using explicit or implicit procedures.

Explicit difference scheme is based upon the values of a variable known at the beginning of a time step. Note that the end of a time step, *n*, is the beginning of the next time step, *n* + 1. The explicit difference expression of $\partial p/\partial x$ and $\partial^2 p/\partial x^2$ are shown below (see Figure C–2):

$$\left(\frac{\partial p}{\partial x}\right)_{i-\frac{1}{2},n} = \frac{P_{i,n} - P_{i-1,n}}{\Delta x} \quad \text{(C–56)}$$

$$\left(\frac{\partial p}{\partial x}\right)_{i+\frac{1}{2},n} = \frac{P_{i+1,n} - P_{i,n}}{\Delta x} \quad \text{(C–57)}$$

$$\left(\frac{\partial^2 p}{\partial x^2}\right)_{i,n} = \frac{P_{i-1,n} - 2P_{i,n} + P_{i+1,n}}{(\Delta x)^2} \quad \text{(C–58)}$$

In implicit difference, the spatial derivatives are evaluated at current time level, *n* + 1 rather than at the previously known time, *n*.

Solutions

The finite difference approximations of the partial differential flow equations ultimately involve a set of simultaneous equations requiring matrix problem solutions. The equations in matrix notation follow:

$$\begin{bmatrix} a_{11} & a_{12} & \cdot & \cdot & a_{1N} \\ a_{21} & a_{22} & \cdot & \cdot & a_{2N} \\ \cdot & & & & \\ \cdot & & & & \\ a_{N1} & a_{N2} & \cdot & \cdot & a_{NN} \end{bmatrix} \begin{bmatrix} p_1 \\ p_2 \\ \cdot \\ \cdot \\ p_n \end{bmatrix} = \begin{bmatrix} b_1 \\ b_2 \\ \cdot \\ \cdot \\ b_N \end{bmatrix} \tag{C–59}$$

where the a and b are known values, and the p are to be determined.

The pressures and saturations can be solved explicitly, implicitly, or by a combination method. The method of solution will affect: (1) stability—the more implicit formulation generally being more stable, (2) accuracy—the truncation error, time step, and grid size affecting the accuracy of the solution, and (3) cost involving the computer storage and run time.

One of the most common formulation methods of numerical simulation is the Implicit Pressure-Explicit Saturation (IMPES) method. This involves solving first implicitly (as required for stability) for the phase pressures at each point and then solving explicitly for the saturations. Its appeal is a result of greatly reduced computing requirements, because it avoids the simultaneous implicit solution for several unknown at each point.

The Implicit method involves solving both the phase pressures and saturations implicitly at each point. It offers a more stable solution at the expense of more computing time. This type of solution is particularly needed for water and/or gas coning problem and the gas percolation problem.

REFERENCES

1. Arps, J. J., et al. "A Statistical Study of Recovery Efficiency," *API Bulletin D14* (1967): 1–33.
2. Doscher, T. M., et al. "Statistical Analysis of Crude Oil Recovery and Recovery Efficiency," *API Bulletin D14* (1984): 5–46.
3. Arps, J. J. "Estimation of Decline Curves," *Trans. AIME* 160 (1945): 228–247.
4. Arps, J. J. "Estimation of Primary Oil Reserves," *Trans. AIME* 207, (1956): 182–191.
5. Havlena, D. and A. S. Odeh. "The Material Balance as an Equation of Straight Line," *J. Pet. Tech.* (August 1963): 896–900.
6. Havlena, D. and A. S. Odeh. "The Material Balance as an Equation of Straight Line—Part II, Field Cases," *J. Pet. Tech.* (July 1964): 815–822.
7. Wang, B. and T. S. Teasdale. GASWAT-PC: A Microcomputer Program for Gas Material Balance with Water Influx. SPE Paper 16484, Petroleum Industry Applications of Microcomputers, Del Lago on Lake Conroe, Texas, June 23–26, 1987.
8. Litvak, B. L. Texaco E&P Technology Department: Personal Contact.

APPENDIX D

▼ ▼ ▼

Fluid Flow Equations

FLUID FLOW IN RESERVOIRS

Fluid flow in reservoirs is characterized by:

1. Flowing phase [i.e., single phase (oil or gas or water), two phase (oil and gas or oil and water or gas and water), and three phase (oil, gas and water)].
2. Flow direction [i.e., linear (x, y or z), radial, and spherical]. One-, two- or three-dimensional flow relates to the number of flowing directions.
3. Fluid state (i.e., compressible or incompressible).
4. Flow condition (i.e., steady state or unsteady-state).

Darcy's law as shown below is the basic equation to describe the flow of fluids through porous media:

$$q = -\frac{kA}{\mu}\left(\frac{dp}{ds} - \frac{\rho g}{1.0133}\frac{dz}{ds} \times 10^{-6}\right) \qquad \text{(D–1)}$$

where:

A = the cross-sectional area of rock and pore, in the direction of flow, cm^2
dp/ds = pressure gradient along the direction of flow, atm/cm
dz/ds = gradient in the vertical direction
g = acceleration due to gravity, cm/sec^2 (= 980.7 cm/sec^2)
k = permeability, darcies
μ = viscosity of flowing fluid, centipoise (cp)
ρ = density of the flowing fluid, g/cm^3
q = flow of fluid, cm^3/sec

STEADY STATE FLOW

When pressure and flow rate at any point in the flowing system do not change, the condition is described as steady state flow. The various forms of the Darcy's equation for steady-state single phase flow are given below:

Linear Flow (Field Units)

1. Incompressible (flow from the point 1 to the point 2)

$$q = \frac{1.127\,kA\left(p_1 - p_2\right)}{\mu L} \tag{D–2}$$

2. Compressible (flow from the point 1 to the point 2)

$$q_{sc} = \frac{3.164\,T_{sc}Ak}{p_{sc}TzL\mu}\left(p_1^2 - p_2^2\right)$$
$$\left(3.164 = \frac{1.127}{2} * 5.615\right) \tag{D–3}$$

Radial Flow (Field Units)

1. Incompressible (flow from the external radius to the wellbore)

$$q = \frac{7.081\,kh\left(p_e - p_w\right)}{\mu \ln\left(r_e/r_w\right)}$$
$$\left(7.081 = 1.127 * 2\pi\right) \tag{D–4}$$

2. Compressible (flow from the external radius to the wellbore)

$$q_{sc} = \frac{19.880\,T_{sc}kh}{p_{sc}Tz\mu \ln\left(r_e / r_w\right)}\left(p_e^2 - p_w^2\right)$$
$$= \frac{703.0\,kh\left(p_e^2 - p_w^2\right)}{\mu zT \ln\left(r_e / r_w\right)} \tag{D–5}$$
$$\left(19.880 = \frac{1.127}{2} * 5.615 * 2\pi\right)$$

where:

A = area, ft^2
h = thickness, ft
k = permeability, darcies

L = length of the system, ft
μ = viscosity, cP
p = pressure, psi
p_e = pressure at radius r_e, psia
p_{sc} = standard pressure, psia
p_w = pressure at well radius r_w, psia
q = flow of fluid, bbl/day
q_{sc} = flow rate at standard conditions, SCF/day
r_e = outer radius of well influence, ft
r_w = wellbore radius, ft
T = reservoir temperature, °R
T_{sc} = standard temperature, °R
z = deviation (compressibility) factor

The steady-state radial flow equation for incompressible fluid is used to calculate the productivity or injectivity index of a well. For example, the oil productivity index of a well is given by

$$J = \frac{q_o}{(p_e - p_w)} = C\left(\frac{k_o}{\mu_o B_o}\right) \tag{D–6}$$

where:

$$C = \frac{7.081h}{\ln(r_e / r_w)} \tag{D–7}$$

B_o = oil formation volume factor
J = productivity index, PI, STBO/day/psi
k_o = effective permeability to oil, darcy
q_o = oil flow rate, STBO/day
μ_o = oil viscosity, cp

The productive capacity of a gas well under open-flow conditions (i.e., open-flow potential) is calculated using the steady-state radial flow equation for compressible fluid, as follows:

$$q_{sc} = c\left(p_f^2 - p_w^2\right)^n \tag{D–8}$$

where:

$$c = \frac{703k_g h}{\mu_g z T \ln(r_e / r_w)} \tag{D–9}$$

q_{sc} = gas flow rate, scf/day
k_g = effective permeability to gas, darcy
μ_g = gas viscosity, cp
p_f = formation pressure, psia

n = 1 for completely laminar steady-state flow
n = 0.5 for completely turbulent steady-state flow

UNSTEADY STATE FLOW

In practice, the flow of fluids is transient or unsteady state, (i.e., the rates and pressures change with time and position). Based upon (1) law of conservation of mass, (2) Darcy's fluid flow law, and (3) equation of state of fluid (slightly compressible fluid), the following equation results:

$$\frac{\partial^2 p}{\partial r^2} + \frac{1}{r}\frac{\partial p}{\partial r} = \frac{1}{\eta}\frac{\partial p}{\partial t} \quad \text{(D–10)}$$

where:

$$\eta = \frac{k}{\phi \mu c}$$

ϕ = porosity, fraction
c = compressibility, vol/vol/psi

This equation is called the "diffusivity equation." Analytical solutions of this equation for specified initial and boundary conditions are particularly very useful for well-pressure test analysis, and natural-water influx calculations.

WATER INFLUX

The quantitative evaluation of the cumulative water encroachment, W_e, into a reservoir is one of the important problems of primary production analysis. Since it is not directly amenable to measurement, its evaluation must necessarily be deduced from indirect estimates. In fact, to calculate this influx, the engineer confronts what is inherently the greatest uncertainty in the whole subject of reservoir engineering. Calculations of W_e require a mathematical model which, in turn, relies on the properties of the aquifer (i.e., fluid properties, permeability, thickness, geometrical configuration, etc.), yet these are seldom known since wells are not deliberately drilled into the aquifer to obtain these data.

The general water influx equation is expressed as

$$W_e = US(p,t) \quad \text{(D–11)}$$

where U is an aquifer constant and $S(p,t)$ is aquifer function, which is defined separately for different aquifer types.

Wang and Teasdale presented a list of theoretical aquifer functions and aquifer constants for small, Schilthuis steady-state, Hurst simplified steady-state, Van Everdingen Hurst infinite linear and radial unsteady-state aquifers.[1–7]

Because of the many uncertainties in the model (i.e., steady state, unsteady state, geometry, dimensions, and properties of the aquifer), direct calculation of water influx, even though it is possible, is somewhat unreliable. For best accuracy, water influx calculations are made in conjunction with the overall material balance of the reservoir.

IMMISCIBLE DISPLACEMENT

Displacement of oil from a porous medium by immiscible fluids, water, or gas can be described by the fractional flow equation and the frontal advance theory.[8] In practical units, considering viscous, gravitational, and capillary effects the fractional flow equation for water displacing oil is

$$f_w = \frac{1 + 0.001127 \dfrac{k k_{ro}}{\mu_o} \dfrac{A}{q_t} \left[\dfrac{\partial p_c}{\partial L} - 0.433 \Delta\rho \sin\alpha_d \right]}{1 + \dfrac{\mu_w}{\mu_o} \dfrac{k_{ro}}{k_{rw}}} \qquad \text{(D–12)}$$

where:

A = area, sq. foot
f_w = fraction of water flowing
k = absolute permeability, millidarcy
k_{ro} = relative permeability to oil
k_{rw} = relative permeability to water
μ_o = oil viscosity, cp
μ_w = water viscosity, cp
L = distance along direction of flow, ft
p_c = capillary pressure = $p_o - p_w$, psi
q_t = total flow rate = $q_o + q_w$, B/day
$\Delta\rho$ = water-oil density difference = $p_w - p_o$, gm/cc
α_d = angle of formation dip to the horizon, degree

The fractional flow of water for given rock and fluid properties, and flooding conditions is a function of water saturation only because the relative permeability and capillary pressure are functions of saturation only.

Neglecting gravity and capillary effects, the above fractional flow equation is reduced to

$$f_w = \frac{1}{1+\frac{\mu_w}{\mu_o}\frac{k_{ro}}{k_{rw}}} \tag{D–13}$$

Using the oil-water relative permeability data shown in Figure D–1 and an oil-water viscosity ratio of 2, calculated fractional flow curve is shown in Figure D–2.

The linear frontal advance equation for water, based upon conservation of mass and assuming incompressible fluids, is given by

$$\left(\frac{\partial x}{\partial t}\right)_{S_w} = \frac{q_t}{A\phi}\left(\frac{\partial f_w}{\partial S_w}\right)_t \tag{D–14}$$

This equation states that the rate of advance of a plane of a fixed water saturation, S_w, at a time, t, is equal to the total fluid velocity multiplied by the change in composition of the flowing stream caused by a small change in the saturation of the displacing fluid.

FIGURE D–1. Oil-Water Relative Permeabilities

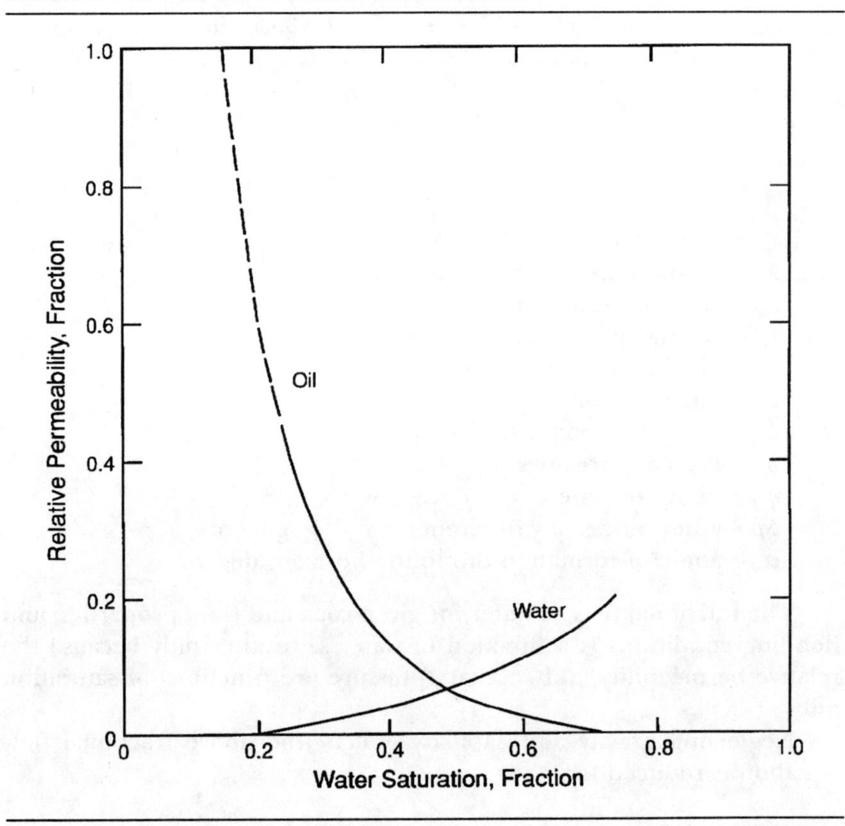

FIGURE D–2. Fractional Water Flow

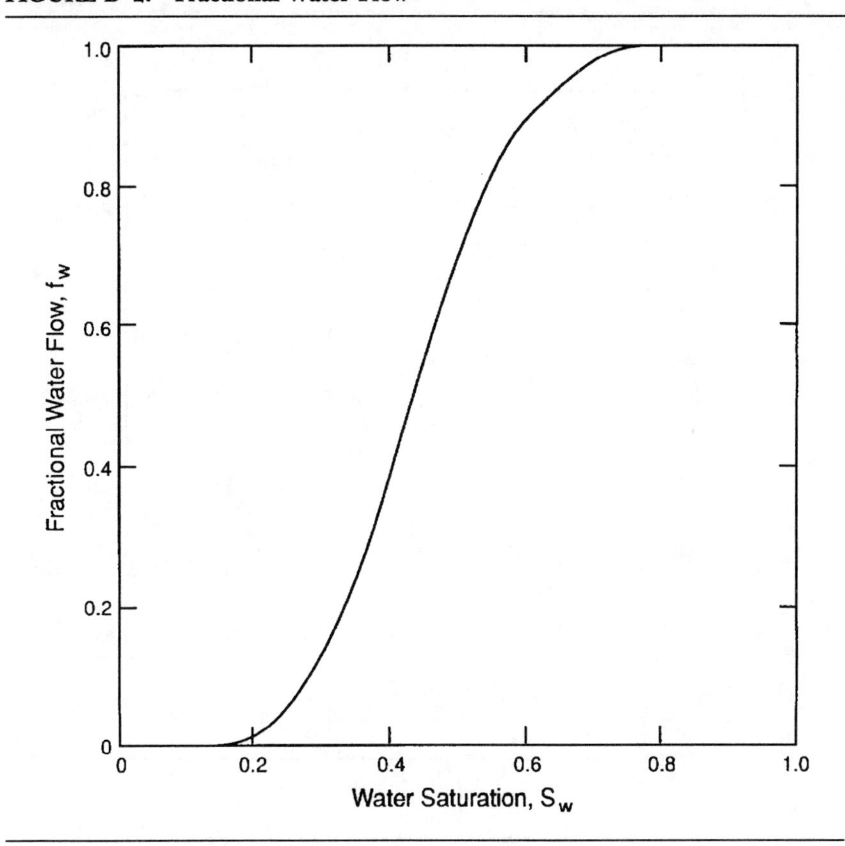

The frontal advance equation can be used to derive the expressions for average water saturation as follows:

$$\textit{At breakthrough}: \overline{S}_{wbt} - S_{wc} = \left(\frac{\partial S_w}{\partial f_w}\right)_f = \frac{S_{wf} - S_{wc}}{f_{wf}} \tag{D–15}$$

$$\textit{After breakthrough}: \overline{S}_w - S_{w2} = \frac{1 - f_{w2}}{\left(\frac{\partial f_w}{\partial S_w}\right) S_{w2}} \tag{D–16}$$

where:

f_{wf} = fraction of water flowing at the flood front
f_{w2} = fraction of water flowing at the producing end of the system
$\overline{S}_w$ = average water saturation after breakthrough, fraction
S_{wf} = water saturation at the flood front, fraction

FIGURE D–3. Determination of Average Water Saturation at Breakthrough

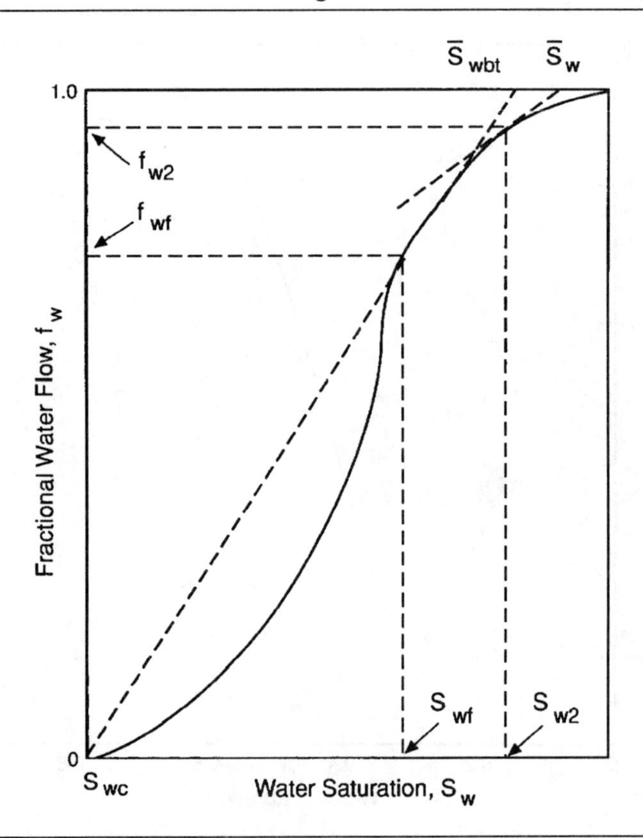

$\bar{S}_{wbt}$ = average water saturation at breakthrough, fraction
S_{wc} = connate water saturation, fraction
S_{w2} = water saturation at the producing end of the system, fraction

Figure D–3 presents graphical solutions for average water saturations at and after water breakthrough. The average water saturations can be used to calculate displacement efficiencies before and after water breakthrough.

Displacement efficiency that is governed by rock and fluid properties is given by:

$$E_D = \frac{\dfrac{S_{oi}}{B_{oi}} - \dfrac{S_{or}}{B_{or}}}{\dfrac{S_{oi}}{B_{oi}}} \qquad \text{(D–17)}$$

where:

S_o = oil saturation, fraction
B_o = oil formation volume factor, RBO/STBO
i,r = subscripts denoting initial (before flooding) and residual (after flooding) condition, respectively

If oil and water are the only fluids present in the formation

$$S_o = 1 - S_w$$

Then oil displacement efficiency can be re-expressed as

$$E_D = \frac{S_{wor} - S_{wi}}{1 - S_{wi}} \tag{D–18}$$

when:

B_{oi} = B_{or}
S_{wor} = water saturation at the residual oil saturation which can be determined from the fractional flow curve for a given fractional water flow (Figure D–1).

REFERENCES

1. Wang, B. and T. S. Teasdale. GASWAT PC: A Microcomputer Program for Gas Material Balance with Water Influx. SPE Paper 16484, Petroleum Industry Applications of Microcomputers, Del Lago on Lake Conroe, Texas, June 23–26, 1987.
2. Craft, B. C. and M. F. Hawkins. *Applied Petroleum Reservoir Engineering.* Englewood Cliffs, N.J: Prentice-Hall, Inc., 1959, 205–208.
3. Dake, L. P. *Fundamentals of Reservoir Engineering,* New York, N.Y.: Elsevier Scientific Publishing Company, 1978, 303–341.
4. Schilthuis, R. J. "Active Oil and Reservoir Energy," *Trans. AIME* 118 (1936): 33–52.
5. Hurst, W. "Water Influx into a Reservoir and its Application to the Equation of Volumetric Balance," *Trans. AIME* 151 (1943): 57–72.
6. Hurst, W. "The Simplication of the Material Balance Formulas by the Lap Lace Transformation," *Trans. AIME* 213 (1958): 292–303.
7. Van Everdingen, A. F. and W. Hurst. "The Application of the Lap Lace Transformation to Flow Problems in Reservoirs," *Trans. AIME* 186 (1949): 305–324.
8. Buckley, S. E. and M. C. Leverett. "Mechanism of Fluid Displacement in Sands," *Trans. AIME* 146 (1942): 107–116.

APPENDIX E

▼ ▼ ▼

North Ward Estes Field

To illustrate the importance and value of the effective presentation of performance analyses (primary plus secondary) and design of an EOR project, the North Ward Estes field, a mature field located in Ward and Winkler Counties, Texas, has been considered.[1,2]

INTRODUCTION

The North Ward Texas Estes (NWE) field, located in Ward and Winkler Counties, Texas (see Figure E–1), was discovered in 1929.[1] Cumulative oil produced is more than 320 million bbl (25% OOIP). The field has been waterflooded since 1955.

Geologically, the NWE field resides on the western flank of the Central Basin Platform. Yates, the dominant producing formation, includes up to seven major reservoirs and is composed of very fine-grained sandstones to siltstones separated by dense dolomite beds.[1,2] Within the 3,840-acre project area, average depth is 2,600 ft. Porosity and permeability average 16% PV and 37 MD, respectively. Reservoir temperature is 83°F. The flood patterns are 20-acre, five-spots, and line drives.

CO_2 flooding was implemented in early 1989 in a six-section project area located in the better part of the field in terms of cumulative oil production and reservoir rock quality.

FIELD HISTORY AND DEVELOPMENT

Except for the most productive parts, which were drilled on 10-acre spacing, the field was initially developed on 20-acre spacing.

FIGURE E–1. NWE Field *(Copyright © 1991, SPE, from* SPE Reservoir Engineering, *February 1991*[1]*)*

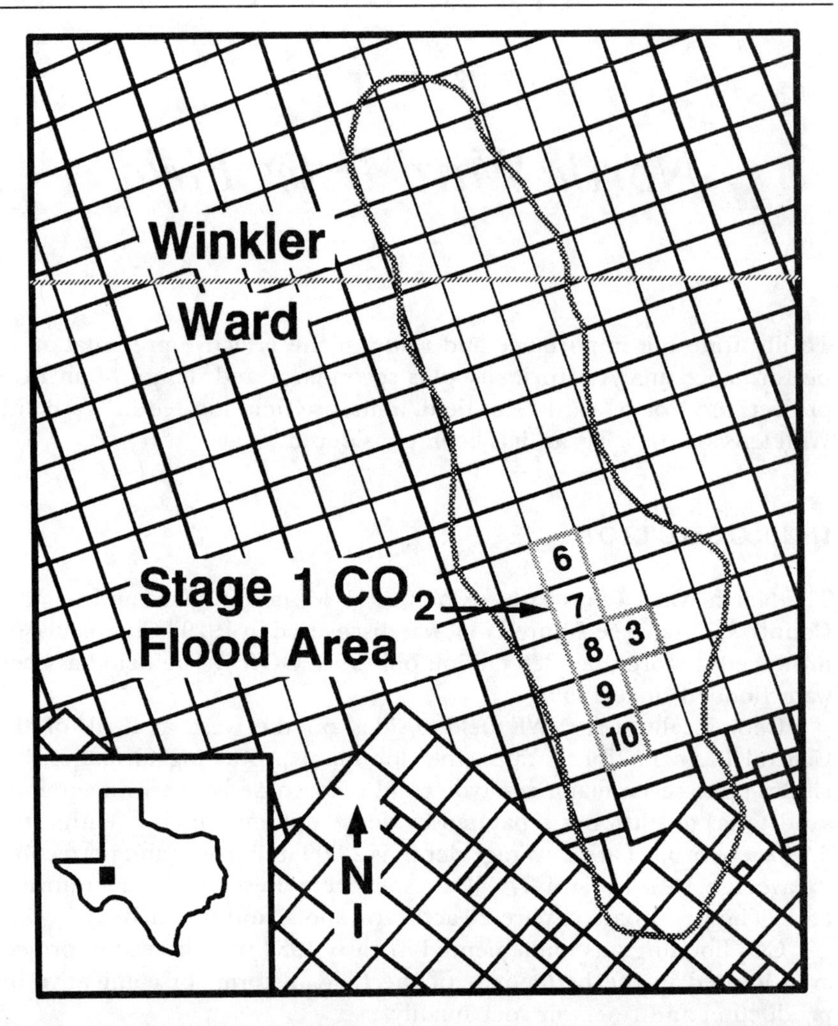

Until the early 1950s, a typical completion consisted of drilling to the top of the Yates, drilling ahead and checking for gas caps, setting casing through the gas sands, drilling to total depth, shooting the producing section with nitroglycerine, cleaning out the hole, and hanging a perforated liner from the casing. Practices changed in the early 1950s to cased-hole completions, hydraulic fracturing, and acidizing. About one-half of the current injectors are shot, open-hole completions. Vertical sweep has been adversely affected because of the inability to measure and control the injection profiles.

Figure E–2 shows the production and injection history of the project area. Primary production peaked in 1944 and was approaching the economic limit in the mid-1950s. A 960-acre pilot waterflood began in 1954. Oil production responded quickly, and the flood was expanded to the rest of the project area during the next two years. The prevailing flood patterns were 40-acre, five-spots.

Oil production increased steadily after 1954, reached a peak in 1960, and then declined at 11%/yr until 1979, when it began to stabilize as a result of drilling infill and replacement wells, injection-profile modifications by means of polymer treatments, and pattern tightening and realignment (Section 3 and 6 through 8 were converted to 20-acre, five-spot patterns and Section 9 and 10 to 20-acre, line drive patterns).

By the end of 1988, the six sections had produced 29% of the OOIP. Waterflooding the Yates has been very successful, as evidenced by the 2.3 ratio of ultimate secondary to ultimate primary production from wells existing at the beginning of waterflooding. The favorable mobility ratio in these reservoirs indicates good areal sweep efficiency. Because of the high Dykstra-Parsons coefficient (0.85) and permeability contrast among the major sands, the vertical conformance has been poor. Even after injection of 2.6 waterflood-moveable PV, less than 50% of the oil recoverable by waterflooding has been produced.

FIGURE E–2. Production and Injection History, Six-Section Project Area *(Copyright © 1991, SPE, from* SPE Reservoir Engineering, *February 1991*[1]*)*

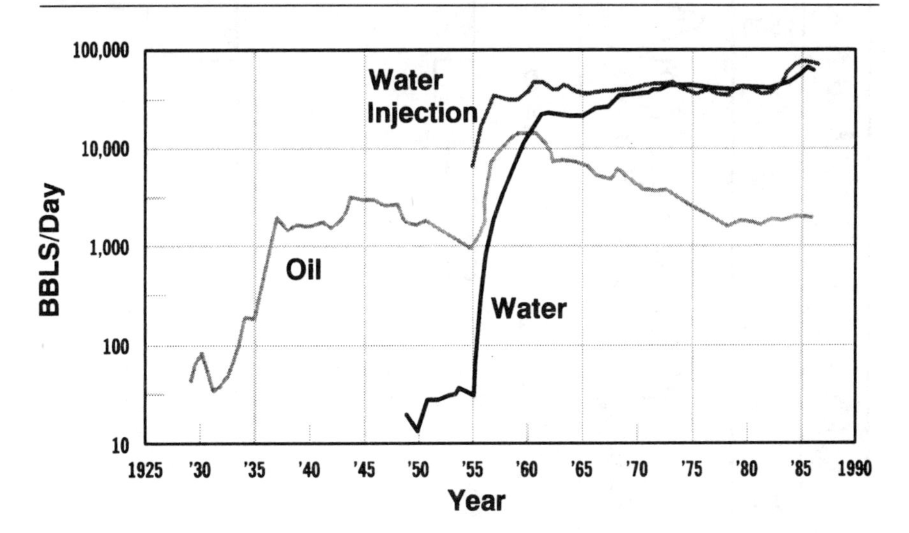

RESERVOIR GEOLOGY AND PROPERTIES

A comprehensive geologic study and reservoir characterization was conducted to characterize the individual reservoirs of the Yates, which consist of very fine-grained sandstones to siltstones separated by dense dolomite beds. In descending order, these sands are Sands BC, D, E, Strays, J_1, and J_2 (see Figure E–3). The general depositional environment was a tidal-flat-to-lagoonal setting situated to the east of and behind the shelf margin. The reservoirs were deposited as sand and silt in the subtidal-to-beach environment and silt-to-clay in the supratidal environment. Depositional strike was parallel to the shelf margin, which is parallel to the present northwest/southeast section lines.

Sand BC is a siltstone to fine-grained sandstone with detrital clay. The depositional environment was that of a shallow-water tidal flat with an abundant amount of windblown sediments. A zone of low porosity and permeability trends northwest/southeast through the middle of the project area. Most of Sand BC was in the original gas cap. Sands D and E are similar to Sand BC, but their porosities and permeabilities are more

FIGURE E–3. Injection-Profile Surveys *(Copyright © 1991, SPE, from* SPE Reservoir Engineering, *February 1991*[1]*)*

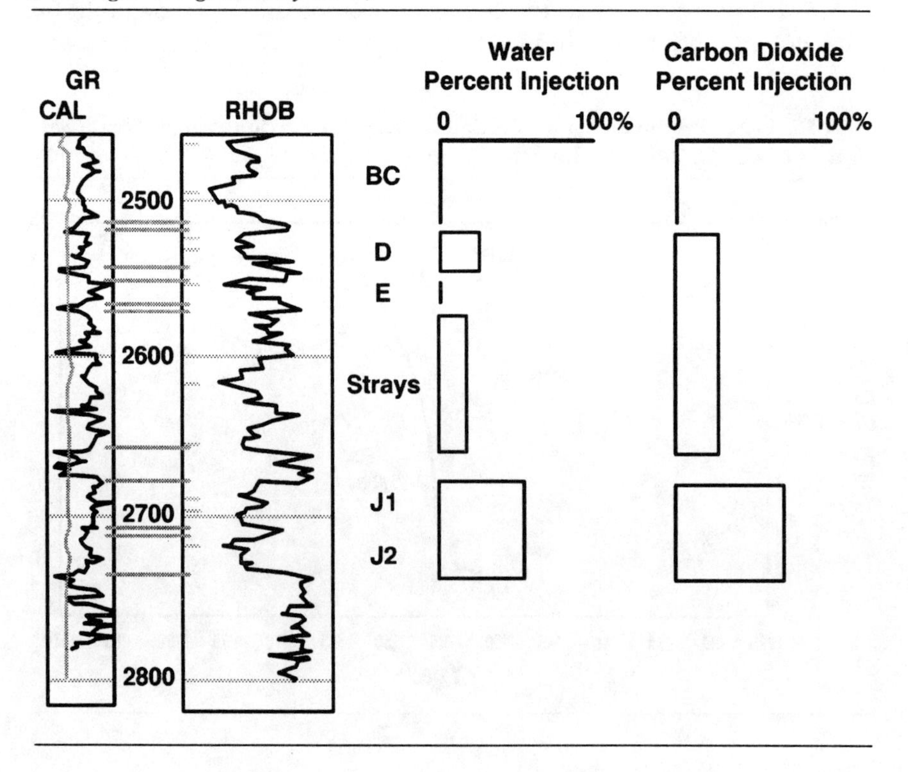

TABLE E–1. Yates Reservoir Properties (CO_2 Project Area) *(Copyright © 1991, SPE, from* SPE Reservoir Engineering, *February 1991[1])*

Formation	Yates
Lithology	Sandstone
Depth, ft	2,500
Reservoir temperature, °F	83
Porosity, % PV	16
Permeability to air, md	37
Dykstra-Parsons coefficient	0.85
Initial conditions	
Water saturation, % PV	50
Reservoir pressure at gas/oil contact, psia	1,400
Saturation pressure, psia	1,400
Oil FVF, RB/STB	1.2
Oil viscosity, cp	1.4
Solution GOR, scf/STB	500
Oil gravity, °API	37
S_{orw}, % PV	25
MMP (pure CO_2), psia	937
Flood patterns (Sections 3 and 6 to 8)	20-acre five-spots
Flood patterns (Sections 9 and 10)	20-acre line drives

variable. The Strays sand is composed of thin-bedded, lenticular, and intertidal to subtidal siltstones and fine-grained sandstones with the highest clay content of any Yates interval. Because of this, permeability and reservoir continuity suffer while porosity remains high. Sands J_1 and J_2 are composed of coarser sands with much less clay content and, therefore, have higher effective porosities and permeabilities. The depositional environment was a beach to near-shore marine where turbulence winnowed finer silts and clays out of the strike-oriented sand deposits. Table E–1 lists average reservoir properties for the Yates.

LABORATORY WORK

Extensive laboratory work was conducted to support the evaluation of CO_2 flooding in the NWE field.

- Black-oil PVT and oil/CO_2 phase-behavior studies of recombined separator oil and gas samples (see Table E–2) determined oil swelling, viscosity reduction, and phase transition pressures vs. mole percent CO_2. The PVT data show the typical complex phase

TABLE E–2. Analysis of Separator and Reservoir Fluids *(Copyright © 1991, SPE, from* SPE Reservoir Engineering, *February 1991*[1]*)*

	Separator		Reservoir
	Gas (mol %)	Liquid (mol %)	Fluid (mol %)
N_2	0.44	0.00	0.19
CO_2	1.25	0.03	0.44
H_2S	1.89	0.04	0.14
Methane	56.89	0.34	20.25
Ethane	15.47	1.01	5.91
Propane	13.44	2.82	5.89
Iso-butane	2.23	1.81	1.46
n-butane	4.88	5.68	4.12
Iso-pentane	1.41	3.97	2.53
n-pentane	1.08	3.67	2.38
Hexanes	0.75	6.40	4.35
Heptanes-plus	0.27	74.23	52.34
	100.00	100.00	100.00

Other Physical Properties

Bubblepoint pressure, psia	870
Solution gas at 870 psia, scf/STB	288
Temperature, °F	83
Oil viscosity at 870 psia, cp	1.60
Molecular weight heptanes-plus	229
Density at 60°F (heptanes-plus)	0.87

behavior exhibited by CO_2/light-crude-oil systems at low reservoir temperature (see Figure E–4).

- Slim-tube experiments determined minimum miscibility pressure (MMP). Figure E–5 shows the results of the displacement of reconstituted reservoir fluid by pure CO_2 in a packed column at different pressures. Additional displacement tests were conducted with five different CO_2/hydrocarbon-gas mixtures. The MMP ranged from 1,010 to 1,350 psia vs. 937 psia for pure CO_2. No significant changes in ultimate slim-tube oil recovery were observed. These tests verified that published correlations adequately estimate the MMP for NWE oil and impure CO_2.
- CO_2 flooding of restored-state composite cores determined the mobilization and recovery of the waterflood residual oil satura-

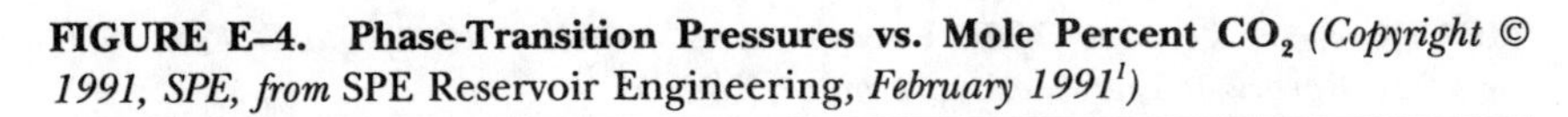

FIGURE E–4. Phase-Transition Pressures vs. Mole Percent CO_2 *(Copyright © 1991, SPE, from* SPE Reservoir Engineering, *February 1991*[1]*)*

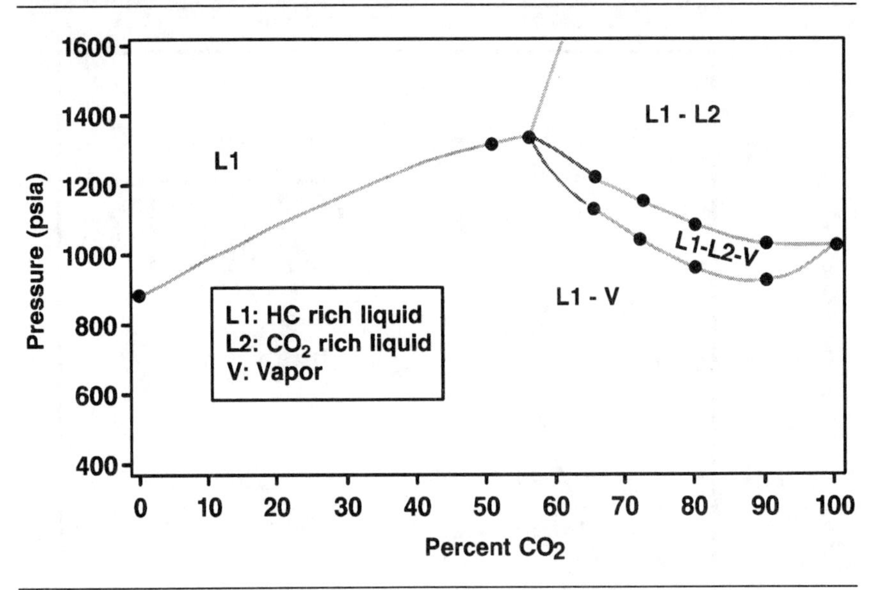

tions, S_{orw}. The core assembly (see Figure E–6) was constructed from l-in.-diameter plugs drilled from NWE cores epoxied into confining stainless steel sleeves. Capillary contact between segments was maintained with sieved core material. The displacement tests were preceded by cleaning the core assembly with toluene, methanol, and CO_2; injecting brine into the evacuated core; displacing the mobile brine with reconstituted reservoir fluid; and waterflooding to S_{orw}.

Table E–3 lists the residual oil saturations to miscible flooding, S_{orm}, determined at reservoir temperature and at pressures a few hundred psi above the MMP with pure and impure CO_2 and with different water-alternating-gas (WAG) injection ratios. These values should be obtained in the reservoir if levels of physical dispersion in the corefloods are comparable with those obtained in the field and if only a small part of the long core was needed to develop multiple-contact miscibility (MCM). No field data are available for NWE; however, where coreflood and field S_{orm} values have been compared, good agreement has been found.

- Amott tests determined the wettability of wettability-preserved cores. The Amott index to oil was zero for all tests, suggesting that the Yates sands are water-wet.

FIGURE E–5. Percent OOIP Recovered vs. Pressure *(Copyright © 1991, SPE, from* SPE Reservoir Engineering, *February 1991*[1]*)*

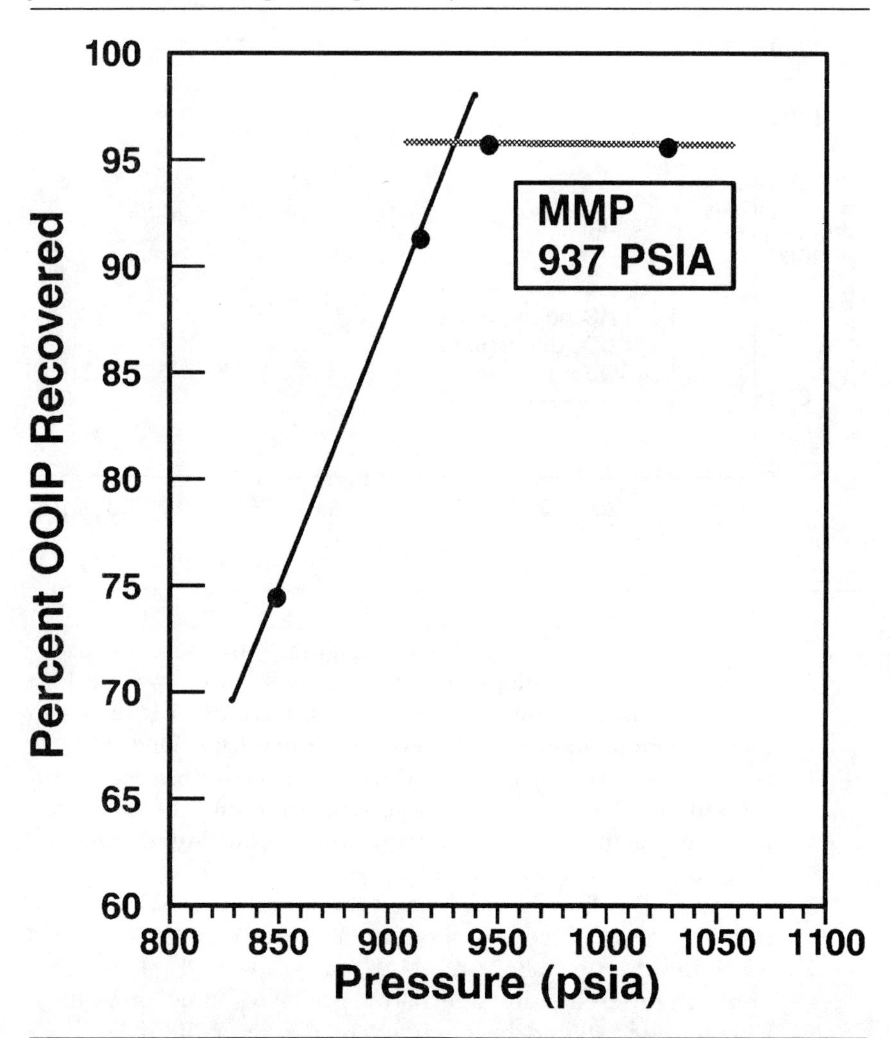

SIMULATION APPROACH

The initial CO_2 flood design called for the selection of typical patterns from the project area, a detailed reservoir characterization of each such pattern, history matches of the waterflood performance, predictions for continuation of the waterflood, predictions for CO_2 flooding, and scale-up of the predictions for these typical patterns to the whole project area.

FIGURE E–6. Long-core Properties *(Copyright © 1991, SPE, from* SPE Reservoir Engineering, *February 1991*[1]*)*

1	2	3	4	5	6	7	8	9	10	11	12
2.85	2.45	2.29	2.26	2.50	2.64	2.69	1.96	2.67	2.23	2.93	2.52

(Individual Lengths in Inches)

Total Length = 30 Inches

1	2	3	4	5	6	7	8	9	10	11	12
14.4	14.6	11.6	17.8	11.2	17.7	10.2	42.8	8.5	56.6	6.9	92.3

(Individual Permeabilities in MD)

Composite Permeability = 12.20 MD (Experimental)

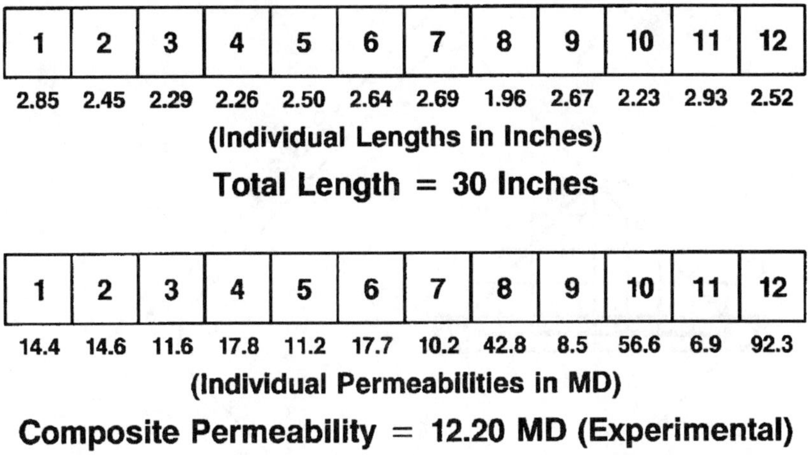

TABLE E–3. CO_2 Corefloods (83°F, 1,000 to 1,200 psig outlet pressure) *(Copyright © 1991, SPE, from* SPE Reservoir Engineering, *February 1991*[1]*)*

				Injection Gas Composition	
Wag Ratio	S_{oi} (% PV)	S_{orw} (% PV)	S_{orm} (% PV)	CO_2 (%)	Hydrocarbon Gas (%)
0:1	58	29	14	100	0
1:1	58	29	11	100	0
1:1	58	28	12	100	0
1:1	59	30	14	88	12
2:1	58	30	17	100	0

However, time constraints, computer cost, and concerns about data availability and quality dictated a change to the simpler approach of average patterns. Three-dimensional-pattern models were developed for four of the six sections. Ten to twelve layers were necessary to characterize the seven major sand bodies of the Yates. An areal view of the model and the layer properties for one of the models are shown in Figure E–7 and Table E–4, respectively. Net pay and porosity for each major sand body are averages developed from geological maps. The permeability stratification within and among the major sand bodies was developed from core

FIGURE E–7. Areal View of 1/8 Five-Spot Pattern *(Copyright © 1991, SPE, from* SPE Reservoir Engineering, *February 1991*[1]*)*

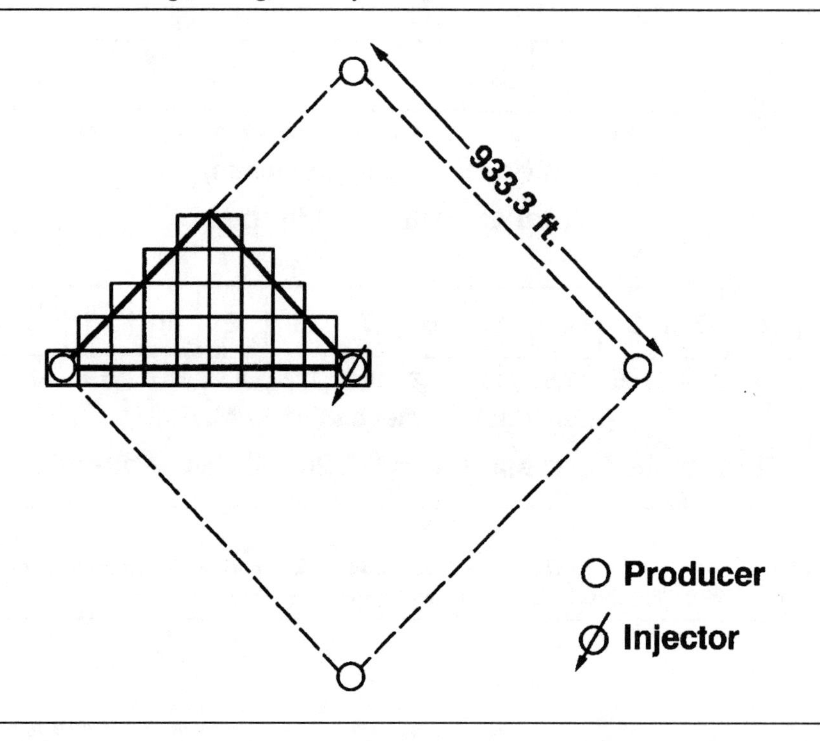

data and injection profiles. The Dykstra-Parsons coefficient for the layered model agreed with that calculated from core data.

A finite-difference, four-component, modified black-oil simulator was selected for history matching and waterflood and CO_2 flood predictions. This simulator is suitable for first-contact miscibility or for multicontact miscibility if the miscibility occurs within a mixing zone having a length that is small compared with the length of the imposed grid.

CO_2 INJECTIVITY TEST

A CO_2 injectivity test was conducted to investigate injectivity losses during CO_2 and water injection cycles. Potential injectivity loss was a concern because of the sensitivity of project economics to injection rates. An injector in good mechanical condition and with no hydraulic fracturing was selected for this purpose. Before, during, and after CO_2 injection,

TABLE E–4. Layer Properties of an Average Pattern *(Copyright © 1991, SPE, from* SPE Reservoir Engineering, *February 1991*[1]*)*

Layer	Sand	S_{oi} (% PV)	S_{gi} (% PV)	Net Pay (ft)	Porosity (%)	$k_{x\text{-}y}$ (md)	k_z * (md)
1	BC (gas)	10	41	10.0	16.6	11.0	0.11
2	BC	51	0	4.8	16.3	11.0	0.11
3	BC	51	0	1.3	16.3	36.8	0.20
4	BC	51	0	8.8	16.3	1.0	0.01
5	D and E (gas)	10	41	5.0	16.8	10.0	0.10 **
6	D and E	51	0	1.6	16.8	54.0	0.20
7	D and E	51	0	19.5	16.8	4.0	0.04
8	Strays	51	0	2.0	15.3	15.0	0.15 **
9	Strays	51	0	11.9	15.3	2.0	0.02
10	J_1 and J_2	51	0	21.5	16.3	26.0	0.20 **
11	J_1 and J_2	51	0	15	16.3	66.0	0.20
12	J_1 and J_2	51	0	11.0	16.3	1.5	0.02

* One percent of $k_{x\text{-}y}$ with 0.2-md maximum.
** Zero transmissibility between Layers 4 and 5, 7 and 8, 9 and 10.

step-rate tests, injection-profile surveys (Figure E–3), and pressure-transient falloff tests were run. After injection of 30 MMscf (1.3% HCPV) of CO_2, the well was returned to water injection. The major conclusions were as follows:

- No reduction in injection rates was observed during or after CO_2 injection. The CO_2 injection rate (expressed in terms of reservoir barrels) was about 20% higher than the water injection rate at the same flowing bottomhole injection pressures.
- No significant change in injection profile was observed during and after CO_2 injection.
- The CO_2 falloff data were used to estimate such parameters as mobility ratio, swept volume, and average CO_2 saturation in the swept region. These values were in agreement with laboratory measurements from CO_2 corefloods.

There is some uncertainty about whether enough CO_2 was injected to detect potential losses in injectivity. Because reductions in injectivity generally are not associated with water-wet systems and because no changes in injectivity were observed during the corefloods (for two of the corefloods, a chase-water injection state was added for measuring injectivity changes), additional expenditures for a prolonged field injectivity test could not be justified.

HISTORY MATCHING

History matching was conducted by entering the scaled oil production and water injection rates for the years 1929 to 1986 and letting the simulator calculate the gas and water production rates and reservoir pressures. Because of limited GOR and pressure data, history matching consisted mostly of matching water production rates. The matches were obtained largely by adjusting layer permeabilities and, to a lesser degree, the oil and water relative permeability curves (see Figure E–8).

To improve the prediction of when CO_2 will breakthrough at the producers, particular attention was paid to matching the water breakthrough time after waterflood initiation. In developing the average pattern models, most of the oil response and water breakthrough observed in the field between 1955 and 1962 were assumed to come from high-permeability zones. This assumption apparently is supported by the good correlation between cumulative oil and cumulative water production for individual wells during 1955–1962 (see Figure E–9). Wells with the highest cumulative water production during this period also had the highest cumulative oil production.

FIGURE E–8. Water/Oil Relative Permeability *(Copyright © 1991, SPE, from* SPE Reservoir Engineering, *February 1991*[1]*)*

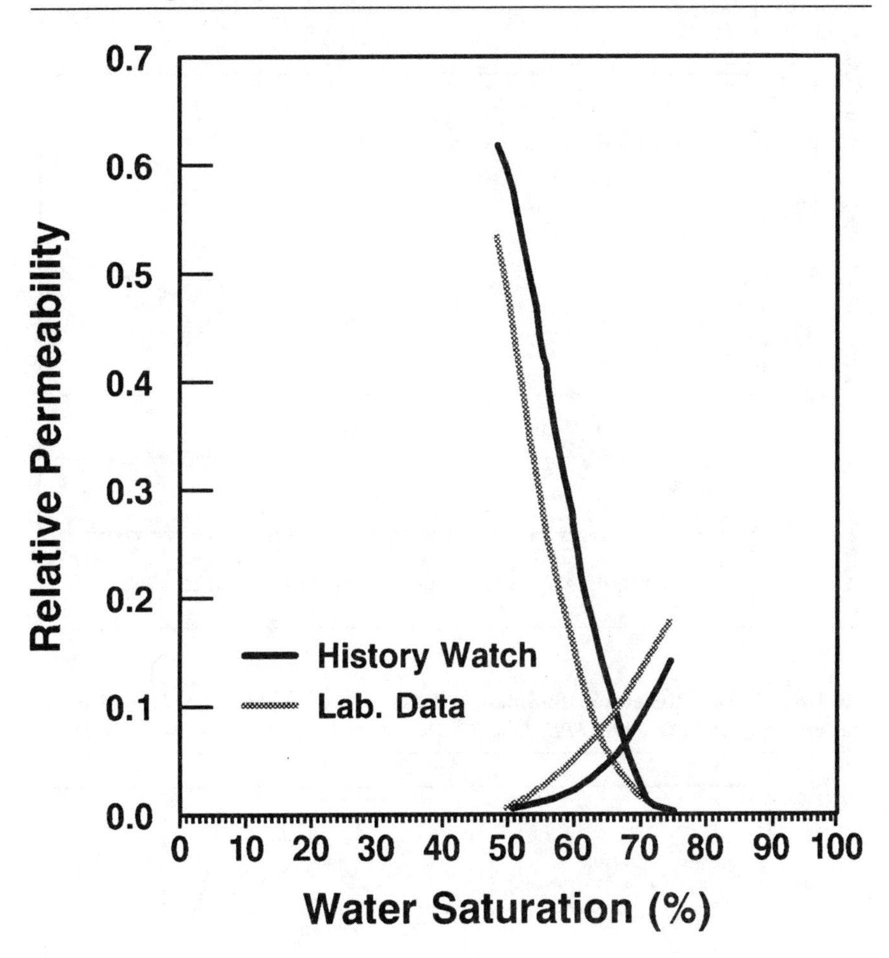

PERFORMANCE PREDICTIONS-PATTERNS

The history matches were followed by prediction runs for continuation of the waterflood and for CO_2 flooding. Figure E–10 shows the simulation results (history match and waterflood and CO_2 flood predictions) for one of the average patterns. Simulator input parameters specific to the CO_2 flood predictions were as follows.

FIGURE E–9. Correlation for Wells on 10-Acre Spacing *(Copyright © 1991, SPE, from* SPE Reservoir Engineering, *February 1991*[1]*)*

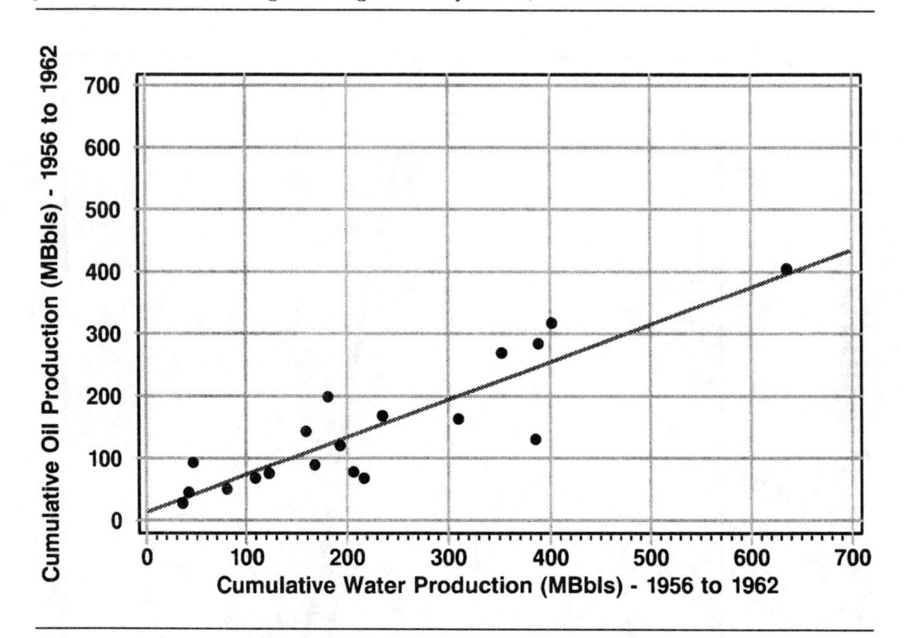

FIGURE E–10. Reservoir Simulation Results for an Average 1/8 Five-spot Pattern *(Copyright © 1991, SPE, from* SPE Reservoir Engineering, *February 1991*[1]*)*

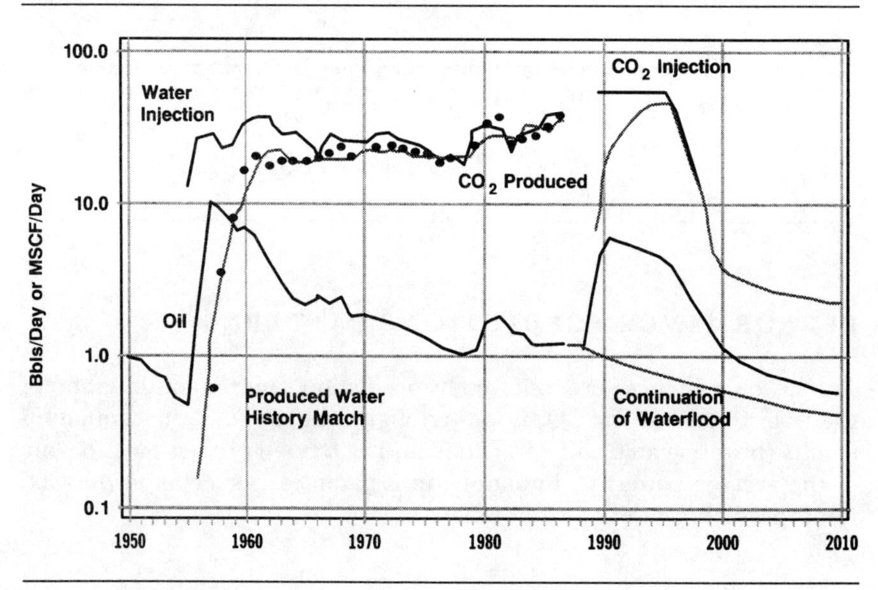

- WAG—Because corefloods implied that a WAG ratio of 1:1 was optimal, CO_2 flood predictions were run at that WAG ratio, injecting 2.5% HCPV per WAG cycle.
- Injection rate—CO_2 injection rates (in terms of reservoir barrels) were increased 20% above the average water injection rates. The results of the field injectivity test justified this increase.
- Slug size—a 38% HCPV CO_2 slug was injected over a 10-year period. As discussed below, this slug size was selected on economic considerations.
- S_{orm}—Predictions were run with 12% PV; the value was determined from CO_2 corefloods conducted at a 1:1 WAG ratio. No additional waterblocking over that already reflected in the experimental S_{orm} was introduced.
- CO_2/Oil Mixing Parameter—a mixing parameter of 0.67 was used in a modified black-oil simulator to approximate the influence of viscous fingering on sweep efficiency in coarsely gridded simulations.

Sensitivity studies were conducted to examine the effects of changes in WAG ratio, S_{orm}, CO_2/oil mixing parameter, and vertical permeability on oil recovery. Continuous CO_2 injection (zero WAG ratio) recovered only 7.1% OOIP, compared with 9.8% OOIP with a WAG ratio of 1:1, mostly because of excessive CO_2 channeling through high-permeability layers. At a WAG ratio of 2:1, peak oil production rates were maintained for a longer period. Incremental recovery, however, decreased to 7% OOIP because of higher S_{orm} range (see Table E–3) and variations in the CO_2/oil-mixing parameter between 0.5 and 0.75, the incremental oil recovery ranged from 6.3 to 10.5% OOIP. Changes in vertical permeability from 0 to 10% of horizontal permeability had a negligible effect on incremental oil recovery. Analytical models also predict that gravity override (the major source of CO_2 flood oil), the distance the solvent will travel from the injector until it is concentrated at the top of the layer, is greater than the distance between injectors and producers for plausible vertical permeabilities.

OPTIMUM ECONOMIC CO_2 SLUG SIZE

Oil recovery predictions were made for seven CO_2 slug sizes (15 to 75% HCPV). The economically optimum slug size was found by balancing the increase in revenues from additional oil production with the cost of purchasing additional CO_2 and the increasing capital and operating costs to process larger volumes of produced CO_2. In terms of rate of return, the optimum slug size was found to be between 38 and 60% HCPV of CO_2 injected. All predictions were run at a 38% HCPV slug size, requiring a CO_2 recycle plant capacity of 65 MMscf/D for the project.

PERFORMANCE PREDICTIONS—PROJECT AREA

The CO_2 flood prediction for the entire project area (see Table E–5) is based on the scale-up of the average pattern simulation results. The scale-up of the four sections for which average pattern simulations were performed is straightforward. The prediction for a given section equals the prediction from the average pattern of that section times the number of patterns to be flooded with CO_2 (i.e., it is simply the reverse of the scale-down step that defined the average patterns).

No pattern simulations were performed for Sections 9 and 10 because of their similarities in waterflood performance with Sections 8 and 7, respectively. Because Section 9 and 10 were converted to line drives in 1979, correction factors had to be developed before the predictions for five-spot patterns could be applied. These correction factors were developed as follows. A line drive model was initialized with the history-matched saturations and pressures from one of the averaged five-spot patterns as of 1979 (the year when pattern realignments and tightening began in the field). The line drive pattern was water-flooded for 10 years (to allow the saturation and pressure distributions from the five-spot to adjust to those found in a line drive) after which a CO_2 flood prediction was made. Annual correction factors to convert five-spot CO_2 flood performance predictions to those expected from line drives were developed from the CO_2 flood prediction for the previously mentioned line drive and the prediction for the five-spot that was used to initialize the line drive.

TABLE E–5. CO_2 Flood Performance Prediction *(Copyright © 1991, SPE, from* SPE Reservoir Engineering, *February 1991[1])*

Recovery to date, % OOIP	29
Ultimate recovery (primary and secondary), % OOIP	31
CO_2 Flood Recovery, % OOIP	8
Slug size, % HCPV	38
CO_2 injection/WAG cycle, % HCPV	2.5
WAG ratio, RB/RB	1:1
CO_2 utilization, Mscf/bbl oil	
Gross	12
Net	4

REFERENCES

1. Winzinger, R. et al. "Design of a Major CO_2 Flood—North Ward Estes Field, Ward County, Texas," *SPERE* (February 1991): 11–16.
2. Thakur, G. C. "Implementation of a Reservoir Management Program." SPE Paper 20748 presented at the 1990 SPE Annual Technical Conference and Exhibition, New Orleans, Sept. 23–26.

APPENDIX F

▼ ▼ ▼

Reservoir Management in the Means San Andres Unit

INTRODUCTION

This case study provides information on the evolution of reservoir management to meet changing economic and technical challenges as the field produced by primary, secondary, and tertiary methods. Reservoir management at Means San Andres Unit has consisted of an ongoing but changing surveillance program supplemented with periodic major reservoir studies to evaluate and make changes to the depletion plan.

Reservoir management techniques began within one year of discovery and became increasingly complex as operations changed from primary to secondary to tertiary. Reservoir description, infill drilling with pattern modification, and reservoir surveillance played key roles in reservoir management. Reservoir description methods evolved from the relatively simple techniques used in the 1930s to the recent use of high-resolution seismic to improve pay correlation between wells.

FIELD DISCOVERY AND DEVELOPMENT

The Means field was discovered in 1934. It is located about 50 miles northwest of Midland, Texas, along the eastern edge of the Central Basin platform (see Figure F–1). Table F–1 lists the reservoir and fluid properties. The field is a north-south-trending anticline separated into a North Dome and a South Dome by a dense structural saddle running east and west near the center of the field (see Figure F–2). Production in this field is from the Grayburg and San Andres formations at depths ranging from 4,200 to 4,800 ft. Figure F–3 is a type log showing the zonation.

The Grayburg is about 400 ft thick with the basal 100 to 200 ft considered gross pay. Production from the Grayburg was by solution-gas

FIGURE F–1. Permian Basin Principal Geologic Provinces *(Copyright © 1992, SPE, from* JPT, *April, 1992*[1]*)*

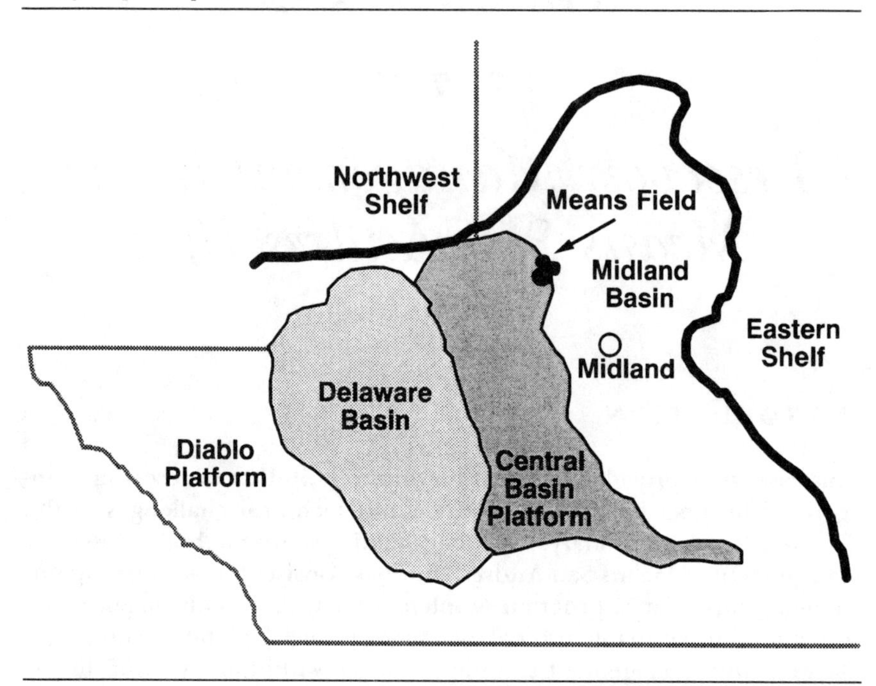

TABLE F–1. Reservoir and Fluid Properties, Means San Andres Unit

Formation name	San Andres
Lithology	Dolomite
Area, acres	14,328
Depth, ft	4,400
Gross thickness, ft	300
Average net pay, ft	54
Average porosity, %	9.0 (up to 25)
Average permeability, md	20.0 (up to 1,000)
Average connate water, %	29
Primary drive	Weak waterdrive
Average original pressure, psig	1,850
Stock-tank gravity, °API	29
Oil viscosity, cp	6
FVF, RB/STB	1.04
Saturation pressure, psi	310

FIGURE F–2. Structural Map *(Copyright © 1992, SPE, from* JPT, *April, 1992[1])*

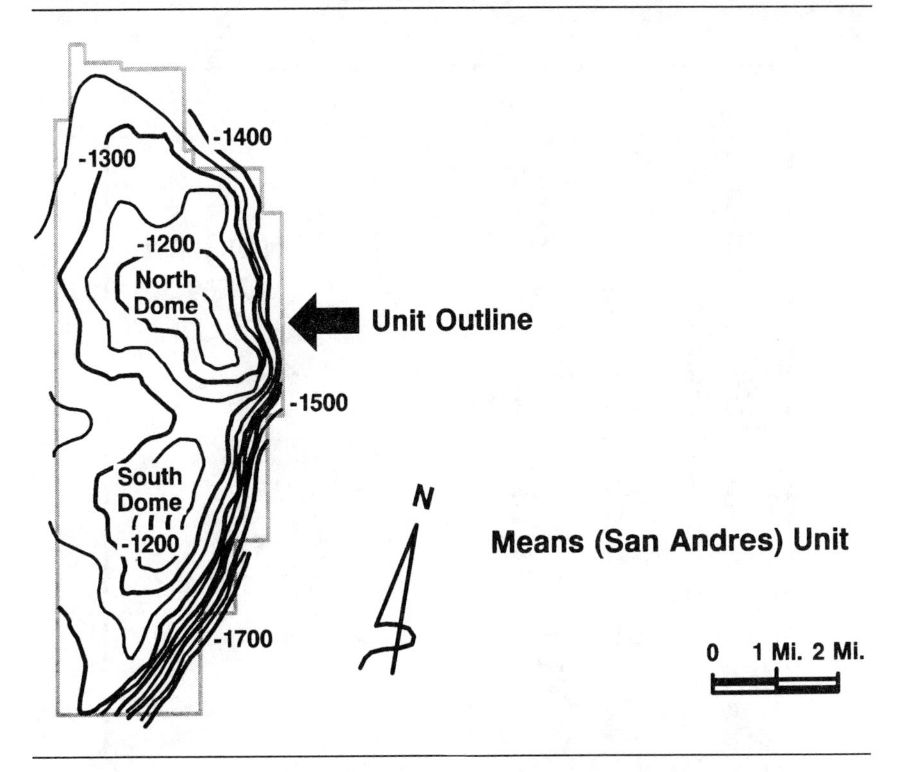

drive with the bubble-point at the original reservoir pressure of 1,850 psi. The Grayburg is of poorer quality, and its production has been minor relative to the San Andres. The San Andres is more than 1,400 ft thick, with the upper 200 to 300 ft being productive. The primary producing mechanism in the San Andres was a combination of fluid expansion and a weak waterdrive.

RESERVOIR MANAGEMENT DURING PRIMARY AND SECONDARY OPERATIONS

The first reservoir study was completed in 1935, and it concentrated on reservoir management related to primary recovery. Later, in 1959, a reservoir study was conducted to evaluate secondary recovery. Highlights of this study included one of Humble's first full-field computer simulations. For

FIGURE F–3. Type Log Means San Andres Unit MSAU 1216 *(Copyright © 1992, SPE, from* JPT, *April, 1992*[1]*)*

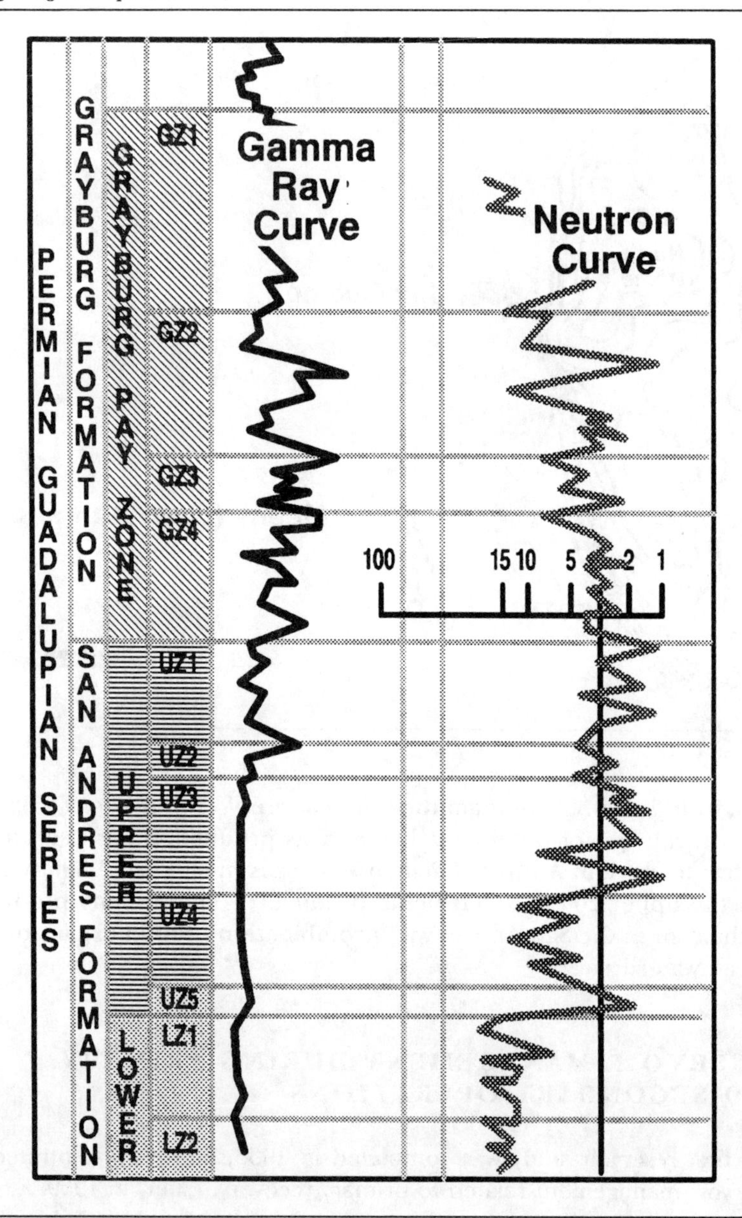

this study, additional data were accumulated, including further logging, fluid sampling, and special core data (e.g., capillary pressures and relative permeabilities).

Cross-sections, like the one shown in Figure F–4, aided in the design of an initial waterflood pattern. In 1963 the field was unitized and a peripheral flood was initiated (see Figure F–5). Twenty-four wells, distributed throughout the unit, were permanently shut-in and maintained as pressure response wells to monitor reservoir pressure. In 1967, as a result of increased allowable, it was realized that the peripheral injection pattern no longer provided sufficient pressure support.

In 1969 a reservoir engineering and geological study was conducted to determine a new depletion plan to offset the pressure decline. The geologic study included a facies study. In the North Dome, pressure data were correlated with the geological data to identify three major San Andres intervals: Upper San Andres, Lower San Andres oil zone, and Lower San Andres aquifer. A permeability barrier was mapped between the Upper and Lower San Andres. Analysis of pressure data from the observation wells indicated that neither North Dome nor South Dome were receiving adequate pressure support. This study recommended interior injection with a 3:1 line drive (see Figure F–6). Following implementation of this program, the unit production increased from 13,000 BOPD in 1970 to greater than 18,000 BOPD in 1972.

After reaching a peak in 1972, oil production again began to decline. A reservoir study conducted in 1975 indicated that all the pay was not being flooded effectively by the 3:1 line-drive pattern. An in depth geological study showed a lack of lateral and vertical distributions of pay. Old gamma-ray/neutron logs were correlated with core data to deter-

FIGURE F–4. West-East Structural Cross Section Means San Andres Unit *(Copyright © 1992, SPE, from* JPT, *April, 1992*[1]*)*

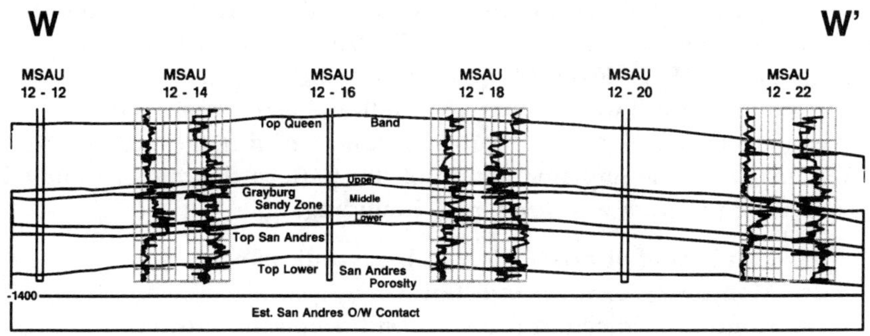

FIGURE F–5. Waterflood Injection Pattern *(Copyright © 1992, SPE, from* JPT, *April, 1992*[1]*)*

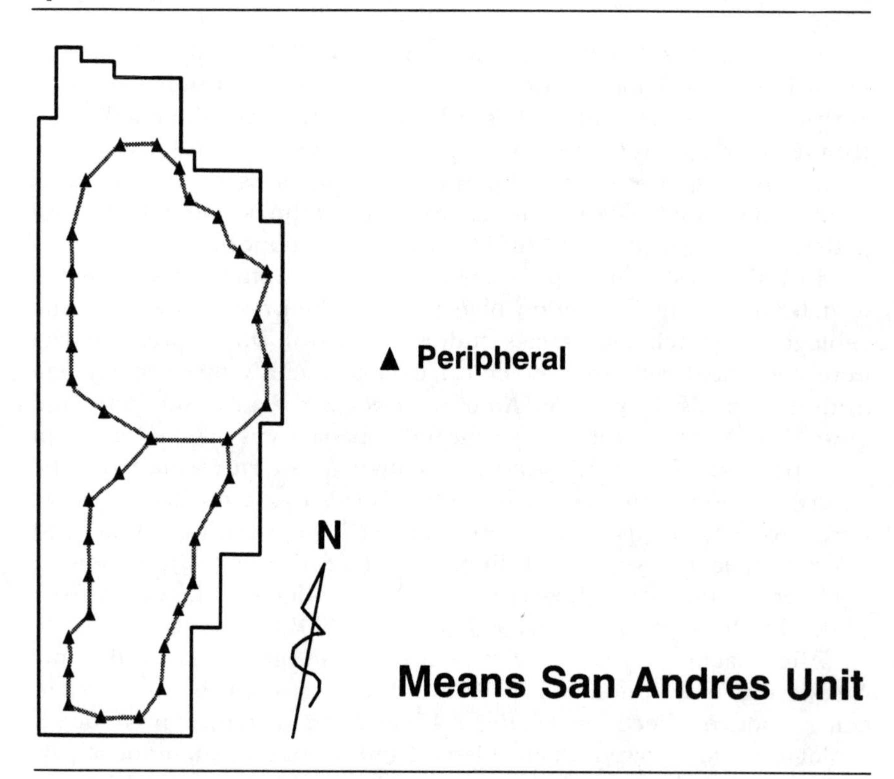

mine porosity-feet. Original-oil-in-place was calculated for up to six zones in each well in the field. This geological work provided the basis for a secondary surveillance program and later for the design and implementation of the CO_2 project.

A major infill drilling program was undertaken based upon potential additional recovery estimated from infill drilling with pattern densification (see Figure F–7). From 141 (20-acre spacing with an 80-acre inverted 9-spot pattern) infill wells, about 15.4 million barrels of incremental oil was estimated (see Figure F–8 for oil production with and without infill program). With the implementation of this pattern flooding, a detailed surveillance program was developed, including:

- Monitoring of production (oil, gas and water).
- Monitoring of water injection.
- Control of injection pressures with step-rate tests.

FIGURE F–6. Waterflood Injection Pattern *(Copyright © 1992, SPE, from* JPT, *April, 1992*[1]*)*

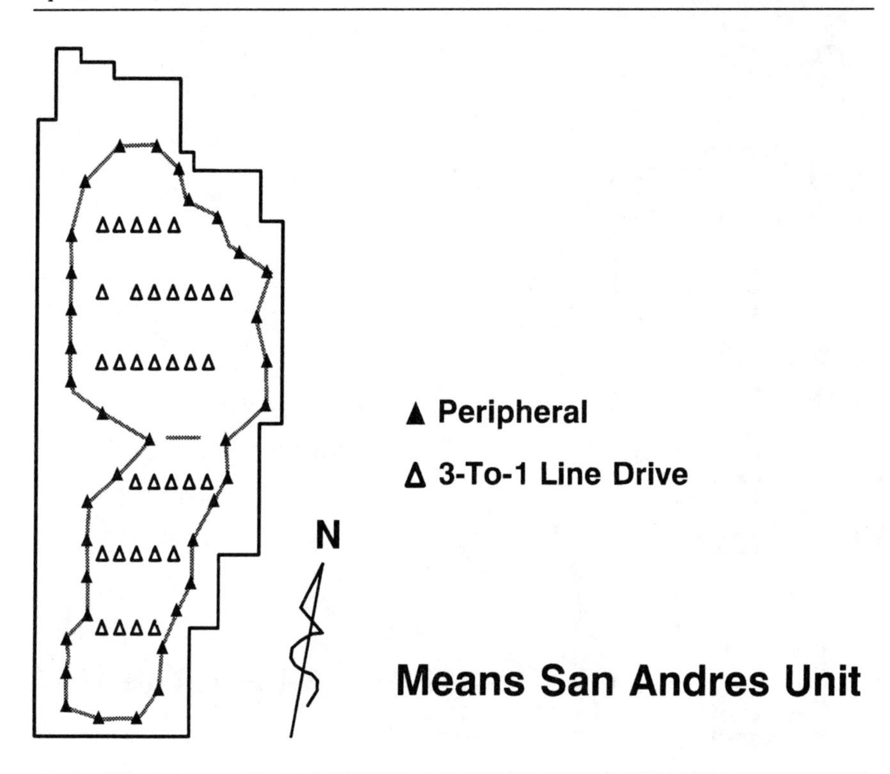

- Pattern balancing with computer balance programs.
- Injection profiles to ensure injection into all pay.
- Specific production profiles.
- Fluid level checks to ensure pump-off of producing wells.

RESERVOIR MANAGEMENT DURING TERTIARY OPERATIONS

In 1981–1982, a CO_2 tertiary-recovery reservoir study was conducted. At that time, several major CO_2 projects had been proposed for San Andres reservoirs, but none had been implemented. Although Means was similar to other San Andres fields in the Permian Basin, some properties (e.g., 6-cp oil viscosity, relatively high minimum miscibility pressure, and low formation parting pressure) made the Means Unit somewhat unique.

FIGURE F–7. Waterflood Injection Pattern *(Copyright © 1992, SPE, from* JPT, *April, 1992*[1]*)*

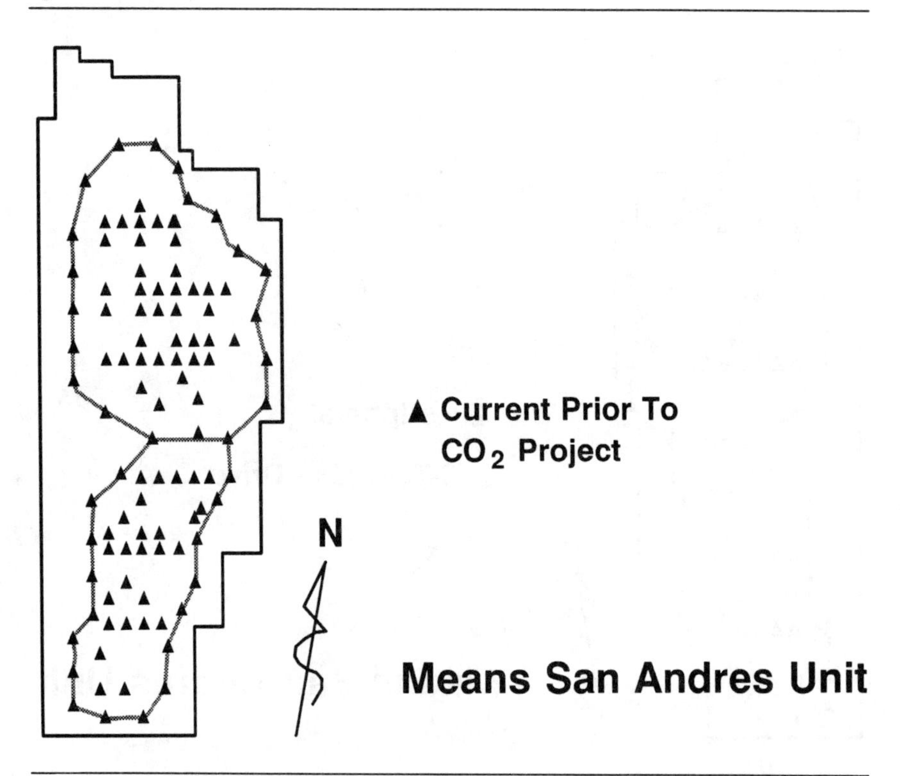

A CO_2 flood pilot along with extensive laboratory and simulation works were initiated. A detailed reservoir description program preceded this work and became the basis for planning the CO_2 tertiary project. Although this reservoir description was the building block for the project, it was continuously updated during the planning and implementation phases of the CO_2 project as more data became available.

Several 10-acre wells (generally injectors) were drilled as a part of the CO_2 project. The project consisted of 167 patterns on 6,700 acres, and it included 67% of the productive acres and 82% of OOIP.

A comprehensive surveillance program had been present during the waterflood. Before developing a similar program for CO_2 flooding, an operating philosophy was created by personnel from engineering, geology, and operations, and it was submitted to management for approval and support. Major operating objectives included:

FIGURE F–8. MSAU Infill Drilling *(Copyright © 1992, SPE, from* JPT, *April, 1992*[1]*)*

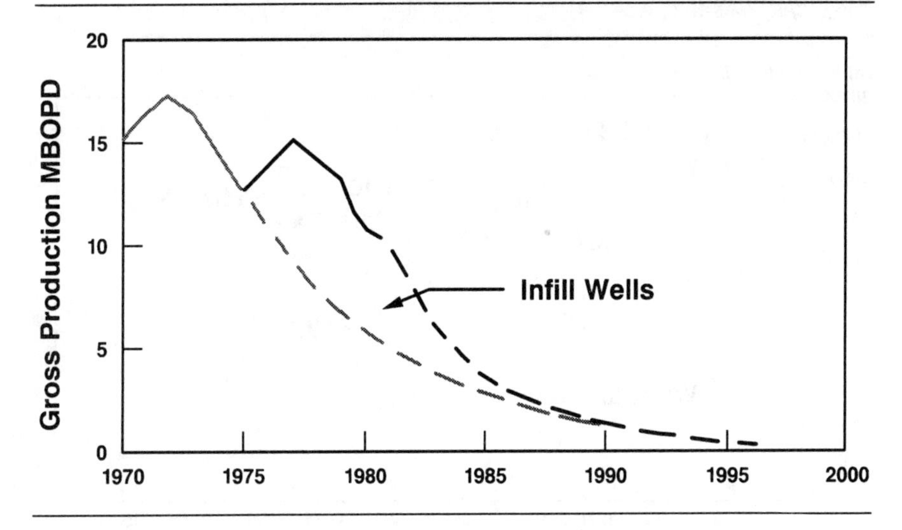

- Completing injectors and producers in all floodable pay.
- Maintaining reservoir pressure near the MMP of 2,000 psi.
- Maximizing injection below fracture pressure.
- Pumping off producers.
- Obtaining good vertical distribution of injection fluids.
- Maintaining balanced injection/withdrawals by pattern.

To satisfy these objectives, a surveillance program was developed involving engineers, geologists, and operations personnel. Major areas of surveillance included:

- Areal flood balancing.
- Vertical conformance monitoring.
- Production monitoring.
- Injection monitoring.
- Data acquisition and management.
- Pattern performance monitoring.
- Optimization.

The objectives of the surveillance program were to maximize oil recovery and flood efficiency and to identify and evaluate new opportunities and technologies. In addition, one of the key objectives was to obtain better reservoir descriptions to understand the reservoir processes. This effort included the use of high resolution seismic to improve pay

FIGURE F–9. Means San Andres Unit Oil Production 1970–1990 *(Copyright © 1992, SPE, from* JPT*, April, 1992*[1]*)*

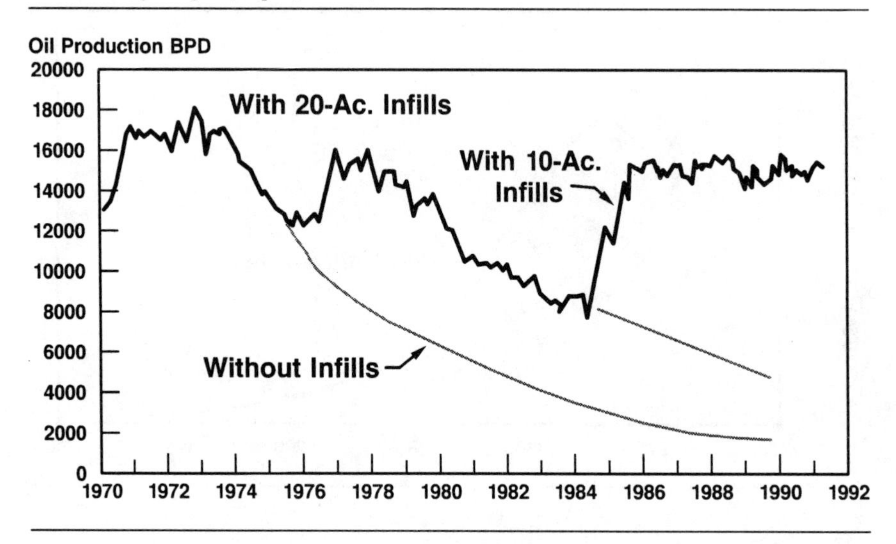

FIGURE F–10. MSAU CO_2 Project Area *(Copyright © 1992, SPE, from* JPT*, April, 1992*[1]*)*

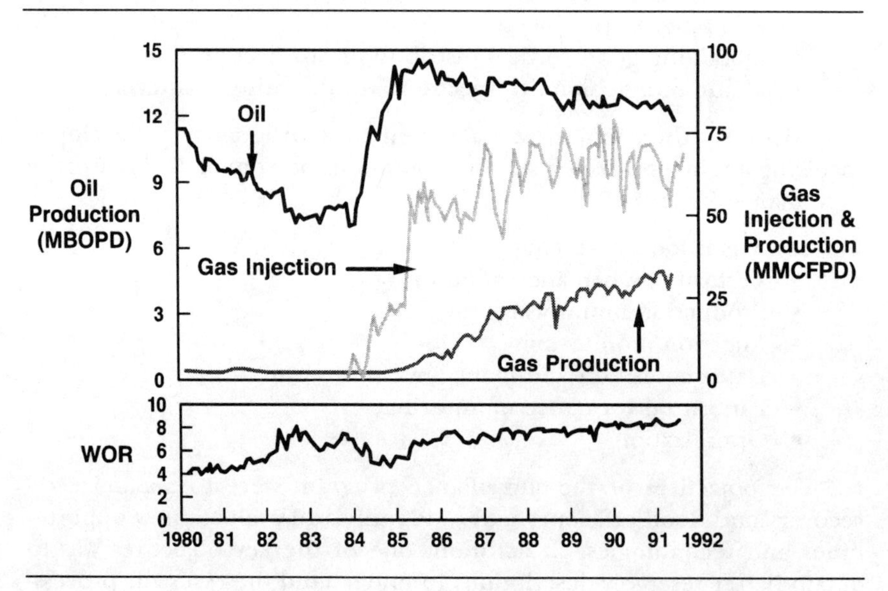

correlation between wells. The application of seismic sequence stratigraphic yielded significant insights into reservoir's complexities. It provided geometric template to constrain basic well-log correlations properly within a chronostratigraphic framework. Seismic-scale stratal geometries, combined with detailed geologic rock descriptions, defined this sequence framework. In addition, the stratigraphic model was used to optimize CO_2 project completions by ensuring that all pays are opened in both producers and injectors based on correlative stratigraphic zones.

As mentioned earlier, the CO_2 project was implemented as part of an integrated reservoir management plan, which included CO_2 injection, infill drilling, pattern changes, and expansion of the Grayburg waterflood outside the project area. Total unit oil production is shown in Figure F–9 with and without the project. Figure F–10 shows the performance of the tertiary project area.

REFERENCE

1. Stiles, L. H. and Magruder, J. B. "Reservoir Management in the Means San Andres Unit," *JPT* (April 1992): 469–475.

Bibliography

Amyx, Bass and Whiting. *Petroleum Reservoir Engineering.* New York: McGraw-Hill Book Company, 1960.

Anderson, J. H. et al. "Brassey Field Miscible Flood Management Program Features Innovative Tracer Injection." Paper 24874 presented at the SPE Annual Technical Conference and Exhibition, Washington, D.C., October 4–7, 1992.

Arps, J. J., et al. "A Statistical Study of Recovery Efficiency," *Bull. D14, API* (1967).

Arps, J. J. "Estimation of Decline Curves," *Trans., AIME* (1945).

Arps, J. J. "Estimation of Primary Oil Reserves," *Trans., AIME* 207, (1956): 182–191.

Beal, C. "The Viscosity of Air, Water, Natural Gas, Crude Oil and its Associated Gases at Oil Field Temperature and Pressure," *Trans., AIME* (1946): 94.

Begg, S. H., R. R. Carter, and P. Dranfield. "Assigning Effective Values to Simulator Gridblock Parameters for Heterogeneous Reservoirs," *SPERE* (November 1989): 455–63.

Beggs, H. D. and J. R. Robinson. "Estimating the Viscosity of Crude Oil Systems," *J. Pet. Tech.* (September 1975): 1140.

Boberg, T. C. Thermal Methods of Oil Recovery. New York: John Wiley & Sons, 1988.

Breitenbach, E. A. "Reservoir Simulation: State of the Art," *J. Pet. Tech.* (September 1991): 1033–36.

Buckley, S. E. and M. C. Leverett. "Mechanisms of Fluid Displacement in Sands," *Trans., AIME* 146, (1942): 107–116.

Burnett, D. B. and M. W. Dann. "Screening Tests for Enhanced Oil Recovery Projects." Paper 9710 presented at the Permian Basin Oil and Gas Recovery Symposium, Midland, TX, March 12–13, 1981.

Calhoun, J. C. "A Definition of Petroleum Engineering," *J. Pet. Tech.* (July 1963).

Campbell, R. A. and J. M. Campbell, Sr. *Mineral Property Economics, vol. 3: Petroleum Property Evaluation,* Campbell Petroleum Series, Norman, Oklahoma (1978).

Chew, J. and C. A. Connally. "A Viscosity Correlation for Gas Saturated Crude Oils," *Trans., AIME* (1959): 20.

Clark, N. J. *Elements of Petroleum Reservoirs,* Henry L. Doherty Series, SPE of AIME, Dallas, Texas (1969): 66–83.

Coats, K. H. "Reservoir Simulation: State of the Art," *J. Pet. Tech.* (August 1982): 1633–42.

Coats, K. H. "Use and Misuse of Reservoir Simulation Models," SPE Reprint Series No. 11: Numerical Simulation, 1973, 183–190.

Coats, K. H. "Use and Misuse of Reservoir Simulation Models," *J. Pet. Tech.* (November 1969): 1391–98.

Coats, K. H. "Reservoir Simulation: State of the Art," *J. Pet. Tech.* (August 1982): 1633–42.

Cole, F. W. *Reservoir Engineering Manual*, Houston, Texas: Gulf Publishing Company (1969): 284–288.

Corey, A. T. *Prod. Monthly* 19 no. 1, (1954): 38.

Craft, B. C. and M. F. Hawkins. *Applied Petroleum Reservoir Engineering.* Englewood Cliffs, NJ.: Prentice-Hall, Inc., 1959.

Craig, F. F., et al. "Optimized Recovery Through Continuing Interdisciplinary Cooperation," *J. Pet. Tech.* (July 1977): 755–760.

Craig, F. F., Jr. *The Reservoir Engineering Aspects of Waterflooding*, SPE Monograph 3, Richardson, TX (1971).

Croes, G. A. and N. Schwarz. "Dimensionally Scaled Experiments and the Theories on the Water-Drive Process," *Trans., AIME* 204 (1955): 35–42.

Dake, L. P. *Fundamentals of Reservoir Engineering.* New York, N.Y.: Elsevier Scientific Publishing Company, 1978, 303–344.

Dandona, A. K., R. B. Alston, and R. W. Braun. "Defining Data Requirements for a Simulation Study." Paper 22357 presented at SPE International Meeting on Petroleum Engineering, Beijing, China, March 24–27, 1992.

Deppe, J. C. "Injection Rates—The Effect of Mobility Ratio, Area Swept, and Pattern," *Soc. Pet. Eng. J.* (June 1961): 81–91.

Dietrich, J. K. and Bondor, P. L. "Three-Phase Oil Relative Permeability Models." Paper 6044 presented at the SPE Annual Meeting in New Orleans, October, 1976.

Doscher, T. M. et al. "Statistical Analysis of Crude Oil Recovery and Recovery Efficiency," *Bull. D14, API* (1984).

Douglas, J., Jr., D. W. Peaceman, and H. H. Rachford, Jr. "A Method for Calculating Multi-Dimensional Immiscible Displacement," *Trans., AIME* 216 (1959): 297–306.

Durrani, A. J., et al. "The Rejuvenation of the 30-year-old McAllen Ranch Field: An Application of Cross-Functional Team Management." Paper 24872 presented at the SPE Annual Technical Conference and Exhibition, Washington, D.C., October 4–7, 1992.

Dyes, A. B., B. H. Caudle, and R. A. Erickson. "Oil Production After Breakthrough as Influenced by Mobility Ratio," *Trans., AIME* 201 (1954): 81–86.

Dykstra, H. and R. L. Parsons. "*The Prediction of Oil Recovery by Waterflooding,*" *Secondary Recovery of Oil in the United States*, 2nd ed., *API* (1950): 160–174.

Emanuel, A. S., et al. "Reservoir Performance Prediction Methods Based on Fractal Geostatistics," *SPERE* (August 1989): 311–18.

Ershaghi, I. and Omoregie O. "A Method for Extrapolation of Cut vs. Recovery Curves," *J. Pet. Tech.* (February 1978): 203–204.

Ershaghi, I. and D. Abdassah. "A Prediction Technique for Immiscible Processes Using Field Performance Data," *J. Pet. Tech.* (April 1984): 664–670.

Essley, P. L. "What is Reservoir Engineering?" J. *Pet. Tech.* (January 1965): 19–25.

Evers, J. F. "How to Use Monte Carlo Simulation in Profitability Analysis." Paper 4401 presented at the SPE Rocky Mountain Regional Meeting, Casper, WY, May 15–16, 1973.

Farouq Ali, S. M. and S. Thomas. "The Promise and Problems of Enhanced Oil Recovery Methods." Paper CIM No. 26 presented at the South Saskatchewan Section Meeting, Regina, Saskatchewan, Sept. 25–27, 1989.

Frank, Jr., J. R., E. Van Reet, and W. D. Jackson. "Combining Data Helps Pinpoint Infill Drilling Targets in Texas Field," Oil *& Gas J.*, May 31, 1993.

Frizzell, D. F., Texaco E&P Technology Department: Personal Contact.

Garb, F. A. "Oil and Gas Reserves Classification, Estimation, and Evaluation," *J. Pet. Tech.* (March 1985): 373–390.

Garb, F. A. "Assessing Risk In Estimating Hydrocarbon Reserves and in Evaluating Hydrocarbon-Producing Properties," *J. Pet. Tech.* (June 1988): 765–768.

Ghauri, W. K., A. F. Osborne, and W. L. Magnuson. "Changing Concepts in Carbonate Waterflooding—West Texas Denver Unit Project—An Illustrative Example," *J. Pet. Tech.* (June 1974): 595–606.

Ghauri, W. K. "Innovative Engineering Boosts Wasson Denver Unit Reserves," *Pet. Eng.* (December 1974): 26–34.

Ghauri, W. K. "Production Technology Experience in a Large Carbonate Waterflood, Denver Unit, Wasson San Andres Field," *J. Pet. Tech.* (September 1980): 1493–1502.

Goolsby, J. L. "The Relation of Geology to Fluid Injection in Permian Carbonate Reservoirs in West Texas," Southwest Pet. Short Course—Lubbock, TX, 1965.

Gordon, S. P. and O. K. Owen. "Surveillance and Performance of an Existing Polymer Flood: A Case History of West Yellow Creek." Paper 8202 presented at the SPE Annual Technical Conference and Exhibition, Las Vegas, NV, Sept. 23–26, 1979.

Guthery, S., K. Landgren, and J. Breedlove. "Data Exchange Standard Smooths E&P Integration," *Oil and Gas J.* (May 31, 1993): 29–35.

Halbouty, M. T. "Synergy is Essential to Maximum Recovery," *J. Pet. Tech.* (July 1977): 750–754.

Haldorsen, H. H. and T. Van Golf-Racht. "Reservoir Management into the Next Century." Paper NMT 890023 presented at the Centennial Symposium at New Mexico Tech., Socorro, NM, Oct. 16–19, 1989.

Haldorsen, H. H. and E. Damsleth. "Stochastic Modeling," *J. Pet. Tech.* (April 1990): 404–412.

Hall, H. N. "How to Analyze Waterflood Injection Well Performance," *World Oil* (October 1963): 128–130.

Harris, D. G. "The Role of Geology in Reservoir Simulation Studies," *JPT* (May 1975): 625–632.

Harris, D. G. and C. H. Hewitt. "Synergism in Reservoir Management—The Geologic Perspective," *J. Pet. Tech.* (July 1977): 761–770.

Harris, D. E. and R. L. Perkins. "A Case Study of Scaling Up 2D Geostatistical Models to a 3D Simulation Model." Paper 22760 presented at the SPE Annual Technical Conference and Exhibition, Dallas, TX, October 6–9, 1991.

Havlena, D. "Interpretation, Averaging and Use of the Basic Geological Engineering Data," *J. Canadian Pet. Tech.*, part 1, v. 5, no. 4 (October–December 1966): 153–164; part 2, v. 7, no. 3 (July–September 1968): 128–144.

Havlena, D. and A. S. Odeh. "The Material Balance as an Equation of Straight Line," *J. Pet. Tech.* (August 1963): 896–900.

Havlena, D. and A. S. Odeh. "The Material Balance as an Equation of Straight Line—Part II, Field Cases," *J. Pet. Tech.* (July 1964): 815–822.

Hewitt, T. A. "Fractal Distributions of Reservoir Heterogeneity and Their Influence on Fluid Transport." Paper 15386 presented at the SPE Annual Technical Conference and Exhibition, New Orleans, LA, Oct. 5–8, 1986.

Hickman, T. S. "The Evaluation of Economic Forecasts and Risk Adjustments in Property Evaluation in the U.S.," *J. Pet. Tech.* (February 1991): 220–25.

Honarpour, M., L. F. Koederitz, and H. A. Harvey. "Empirical Equations for Estimating Two-Phase Relative Permeability in Consolidated Rock," *J. Pet. Tech.* (December 1982): 2905–2908.

Hurst, W. "Water Influx into a Reservoir and its Application to the Equation of Volumetric Balance," *Trans., AIME* 151, (1943): 57.

Hurst, W. "Simplication of the Material Balance Formulas by the Lap Lace Transformation," *Trans., AIME* (1958): 213–292.

Johnson, C. R. and T. A. Jones. "Putting Geology Into Reservoir Simulations: A Three-Dimensional Modeling Approach." Paper 18321 presented at the SPE Annual Technical Conference and Exhibition, Oct. 2–5, 1988.

Johnson, J. P. "POSC Seeking Industry Software Standards, Smooth Data Exchange," *Oil and Gas J.* (Oct. 26, 1992): 64–68.

Jordan, J. K. "Reliable Interpretation of Waterflood Production Data," *J. Pet. Tech.* (August 1955): 18–24.

Journel, A. G. and F. G. Alabert. "New Method for Reservoir Mapping," *J. Pet. Tech.* (February 1990): 212–18.

Klins, M. A. *Carbon Dioxide Flooding.* Boston: IHRDC, 1984.

Langston, E. P. and J. A. Shirer. "Performance of Jay/LEC Fields Unit Under Mature Waterflood and Early Tertiary Operations," *J. Pet. Tech.* (February 1985): 261–68.

Langston, E. P., J. A. Shirer, and D. E. Nelson. "Innovative Reservoir Management —Key to Highly Successful Jay/LEC Waterflood," *J. Pet. Tech.* (May 1981): 783–91.

Lantz, J. R. and N. Ali. "Development of a Mature, Giant Offshore Oil Field," *J. Pet. Tech.* (April 1991): 392–97.

Lantz, J. R. and N. Ali. "Development of a Mature Giant Offshore Oil Field, Teak Field, Trinidad." Paper 6237, 22nd Annual OTC Meeting, Houston, TX, May 7–9, 1990.

Lasater, J. A. "Bubble Point Pressure Correlation," *Trans., AIME* (1958): 379.

Litvak, B. L., Texaco E&P Technology Department: Personal Contact.

Martin, F. D. and J. J. Taber. "Carbon Dioxide Flooding," *J. Pet. Tech.* (April 1992): 396–400.

Mattax, C. C. and R. L. Dalton. *Reservoir Simulation,* SPE Monograph Series, Richardson, Texas (1990): 13.

McCain, W. D. The Properties of Petroleum Fluids. Tulsa, Oklahoma: Petroleum Publishing Co., 1973.

McCune, C. C. "On-Site Testing to Define Injection—Water Quality Requirements," *J. Pet. Tech.* (January 1977): 17–24.

Muskat, M. *Physical Principles of Oil Production.* New York: McGraw-Hill Book Co., Inc., 1950.

Neff, D. B. and T. S. Thrasher. "Technology Enhances Integrated Teams' Use of Physical Resources," *Oil and Gas J.* (May 31, 1993): 29–35.

Nolen-Hoeksema, R. C. "The Future of Geophysics in Reservoir Engineering," *Geophysics: The Leading Edge of Exploration* (December 1990): 89–97.

Odeh, A. S. "Reservoir Simulation—What Is It?" *J. Pet. Tech.* (November 1969): 1383–88.

Pariani, G. J. et al. "An Approach to Optimize Economics in a West Texas CO_2 Flood." Paper 22022 presented at the SPE Hydrocarbon Economics and Evaluation Symposium held in Dallas, TX, April 11–12, 1991.

Patterson, S. and J. Altieri. "Business Modeling Provides Focus for Upstream Integration," *Oil and Gas J.* (May 31, 1993): 43–47.

Prats, M., et al. "Prediction of Injection Rate and Production History for Multifluid Five-Spot Floods," *Trans., AIME* 216 (1959): 98–105.

Prats, M. *Thermal Recovery,* SPE Monograph 7, Richardson, TX, 1982.

Raza, S. H. "Data Acquisition and Analysis: Foundational to Efficient Reservoir Management," *J. Pet. Tech.* (April 1992): 466–468.

Robertson, J. D. "Reservoir Management Using 3D Seismic Data," *J. Pet. Tech.* (July 1989): 663–667.

Robertson, J. D. "Reservoir Management Using 3-D Seismic Data," *Geophysics: The Leading Edge of Exploration* (February 1989): 25–31.

Roebuck, I. F., Jr. and L. L. Crain. "Water Flooding a High Water-Cut Strawn Sand Reservoir," *J. Pet. Tech.* (August 1964): 845–50.

Rose, S. C., J. F. Buckwalter, and R. J. Woodhall. *The Design Engineering Aspects of Waterflooding,* SPE Monograph 11, Richardson, TX, 1989.

Saleri, N. G. and R. M. Toronyi. "Engineering Control in Reservoir Simulation." Paper 18305, SPE 63rd Annual Technical Conference and Exhibition, Oct. 2–5, 1988.

Satter, A., J. E. Varnon, and M. T. Hoang. "Reservoir Management: Technical Perspective." Paper 22350 presented at the SPE International Meeting on Petroleum Engineering held in Beijing, China, March 24–27, 1992.

Satter, A. "Reservoir Management Training: An Integrated Approach." SPE 65th Ann. Tech. Conf. & Exb., Sept. 23–26, 1990, New Orleans, LA.

Satter, A. "Reservoir Management Training—An Integrated Approach." SPE Paper 20752, Reservoir Management Panel Discussion, SPE 65th Annual Technical Conference & Exhibition, Sept. 23–26, 1990, New Orleans, LA.

Satter, A., D. F. Frizzell, and J. E. Varnon. "The Role of Mini-Simulation in Reservoir Management." SPE Paper 22370, International Meeting on Petroleum Engineering, March 24–27, 1992, Beijing, China.

Satter, A., D. F. Frizzell, and J. E. Varnon. "The Role of Mini-Simulation in Reservoir Management," Indonesian Petroleum Association, 1991 Annual Convention, Jakarta, Indonesia.

Schilthuis, R. J. "Active Oil and Reservoir Energy," *Trans., AIME* 118, (1936): 33–52.

Schilthuis, R. J. "Active Oil and Reservoir Energy," *Trans. AIME* 118:37.

Schneider, J. J. "Geologic Factors in the Design and Surveillance of Waterfloods in the Thick Structurally Complex Reservoirs in the Ventura Field, California." Paper 4049 presented at the SPE Annual Meeting, San Antonio, TX, Oct. 8–11, 1972.

Seba, R. D. "Determining Project Profitability," *J. Pet. Tech.* (March 1987): 263–71.

Sessions, K. P. and D. H. Lehman. "Nurturing the Geology Reservoir Engineering Team: Vital for Efficient Oil and Gas Recovery." Paper 19780 presented at the SPE Annual Technical Conference and Exhibition, San Antonio, TX, Oct. 8–11, 1989.

Shirer, J. A. "Jay-LEC Waterflood Pattern Performs Successfully." Paper 5534 presented at the SPE Annual Technical Conference and Exhibition, Dallas, TX, Sept. 28–Oct. 1, 1975.

Sloat, B. F. "Measuring Engineering Oil Recovery," *J. Pet. Tech.* (January 1991): 8–13.

Sneider, R. M. "The Economic Value of a Synergistic Organization," Archie Conference, Houston, TX, Oct. 22–25, 1990.

Staggs, H. M. and E. F. Herbeck. "Reservoir Simulation Models—An Engineering Overview," *J. Pet. Tech.* (December 1971): 1428–36.

Staggs, H. M. "An Objective Approach to Analyzing Waterflood Performance." Paper presented at the Southwestern Petroleum Short Course, Lubbock, TX, 1980.

Stalkup, F. I., Jr. *Miscible Displacement,* SPE Monograph 8, 1983.

Standing M. B. "A Generalized Pressure-Volume-Temperature Correlation For Mixture of California Oils and Gases," Drilling and Production Practice API (1947): 275.

Stanley, R. G., et al. "North Ward Estes Geological Characterization," *AAPG Bulletin,* (1990).

Stermole, F. J. and J. M. Stermole. *Economic Evaluation and Investment Decision Methods,* 6th ed., Golden, CO: Investment Evaluation Corporation, 1987.

Stiles, W. E. "Use of Permeability Distribution in Waterflood Calculations," *Trans., AIME* 186 (1949): 9–13.

Stiles, L. H. "Reservoir Management in the Means San Andres Unit." Paper 20751 presented at the SPE Annual Technical Conference and Exhibition, New Orleans, Sept. 23–26, 1990.

Stiles, L. H. and J. B. Magruder. "Reservoir Management in the Means San Andres Unit," *J. Pet. Tech.* (April 1992): 469–475.

Stone, H. L. "Estimation of Three-Phase Oil Relative Permeability," *J. Pet. Tech.* (1970): 214–218.

Taber, J. J. and F. D. Martin. "Technical Screening Guides for the Enhanced Recovery of Oil." Paper 12069 presented at the SPE Annual Technical Conference and Exhibition, San Francisco, CA, Oct. 5–8, 1983.

Talash, A. W. "An Overview of Waterflood Surveillance and Monitoring," *J. Pet. Tech.* (December 1988): 1539–1543.

Tang, R. W. et al. "Reservoir Studies With Geostatistics To Forecast Performance," *SPE Reservoir Engineering* (May 1991): 253–58.

Thakur, G. C. "Implementation of a Reservoir Management Program." SPE Paper 20748 presented at the SPE Annual Technical Conference and Exhibition, New Orleans, LA, Sept. 23–26, 1990.

Thakur, G. C. "Waterflood Surveillance Techniques—A Reservoir Management Approach," *J. Pet. Tech.* (October 1991): 1180–1188.

Thakur, G. C. "Reservoir Management: A Synergistic Approach." SPE Paper 20138, presented at the Permian Basin Oil and Gas Conference, Midland, TX, March 8–9, 1990.

Thakur, G. C. "Engineering Studies of G-1, G-2, and G-3 Reservoirs, Meren Field, Nigeria," *J. Pet. Tech.* (April 1982): 721–732.

Thomas, G. W. "The Role of Reservoir Simulation in Optimal Reservoir Management." Paper 14129, SPE International Meeting on Petroleum Engineering, Beijing, China, March 17–20, 1986.

Trice, M. L. and B. A. Dawe. "Reservoir Management Practices," *J. Pet. Tech.* (December 1992): 1296–1303 & 1349.

Van Everdingen, A. F. and W. Hurst. "The Application of the Lap Lace Transformation to Flow Problems in Reservoirs," *Trans., AIME* (1949): 186.

Wang, B., Texaco E&P Technology Department: Personal Contact.

Wang, B. and T. S. Teasdale. "GASWAT-PC: A Microcomputer Program for Gas Material Balance with Water Influx." SPE Paper 16484, Petroleum Industry Applications of Microcomputers, Del Lago on Lake Conroe, Texas, June 23–26, 1987.

Welge, H. J. "A Simplified Method for Computing Oil Recovery by Gas or Water Drive," *Trans., AIME* 146 (1942): 107–116.

Wiggins, M. L. and R. A. Startzman. "An Approach to Reservoir Management." Paper 20747, SPE Reservoir Management Panel Discussion, SPE 65th Annual Technical Conference & Exhibition, New Orleans, LA, Sept. 23–26, 1990.

Winzinger, R., et al. "Design of a Major CO_2 Flood—North Ward Estes Field, Ward County, Texas." Paper 19654, presented at the SPE Annual Technical Conference, San Antonio, TX, Oct. 8–11, 1989.

Winzinger, R. et al. "Design of a Major CO_2 Flood—North Ward Estes Field, Ward County, Texas," *SPERE* (February 1991): 11–16.

Woods, E. G. and Osmar Abib. "Integrated Reservoir Management Concepts," Reservoir Management Practices Seminar, SPE Gulf Coast Section, Houston, Texas, May 29, 1992.

Woofer, D. M. and J. MacGillivary. "Brassey Oil Field, British Columbia: Development of an Aeolian Sand—A Team Approach," *SPERE* (May 1992): 165–72.

Wyllie, M. R. J. "Reservoir Mechanics—Stylized Myth or Potential Science?" *J. Pet. Tech.* (June 1962): 583–588.

Wyllie, M. R. J. and G. H. F. Gardner. "Generalized Kozeny—Carmen Equation, parts 1 and 2," *World Oil* 146 no. 4 and no. 5, (1958): March, 121–126; April, 210–228.

"Annual Production Report," *OGJ Special, Oil & Gas J.* (April 20, 1992): 51–79.

Miscible Processes, SPE Reprint Series No. 8, 1971.

Miscible Processes 11, SPE Reprint Series No. 18, 1985.

Reservoir Management Panel Discussion, SPE 65th Ann. Tech. Conf. & Exb., New Orleans, LA, Sept. 23–26, 1990.

Reservoir Management Practices Seminar, SPE Gulf Coast Section, Houston, TX, April 26, 1991.

Reservoir Management Panel Discussion, SPE 66th Ann. Tech. Conf. & Exb., Dallas, TX, Oct. 6–9, 1991.

Reservoir Management Sessions, Int'l. Mtg. on Pet. Engr., Beijing, China, March 24–27, 1992.

Reservoir Management Practices Seminar, SPE Gulf Coast Section, Houston, TX, May 29, 1992.

SPE Forum Series V: *Advances in Reservoir Management and Field Applications*, Mt. Crested Butte, CO, August 13–18, 1989.

SPE Forum Series III: *Application of Reservoir Characterization to Numerical Modeling and Reservoir Management*, Mt. Crested Butte, CO, July 28–Aug. 2, 1991.

"Teamwork, New Technology, and Mature Reservoirs," *J. Pet. Tech.* (January 1992): 38–40.

"Texaco Sets Horizontal Well Marks," *Oil and Gas J.* (July 6, 1992): 30–32.

Thermal Recovery Processes, SPE Reprint Series No. 7, 1985.

Thermal Recovery Techniques, SPE Reprint Series No. 10, 1972.

Index

N

O

P

R

S

T

V

W